Beginner's Guide t

2005

Alejandro Reyes, MSME
Certified SolidWorks Professional and Instructor

ISBN: 1-58503-251-4

Schroff Development Corporation

---

www.schroff.com
www.schroff-europe.com

Trademarks and Disclaimer

SolidWorks and its family of products are registered trademarks of Dassault Systemes. Microsoft Windows and its family products are registered trademarks of the Microsoft Corporation.

Every effort has been made to provide an accurate text. The author and the manufacturers shall not be held liable for any parts developed with this book or held responsible for any inaccuracies or errors that appear in the book.

Acknowledgements

I dedicate this book to my lovely wife Patricia and my kids Liz, Ale and Hector, all of whom were very supportive, patient and comprehensive during the writing of the book. To you, all my love.

I also wish to express my gratitude to the many people who encouraged me to write this book, my first published work, and especially to Rosanne Kramer and Dave Pancoast from SolidWorks Corporation; their input, guidance and encouragement were invaluable for the development of this book.

About the Author

Alejandro Reyes obtained his engineering degree from the Instituto Tecnológico de Ciudad Juárez in Mexico as an electro-mechanical engineer; upon finishing he obtained a scholarship for a Masters Degree in the University of Texas at El Paso in mechanical design, with strong focus in Materials Science and Finite Element Analysis.

Currently Alejandro is the President of MechaniCAD Inc., a SolidWorks Authorized Training, Testing and Support Center in El Paso, Texas and Juarez, Mexico. He has been a Certified SolidWorks Instructor and Support Technician since 1998, Certified SolidWorks Professional and COSMOSWorks Certified Support Technician.

His professional interests include finding alternatives and improvements to existing consumer products, FEA analysis and new technologies. On a personal level, he enjoys spending time with his family and friends.

Notes:

Table of Contents

Notes:

Introduction

The purpose of this book is to help the user learn the basic concepts of SolidWorks and good solid modeling practices in an easy to follow guide. It is intended for the new SolidWorks user, users with experience in different CAD systems, and as a teaching aid in classroom training. At the end of this book, the user will have a good understanding of the SolidWorks interface and the most commonly used commands for part modeling, assembly and detailing by completing a project designing all the components, their 2D drawings and assembly drawing with Bill of Materials. The book is focused on the processes to complete a certain task, instead of focusing on individual operations, which are generally simple enough to learn.

SolidWorks is powerful design software for mechanical design; most of the features covered in this book have advanced options, which will not be covered as they are beyond the scope of this book. This book is meant to be a starting point to help the user prepare for more advanced topics.

SolidWorks is the leading 3D mainstream design software in the world, with hundreds of thousands of users in the industry ranging from one man shops to Fortune 500 companies, and in the educational segment in high schools, vocational schools, and a vast majority of the leading universities in the world. It is estimated that some 100,000 students get some level of experience with SolidWorks every year.

Prerequisites

This book was written assuming you have the following prerequisites:

- The reader is familiar with the Windows operating system.
- Experience in mechanical design and drafting.
- Any experience with other CAD systems is a big plus.

Notes:

The SolidWorks Interface

SolidWorks has three main areas: The toolbars and menus, the graphics area, and the Feature Manager/Property Manager. Toolbars and menus are very much like any Microsoft Windows application; some of the icons and menus are even similar to those of Microsoft Office applications and follow the Microsoft Windows rules, including features like drag and drop, copy/paste, etc.

The graphics area is the main part of SolidWorks; this is where parts, assemblies and drawings are visualized. SolidWorks lets the user Zoom, Pan, Rotate, set Orthogonal Views, etc., as well as to change the view style including Shaded, Shaded with Edges, Hidden Lines, Hidden Lines Visible and Wireframe to name a few. The third area is the **Feature Manager**; this is the graphical browser of features and operations, where features can be edited, modified, deleted, reordered, etc.

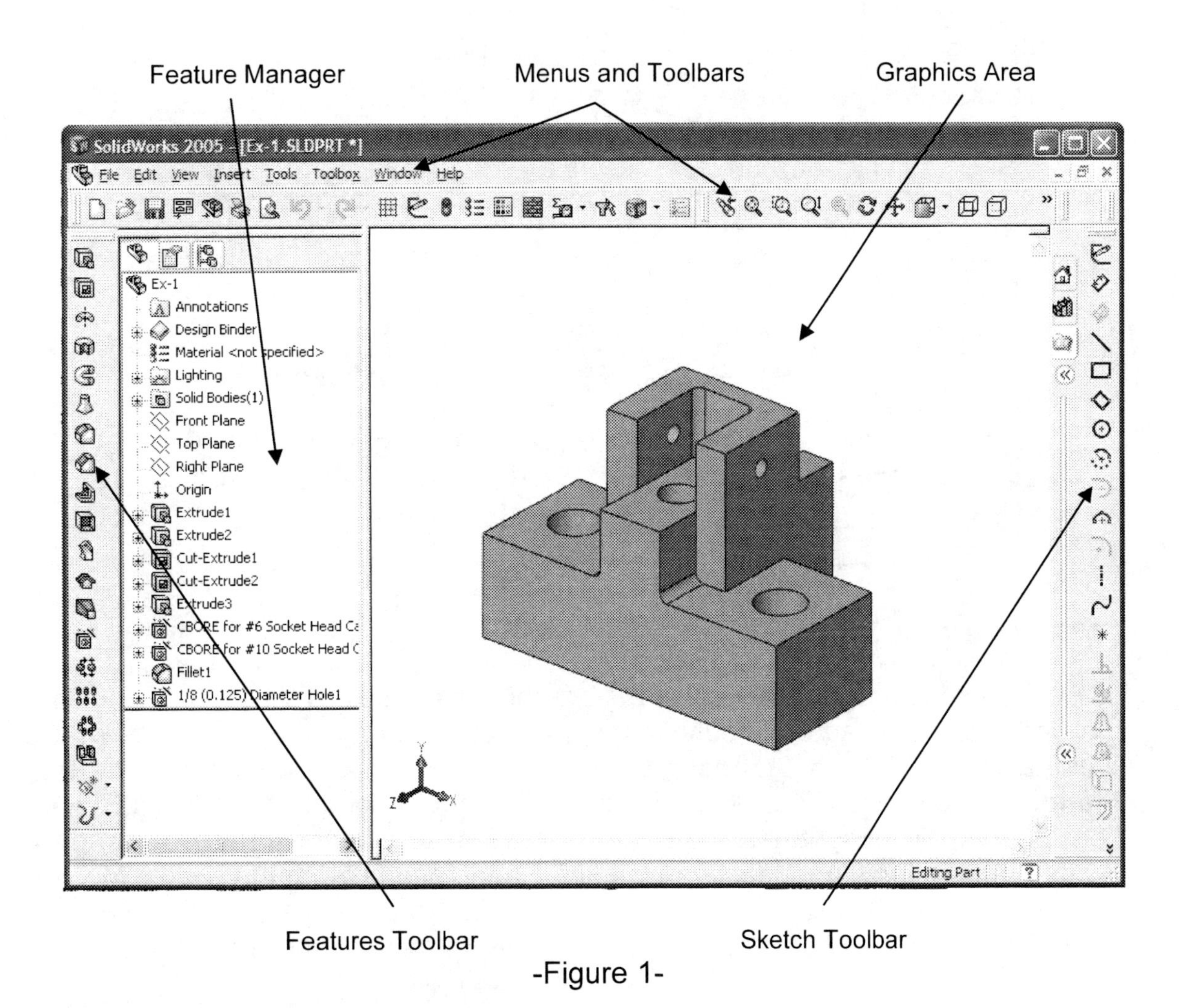

-Figure 1-

The Feature Manager's space is also shared by the **Property Manager**. This is where most of the feature's options are presented to the user; this is also where a selected entity's properties are displayed. The Property Manager will be displayed automatically when needed. The automatic display of the Property Manager can be turned on by selecting the "Tools" menu, "Options, System Options, General" and checking the checkbox "Auto-Show Property Manager". We'll show the user how to view both Feature and Property Managers at the same time later in the book.

In the Property Manager we have a common interface for most commands in SolidWorks, including common Windows Controls such as check boxes, open and closed option boxes, action buttons, etc.

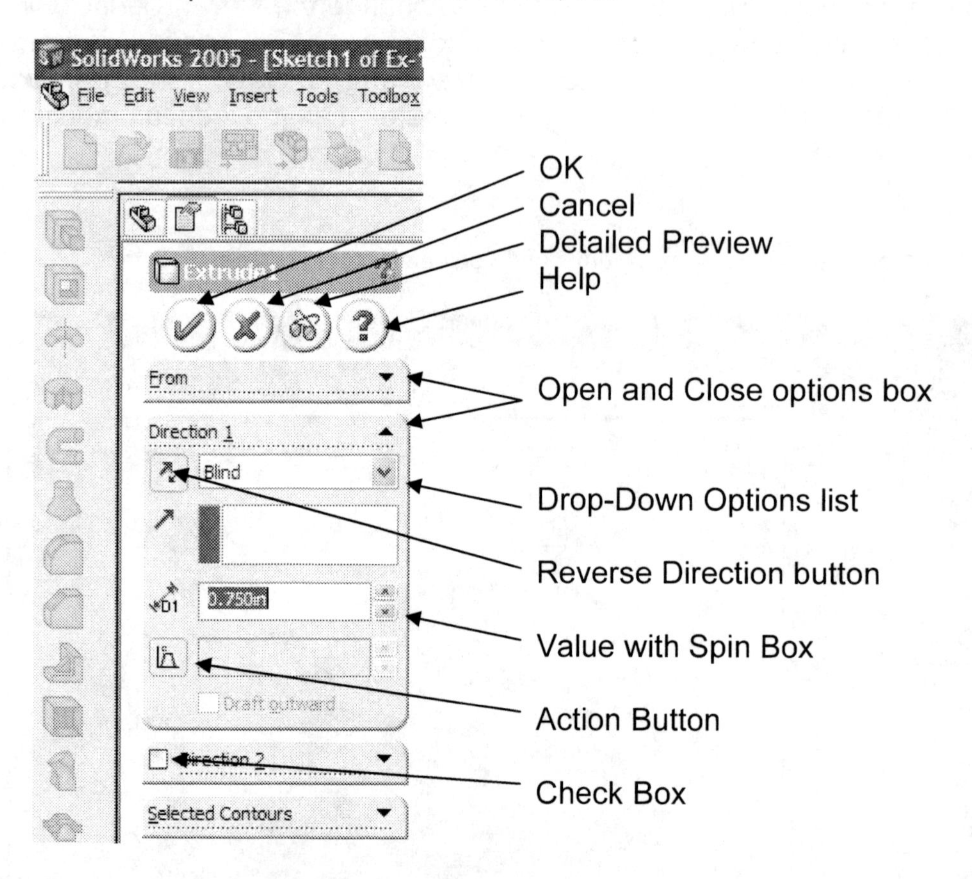

To manipulate the models in the graphics area, select a view manipulation tool, left click and drag the mouse in the graphics area to see its effect; **Previous View**, **Zoom to fit** and **Standard Views** are a single click command. Use the **Standard Views** icon to switch the model to any orthogonal view (Front, Back, Left, Bottom, Top, Right, Isometric, etc). The Mouse wheel can also be used to zoom in and out in the model by rolling the wheel, and clicking the Middle Mouse Button (the wheel) and dragging the mouse **Rotate**s the part in the graphics area. This rotation is automatic about the center of mass of the model.

The view manipulation tools include:

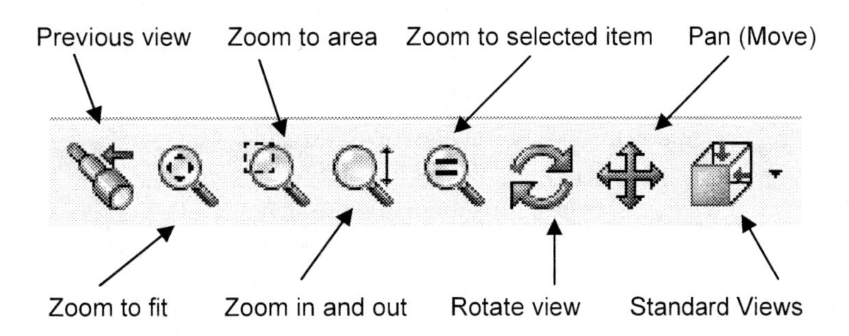

Another way to rotate the models on the screen is with the arrow keys in the keyboard.

To change the **view style**, the corresponding icon can be selected in the View toolbar, the effects will be immediately visible to the user. Feel free to explore them with your first model to become familiar; sometimes it's convenient to switch to a different view style for visibility or selection of entities.

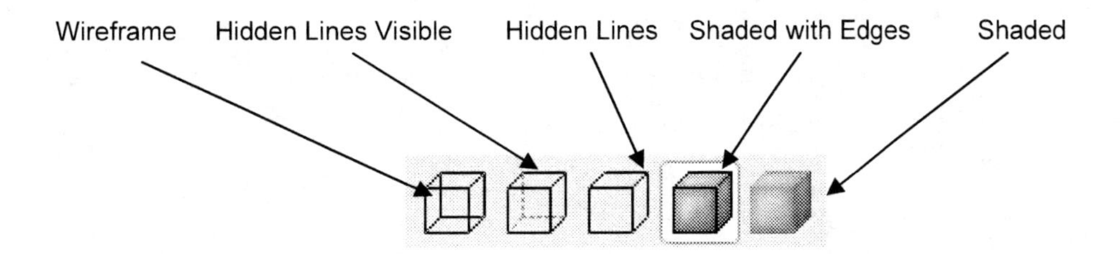

One important thing before we begin: This book is based on the standard toolbar layout of SolidWorks, and will not be using the **Command Manager**. The Command Manager is a very powerful and convenient tool that consolidates many toolbars in a single location, and selecting one toolbar icon displays all of the toolbar's icons. We will not use it to better help the user get familiarized with the SolidWorks interface, and enable them to find the tools always in the same place.

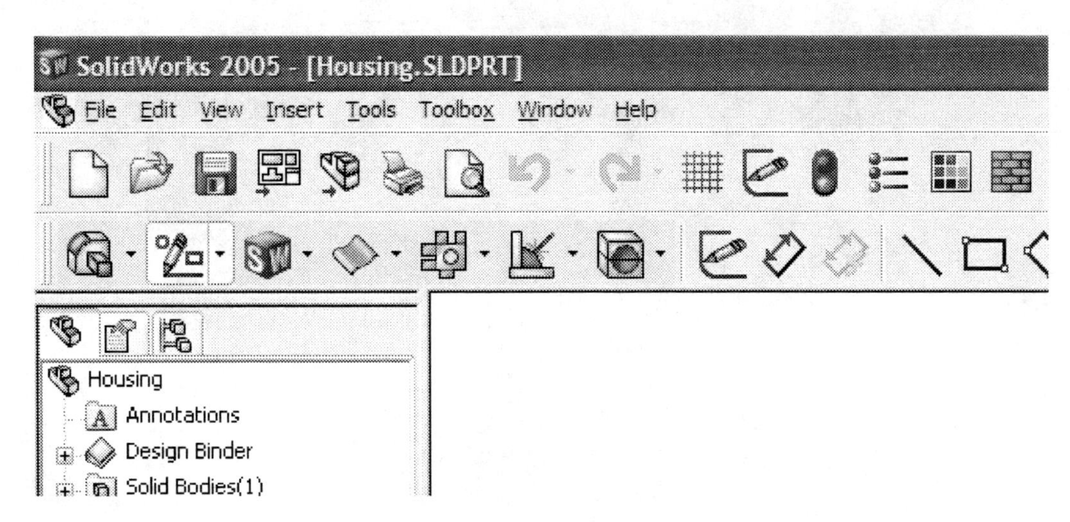

To disable the Command Manager and go back to the standard toolbar layout, select the "View" menu, "Toolbars, Command Manger". You must have a document open to be able to turn the Command Manager on or off and your screen should look like Figure 1.

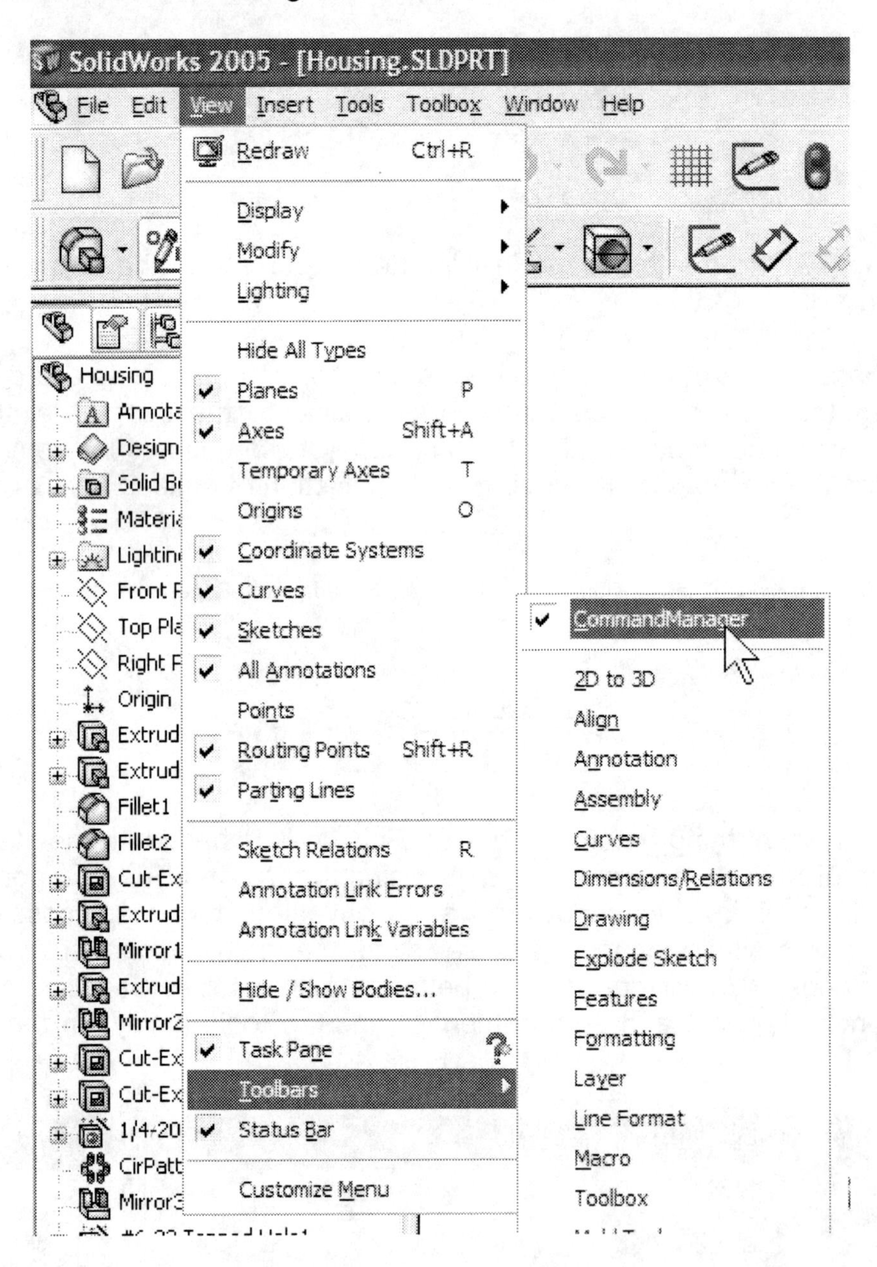

With that said, let's design something…

Part Modeling

The design process in SolidWorks generally starts in the part modeling module, where we create the different pieces that make the design of the product or machine, and are later assembled. In SolidWorks, every component of the design will be modeled separately, and each one is a single file with the extension *.sldprt. SolidWorks is a Feature based software; this means that the models are created by adding features to the model. Features are operations that either add or remove material to a part, for example, extrusions, cuts, rounds, etc. There are also features that do not create geometry, but are used as a construction aid, such as auxiliary planes, axes, etc.

This book will cover many different features to create parts, including the most commonly used tools and options. Some features require a **Sketch** or profile to be created first; these are known as Sketched features. The Sketch is the 2D environment where the sketch or profile is created. It is in the Sketch where most of the parameters and design intelligence is added to the design, including dimensions and geometric relations. Examples of sketched features include Extrusions, Revolved features, Sweeps and Lofts. Extrusions, Cuts and Revolved features will be covered in this book, while Sweeps and Lofts will be included in a future book, as they are generally considered more advanced modeling features.

A 2D Sketch can be created only in Planes or planar (flat) faces. By default, every SolidWorks Part and Assembly has three **default planes** (Front, Top and Right) and an Origin. Most parts can be started in one of these planes. While it is not critical which plane we start our designs in, it is a good idea to plan ahead. The planning that takes place before starting to model a part is called the ***Design Intent***.

SolidWorks is a 3D parametric mechanical design software. **Parametric** design means that the models created are driven by parameters. These parameters include dimensions, geometric relations, equations, etc. When any parameter is modified, the model updates. Good design practices are reflected in how well the Design Intent and model integrity is maintained when parameters are modified.

Notes:

The Housing

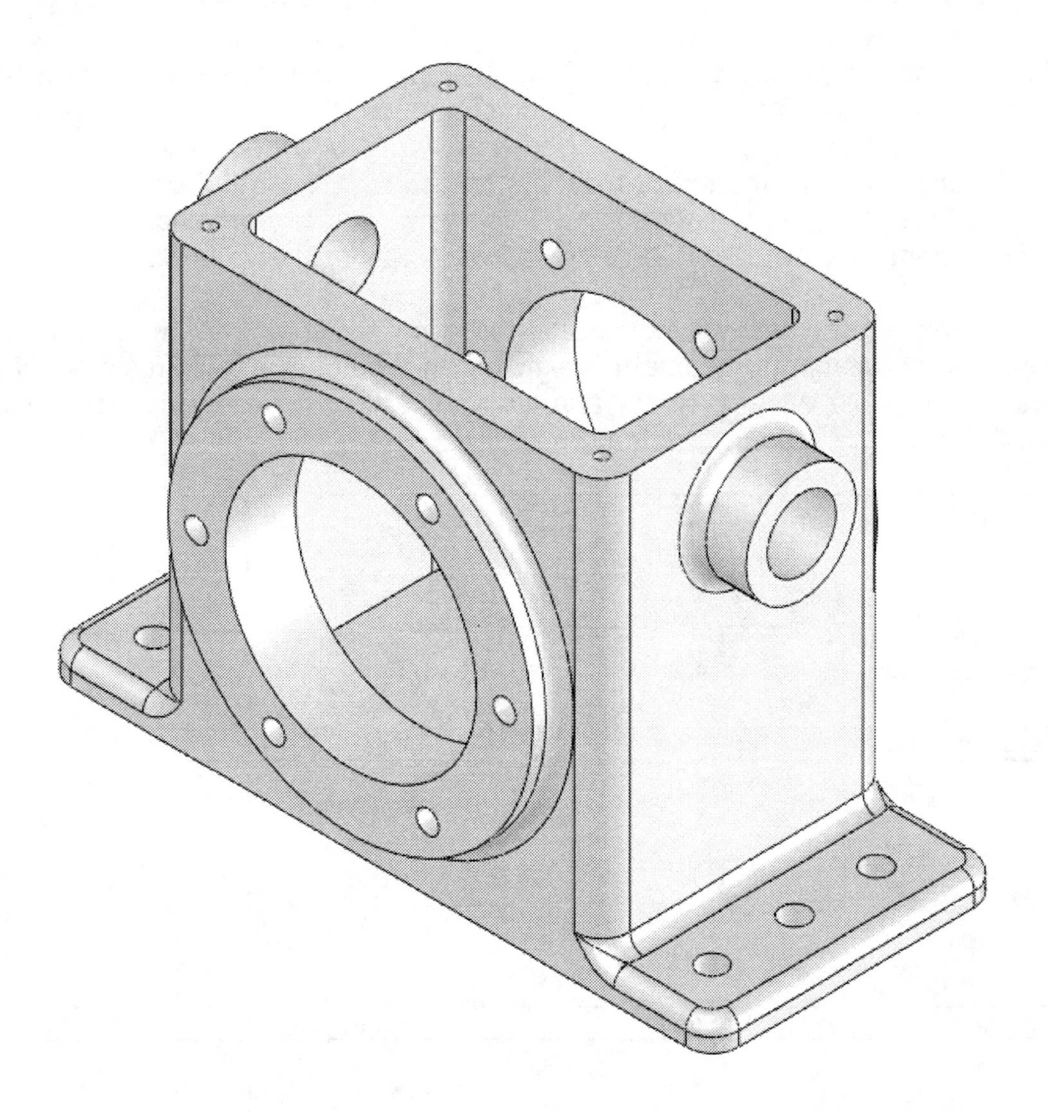

When we start a new design, we have to decide how we are going to make it. Remember that the parts will be made one feature or operation at a time. It takes a little practice to define the optimum sequence to make the features, but this is something that you will master once you learn to think of parts as a sequence of features. To help you understand how to make the Housing part, we'll show a "roadmap" or sequence of features. The order of some of these features can be changed, but remember that we need to make some features before others. For example, we cannot round the corners if there are no corners to round! A sequence will be shown at the beginning of each part, and the dimensional details will be given as we progress.

In this lesson we will cover the following tools and features: creating various sketch elements, geometric relations and dimensions, Extrusions, Cuts, Fillets, Mirror, Hole Wizard, Linear and Circular Patterns. For the Housing, we'll follow the following sequence of features:

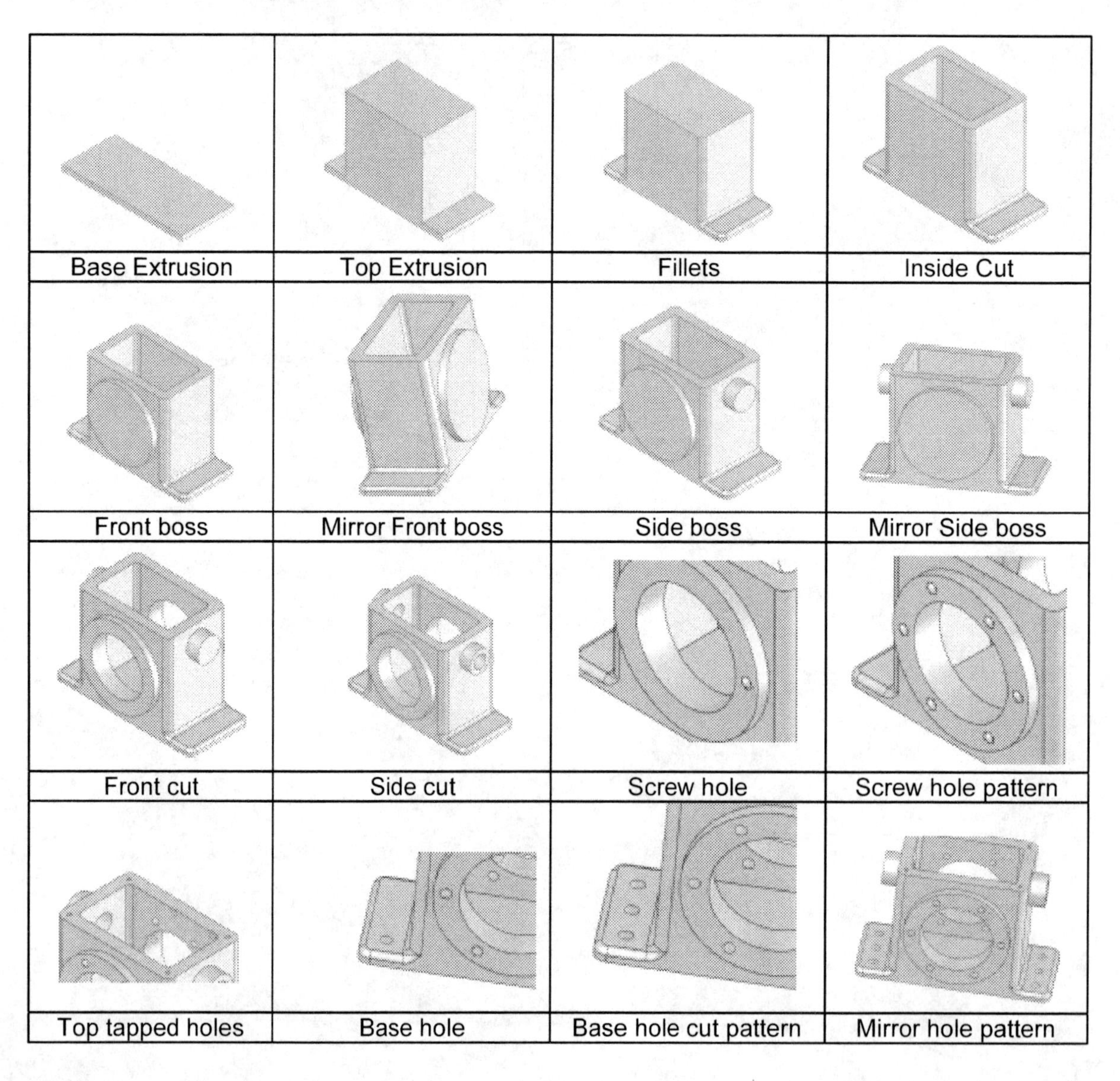

1. - The first thing we need to do after opening SolidWorks, is to make a **New Part** file. Go to the "New" document icon in the main toolbar and select it.

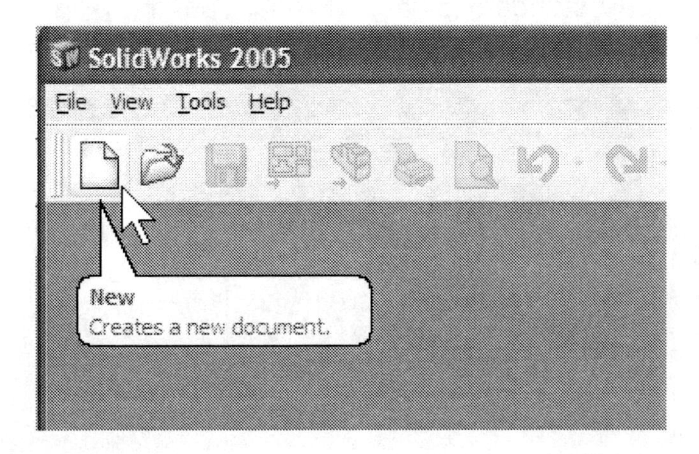

2. - We are now presented with the New Document dialog. If your screen is different than this, click the "Advanced" button in the lower left corner. Now select the "Part" template, and click OK; this way SolidWorks will know that we want to create a Part file. Additional Part templates can be created, with different options and settings, including different units, dimensioning standard, material, color, etc. See Appendix A for information on how to make additional **templates** and change the document **units** to inches or millimeters.

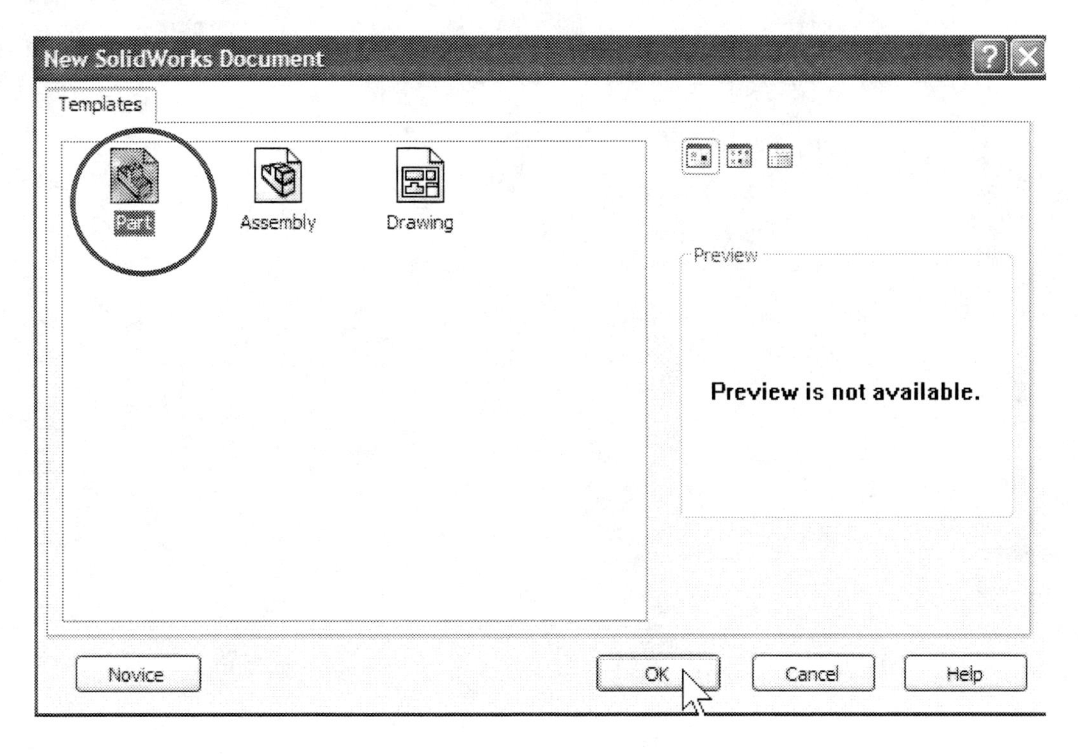

3. - Now that we have an empty Part file, we have to make a new **Sketch** to start modeling the part. Since this is the first sketch of the part, SolidWorks will show us the three standard planes (Front, Top and Right) after selecting the "Sketch" icon for us to select the plane where we want to start.

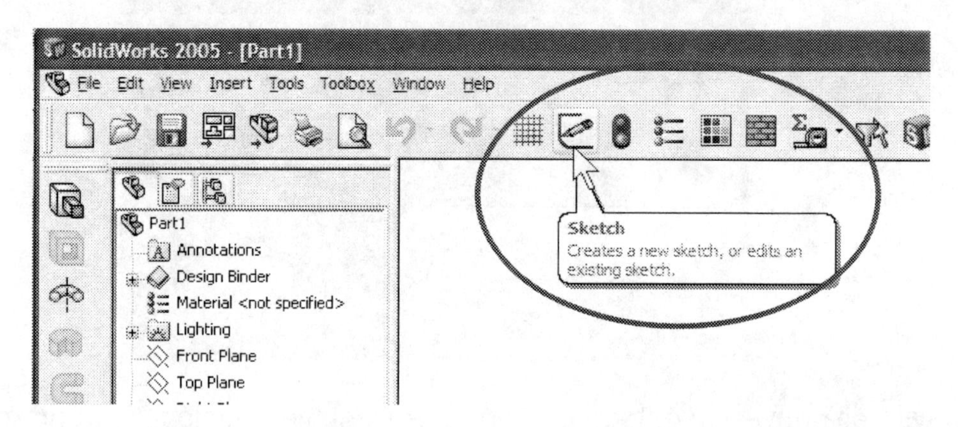

4. - For the Housing, we'll select the "Top Plane" to create the first sketch. We select the Top Plane, because we want to start modeling the part at the base of the Housing and build it up as was shown in the sequence of features. The plane selection will determine the standard view orientations in the part and detail drawings. What we see in the Top view of the model will be the Top View in the drawing.

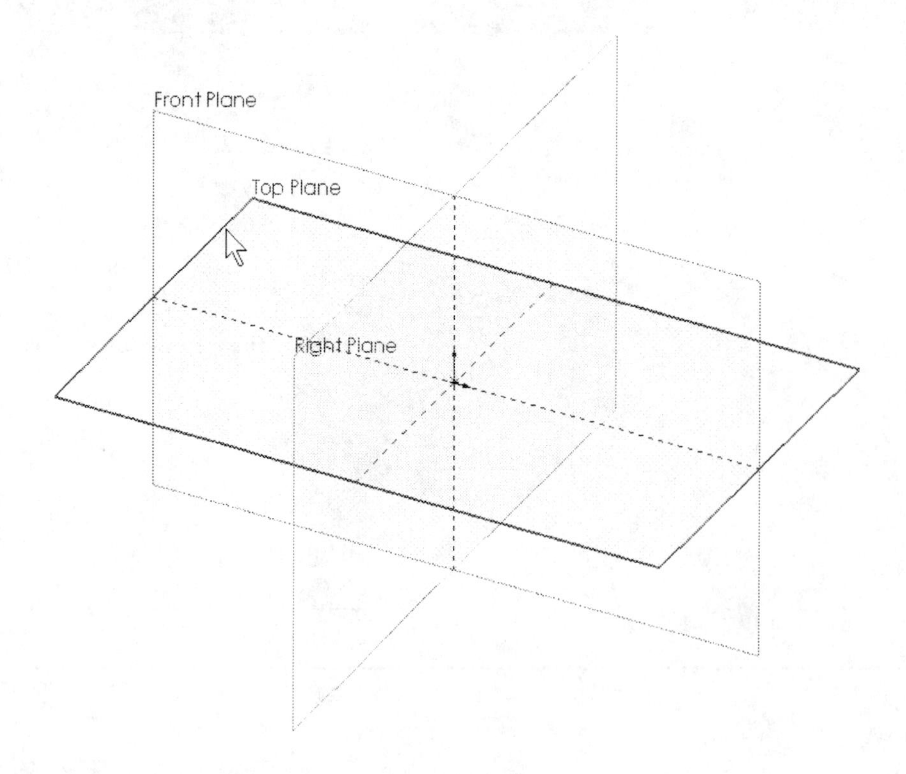

When we make a sketch, we activate the **Sketch** environment. This is where we will create the profiles that will be used to make Extrusions, Cuts, etc. SolidWorks gives us many indications, most of them graphical, to help us know when we are working in a Sketch.

a) The **Confirmation Corner** and the Sketch icon are activated.

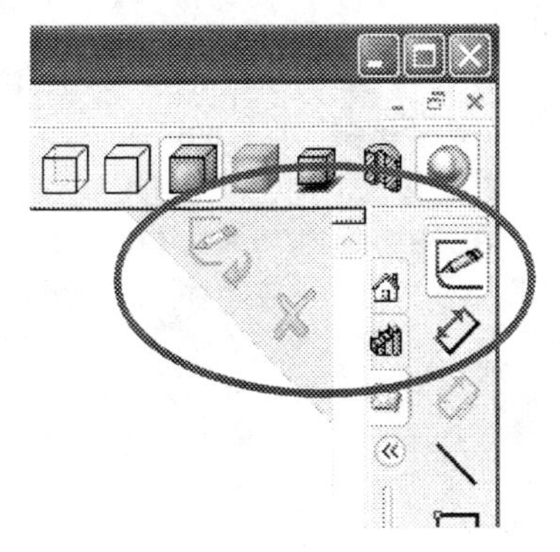

b) The Status bar shows "Editing Sketch".

c) In the Feature Manager "Sketch1" is added at the bottom just under "Origin".

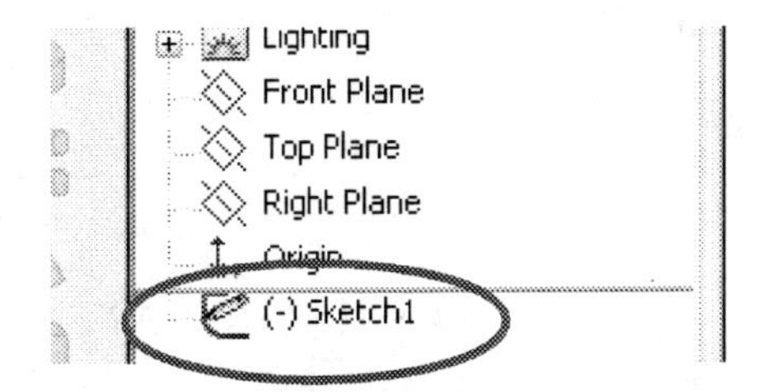

d) The part's Origin is projected in the Sketch plane in red (which in this case coincides with the Origin because we are working in the Top Plane)

e) All the icons in the Sketch toolbar are activated.

f) The title bar of the SolidWorks window shows "Sketch1 of Part1". This is the first sketch of the first part we make in this session of SolidWorks.

As the reader can see, SolidWorks gives us plenty of clues to help us know that we are working in a sketch.

5. - Notice that when we make the first sketch, SolidWorks rotates the view to match the plane that we selected. This is done only in the first sketch to help the user get oriented. In subsequent operations we have to rotate the view ourselves using the view orientation tools or the Middle Mouse Button.

6. - Select the "**Rectangle**" tool from the Sketch Toolbar and click and drag in the graphics area to draw a Rectangle as shown. Don't worry too much about the size; we'll add dimensions later.

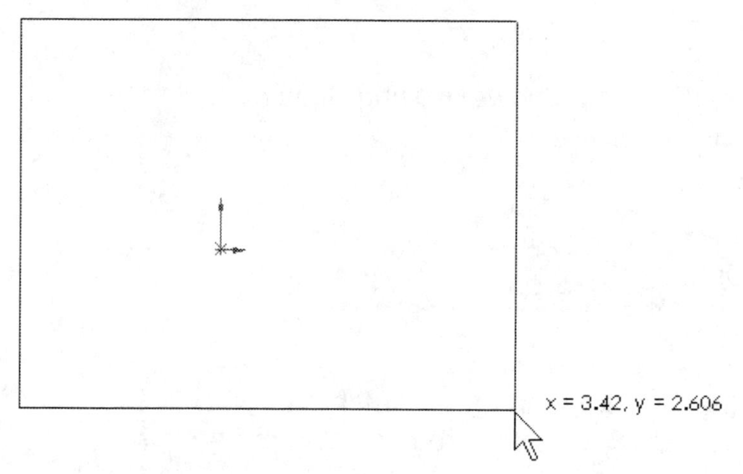

7. - Now we will draw a "**Centerline**" from one corner of the rectangle to the opposite corner. The purpose of this line is to help us center the rectangle about the origin.

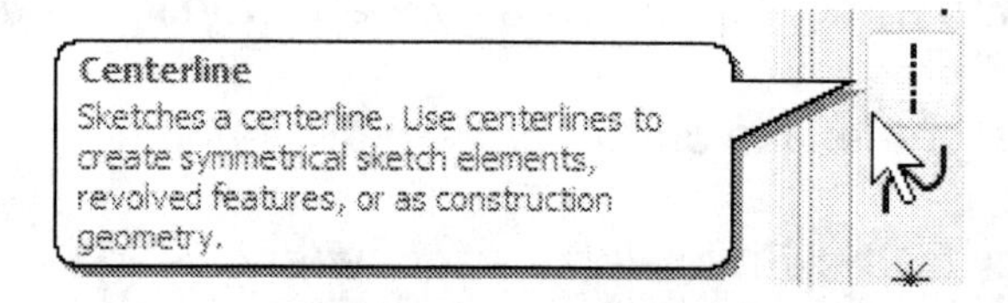

8. - SolidWorks indicates that we will start or finish a line at an existing entity with yellow icons; when the cursor is near an endpoint, line, edge, origin, etc. it will "snap" to it.

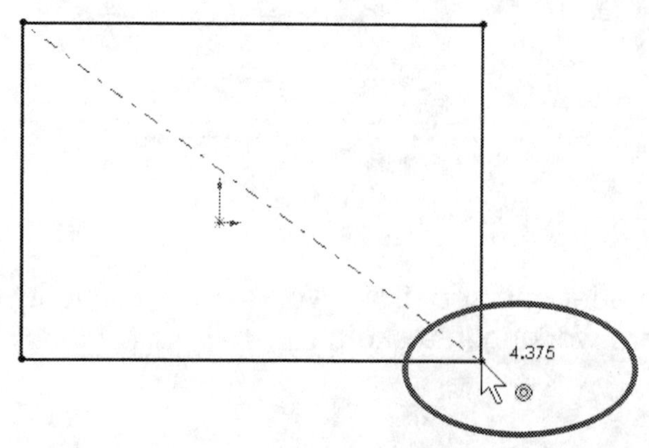

9. - Now we'll select the "**Add Relation**" icon to add a geometric relation between the centerline we just drew and the part's origin in order to center the rectangle about the origin, this way the part will be centered which will be useful in future operations.

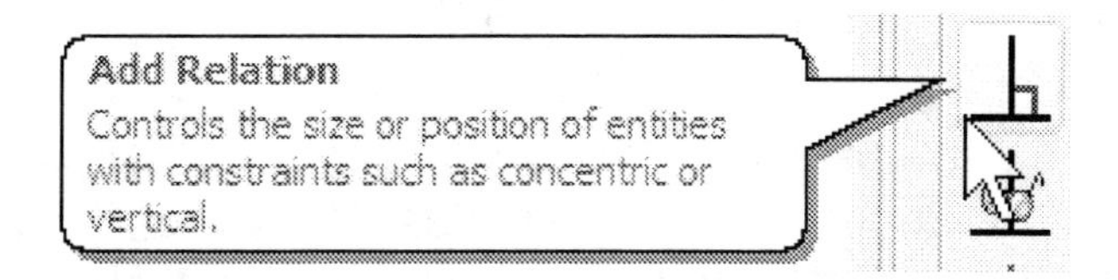

10. - When the icon is selected, the "Add Relations" Property Manager is shown. The **Property Manager** is the area where we will make our selections and choice of options for most commands. Select the previously made centerline and the part origin by clicking on them in the graphics area (notice how they turn green and get listed under the "Selected Entities" box). Click on "**Midpoint**" under the "Add Relations" box to add the relation. Now the center of the line coincides with the Origin. Click on OK (the green checkmark) to finish the command. Click and drag a rectangle corner to see the effect of adding the relation.

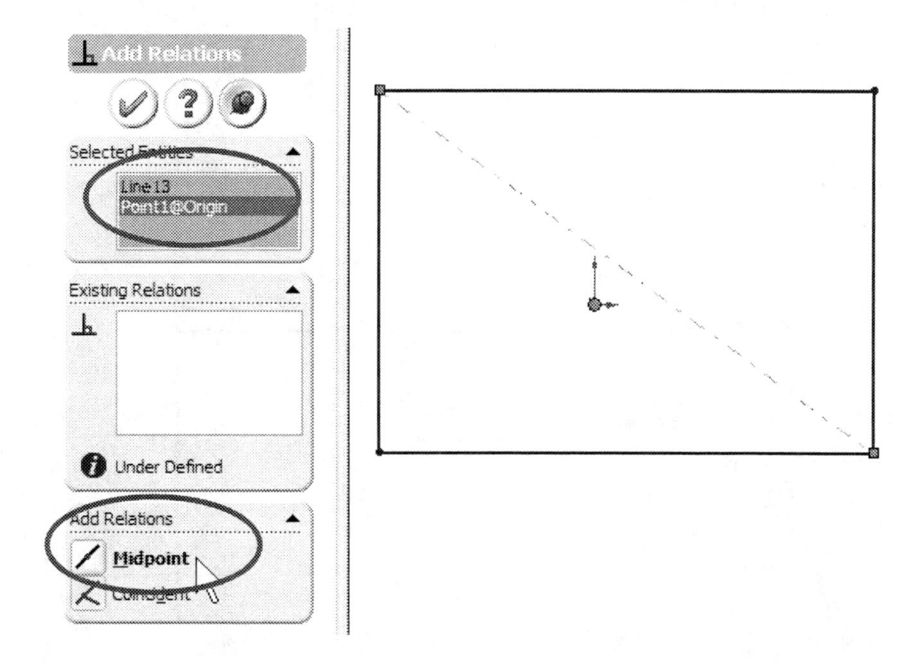

11. - We just added a geometric relation manually, and we also added geometric relations automatically when we drew the rectangle and the centerline in the previous step. SolidWorks allows us to graphically view the existing relations between sketch elements. Go to the "View" menu, "**Sketch Relations**" if not already activated.

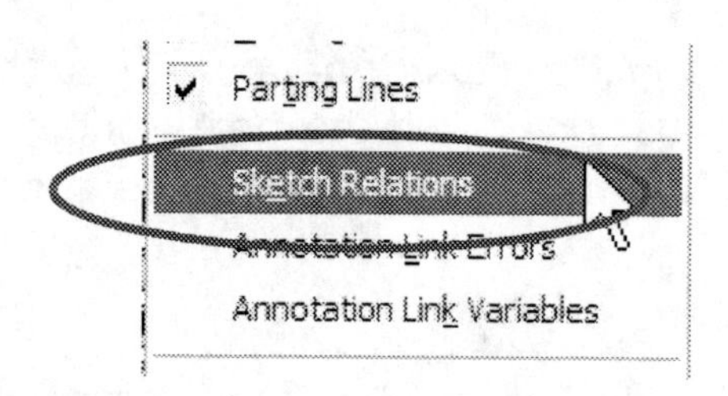

12. - Now we can see the geometric relations graphically represented by small blue icons next to the lines, arcs, etc. Notice that when the cursor rests over a geometric relation icon, the entity or entities that share the relation are highlighted.

NOTE: To delete a geometric relation select the relation icon and press the "Delete" key, or Right Mouse Click on the icon and select "Delete".

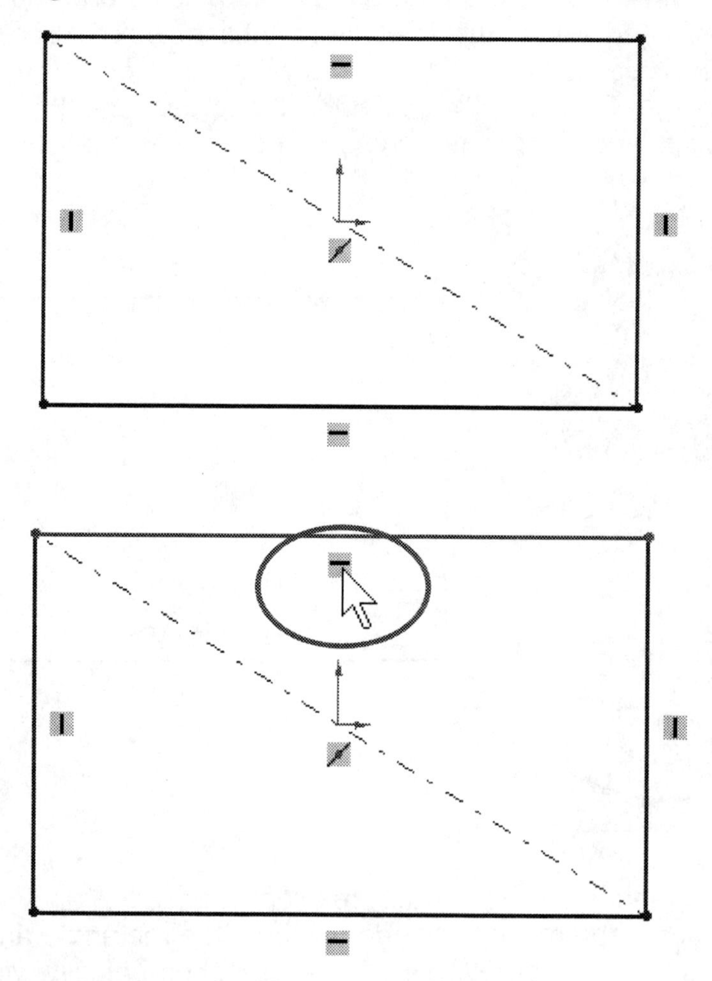

The **Sketch Origin**  defines the local Horizontal (Short red arrow) and local Vertical (Long red arrow) directions, this is important because we may be looking at the part in a different orientation, and vertical may not necessarily mean "up" on the screen. In SolidWorks we can add the following types of geometric relations between sketch entities:

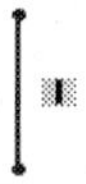	**Vertical** with respect to the sketch vertical direction (Long red arrow in the origin)
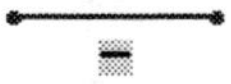	**Horizontal** with respect to the sketch horizontal direction (Short red arrow in the sketch origin)
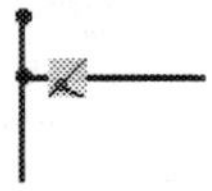	**Coincident** is when an endpoint touches another line, endpoint, edge or circle.
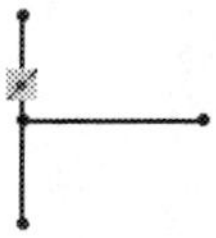	**Midpoint** is when a line endpoint coincides with the middle of another line. The Midpoint relation implies Coincident; it is not necessary to add both relations.
	**Parallel** is when two or more lines have the same inclination.
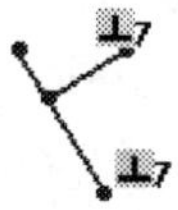	**Perpendicular** is when two lines are 90 degrees from each other.
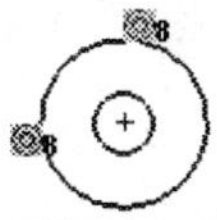	**Concentric** is when two arcs or circles share the same center. Concentric can be also between a point or endpoint and a circle.
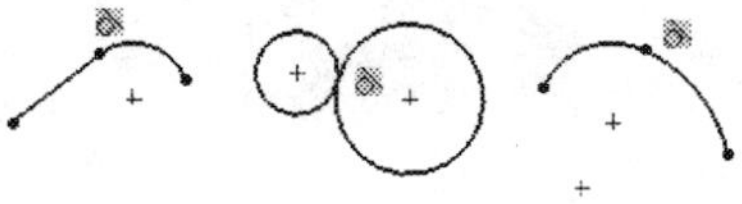	**Tangent** is when a line and an arc or circle, or two arcs or circles are tangent to each other.
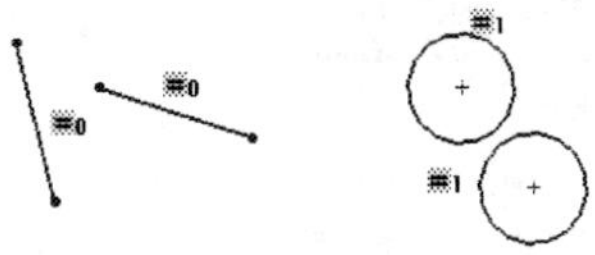	**Equal** is when two or more lines are the same length, or two or more arcs or circles have the same diameter.
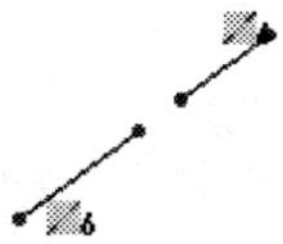	**Collinear** is when two or more lines lie on the same line.

13. - The next step is to add dimensions to the rectangle. Turn off the geometric relations display in the "View" menu "Sketch Relations" to avoid the visual clutter in the screen. Click with the Right Mouse Button in the graphics area and select

"**Smart Dimension**" or select the "Smart Dimension" icon from the Sketch Toolbar. Notice the cursor changes adding a small dimension icon next to it to indicate the Smart Dimension tool is selected.

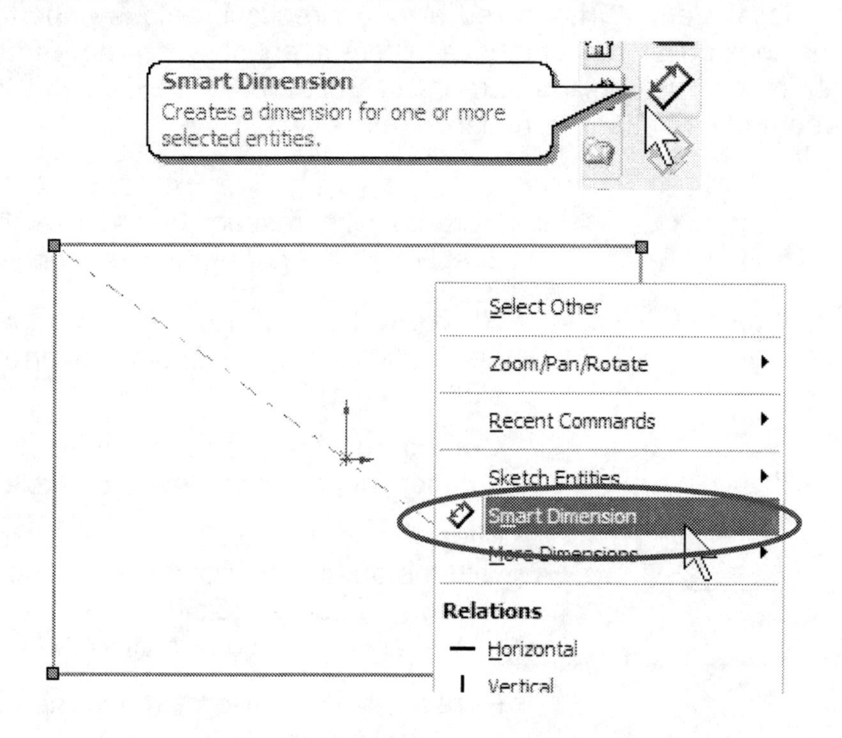

14. - Adding dimensions in SolidWorks is very simple. Click to select the right vertical line and then click just to the right to locate the dimension. SolidWorks will show the "Modify" dialog box, where we can add the 2.625" dimension. Repeat with the top horizontal line and add a 6" dimension.

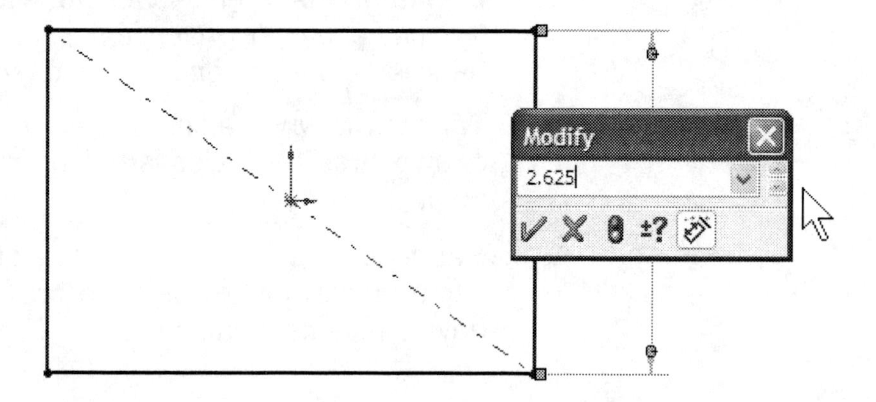

NOTE: View the Appendix if you need to change the **units** from millimeters to inches or vice versa.

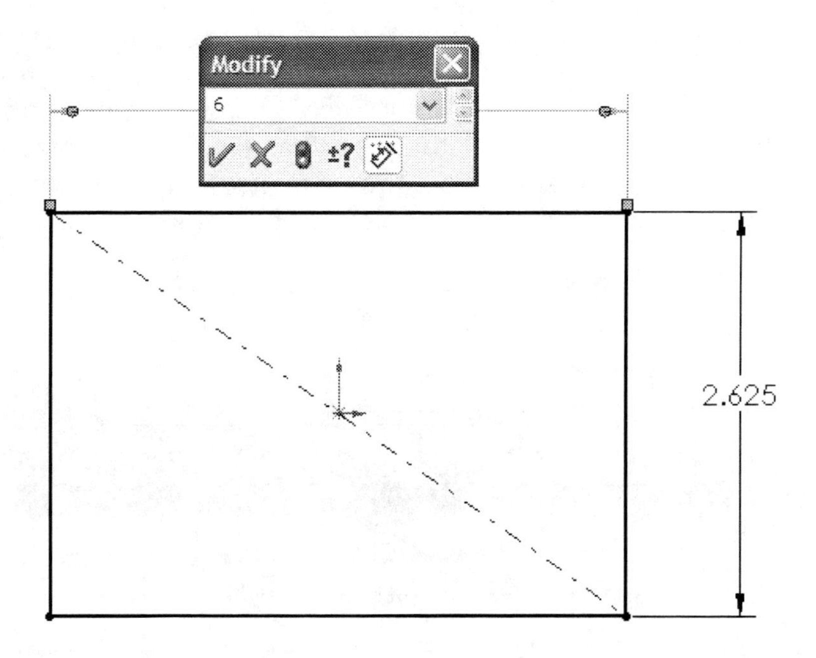

After dimensioning the lines, notice the lines change from Blue to Black. This is the way SolidWorks indicates that the geometry is defined, meaning that we have added enough information (dimensions and/or geometric relations) to define the geometry in the sketch. The status bar also shows "Fully Defined". This is the preferred state before creating a feature, since there is no information missing and the geometry can be accurately described.

A sketch can be in several states; the three main ones are:

- **Under Defined**: (BLUE) Not enough dimensions and/or geometric relations have been provided to define the sketch. Sketch geometry is blue and can be dragged with the mouse.
- **Fully Defined**: (BLACK) The Sketch has all the necessary dimensions and/or geometric relations to completely define it. This is the desired state. Fully defined geometry is black.
- **Over Defined**: (RED) Redundant and/or conflicting dimensions and/or geometric relations have been added to the sketch. If an over-defining dimension is added, SolidWorks will warn the user and offer not to add the dimension. If an over-defining geometric relation is added, delete it or use the "Edit" menu, "**Undo**" command or the "Undo" icon .

15. - Now that the sketch is fully defined, we will create the first feature of the housing; this is where we go from the 2D Sketch to a 3D feature. Select the "**Extrude**" icon from the Features Toolbar, and change the options indicated below to extrude it 0.25". Notice that the first time we create an extrusion, SolidWorks changes to an Isometric view, and gives us a preview of what the feature will look like when finished. To finish the command, select the OK button.

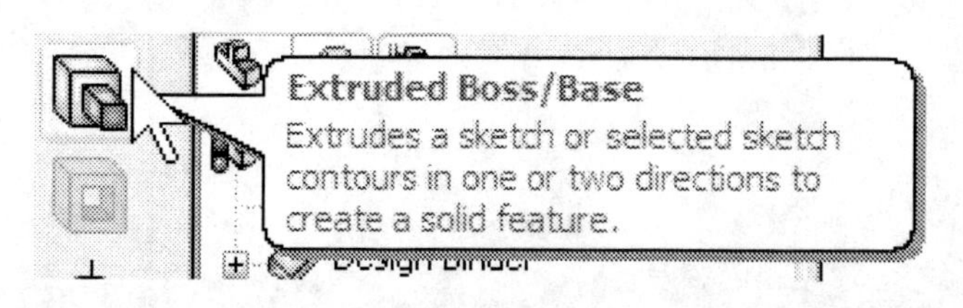

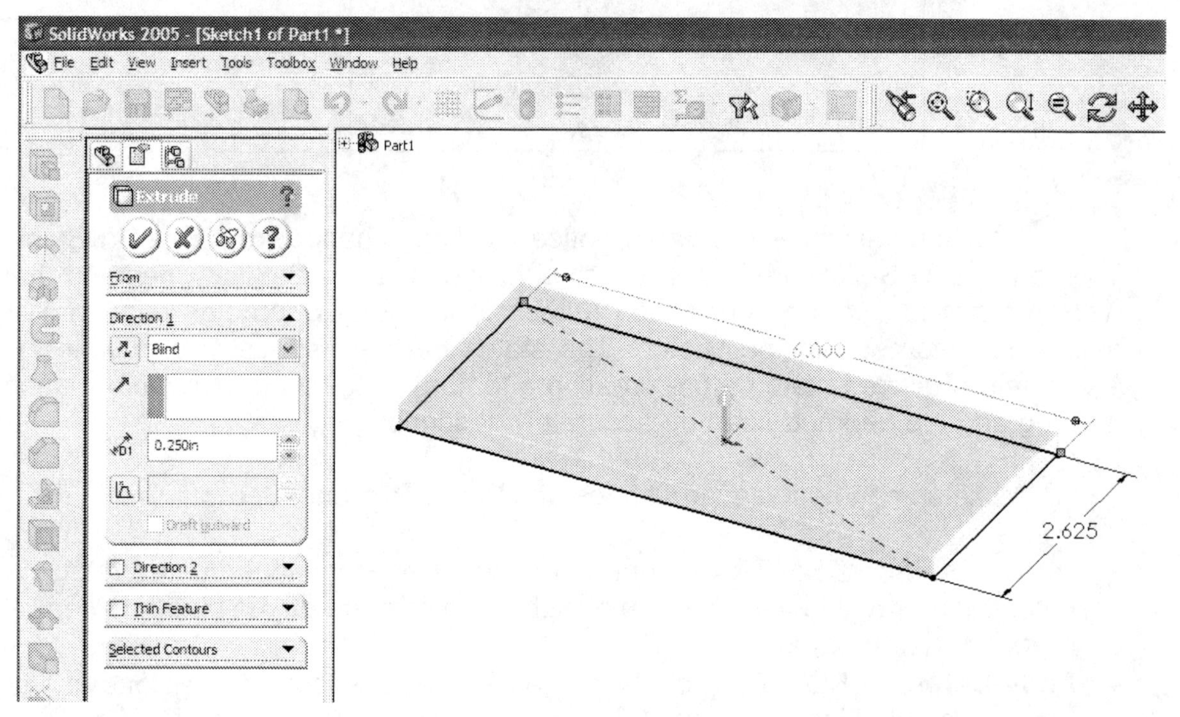

16. - After the first extrusion, notice that "Extrude1" has been added to the Feature Manager. The confirmation corner is no longer active. The status bar now reads "Editing Part" to alert us that we are now editing the part and not the sketch. If we expand the Extrude1 feature in the Feature Manager by clicking on the "+" next to it, we see that "Sketch1" has been absorbed by the "Extrude1" feature.

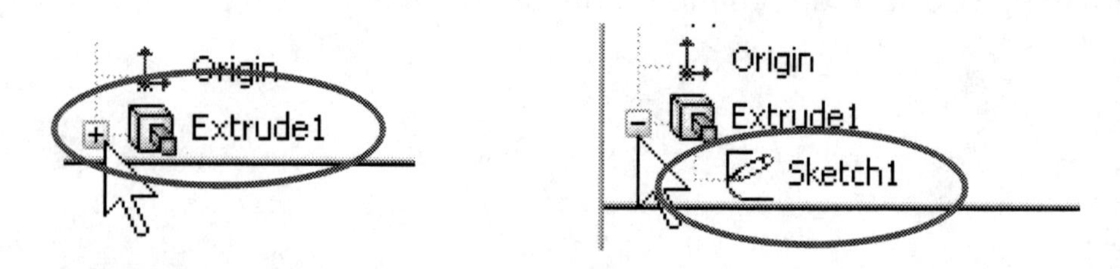

17. - The second feature will be similar to the first one but with different dimensions. To create the second extrusion, we need to make a new sketch. When we select the Sketch icon, SolidWorks gives us a yellow message asking us to select a Plane or a planar face. We'll select the top face of the previous extrusion for the next sketch.

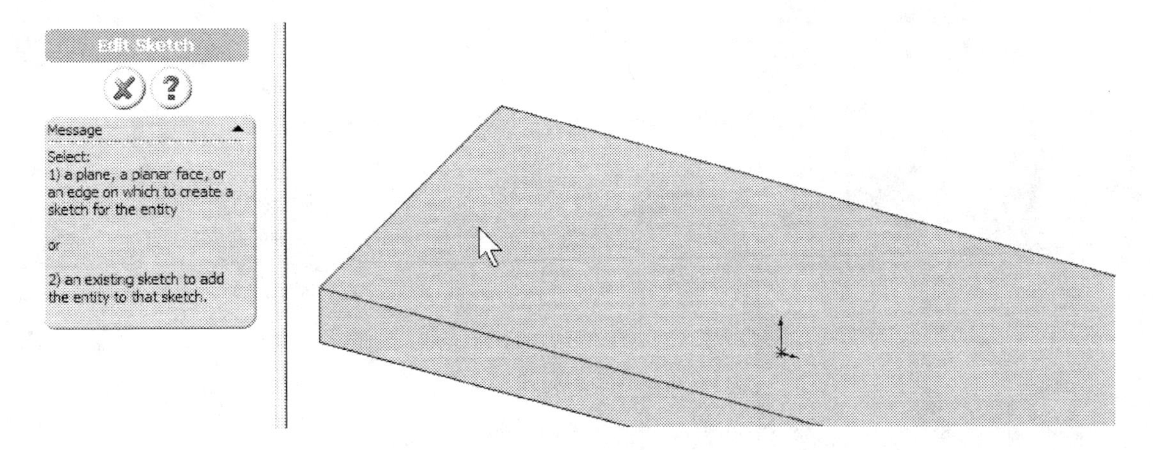

18. - To help us get oriented, we will switch to a **Top View** to see the part from the top. In SolidWorks the user is free to work in any orientation, as long as he/she is able to see what they are doing. Re-orienting the part helps the new user get used to 3D in a more familiar way by looking at it in 2D.

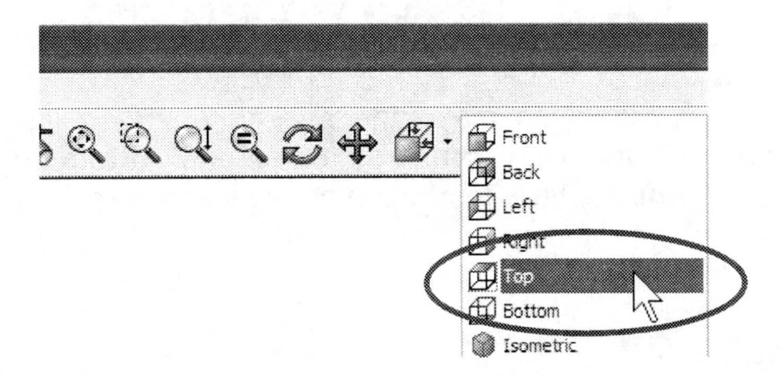

19. - For the second extrusion, repeat steps 6 through 14 and dimension the rectangle 4" wide as shown.

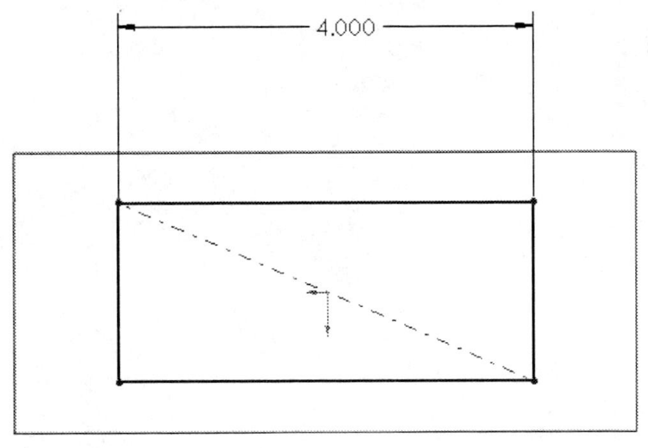

20. - To define the sketch, we'll add a **Colinear** relation between one of the blue lines and the edge of the previous extrusion. Select the "Add Relations" icon from the Sketch Toolbar and select the line and edge indicated. Then click in the "Collinear" relation and hit OK. This relation will fully define our sketch.

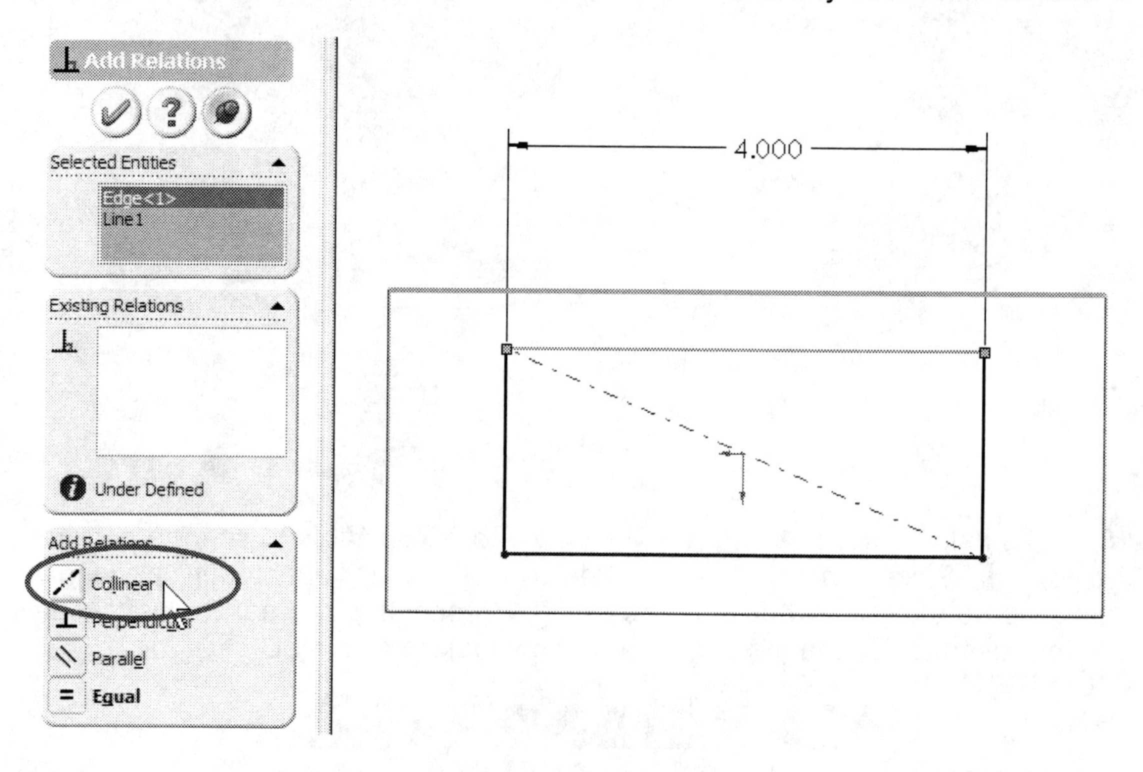

21. - We are now ready to make the second extruded feature. Select the Extrude icon as we did in step 15 and extrude 3.625". From the Standard Views icon, select the **Isometric** view to see the preview of the second extrusion.

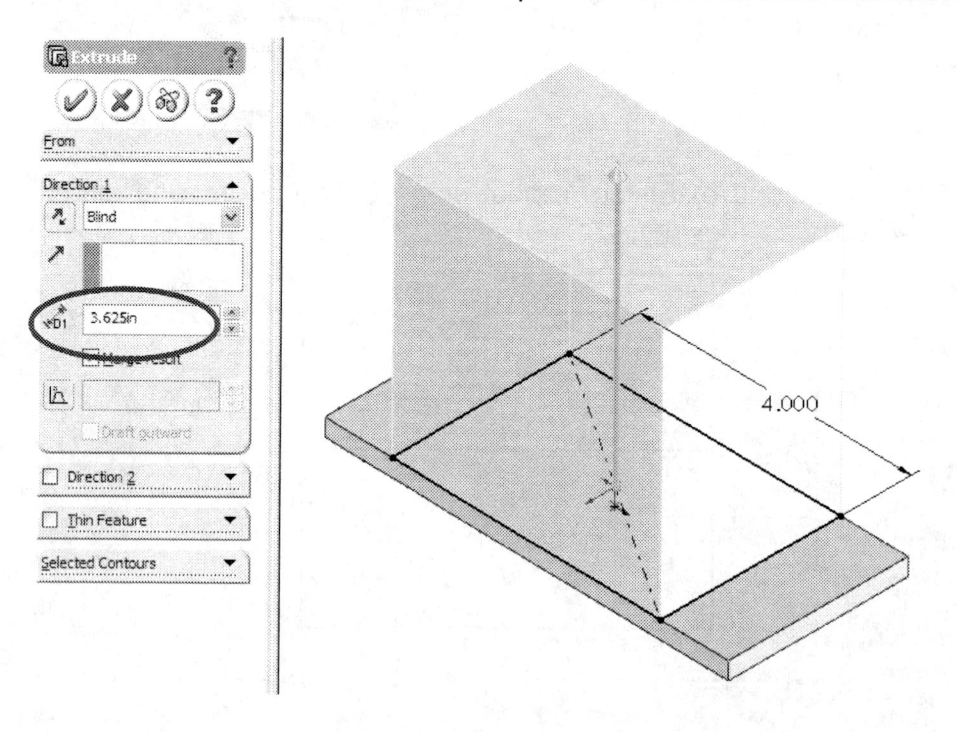

22. - The next step is to round the edges of the two extrusions. To do this, we will select the **Fillet** command. This is what's called an "Applied Feature"; we don't need a sketch to create it, and it's applied directly to the solid model. Select the "Fillet" icon from the Features Toolbar and add 0.25" fillets to the corners indicated. SolidWorks highlights the model edges when we place the cursor on top of them to let us know that we'll select them if we click on them. If an edge or face is mistakenly selected, simply click on it again to de-select it.

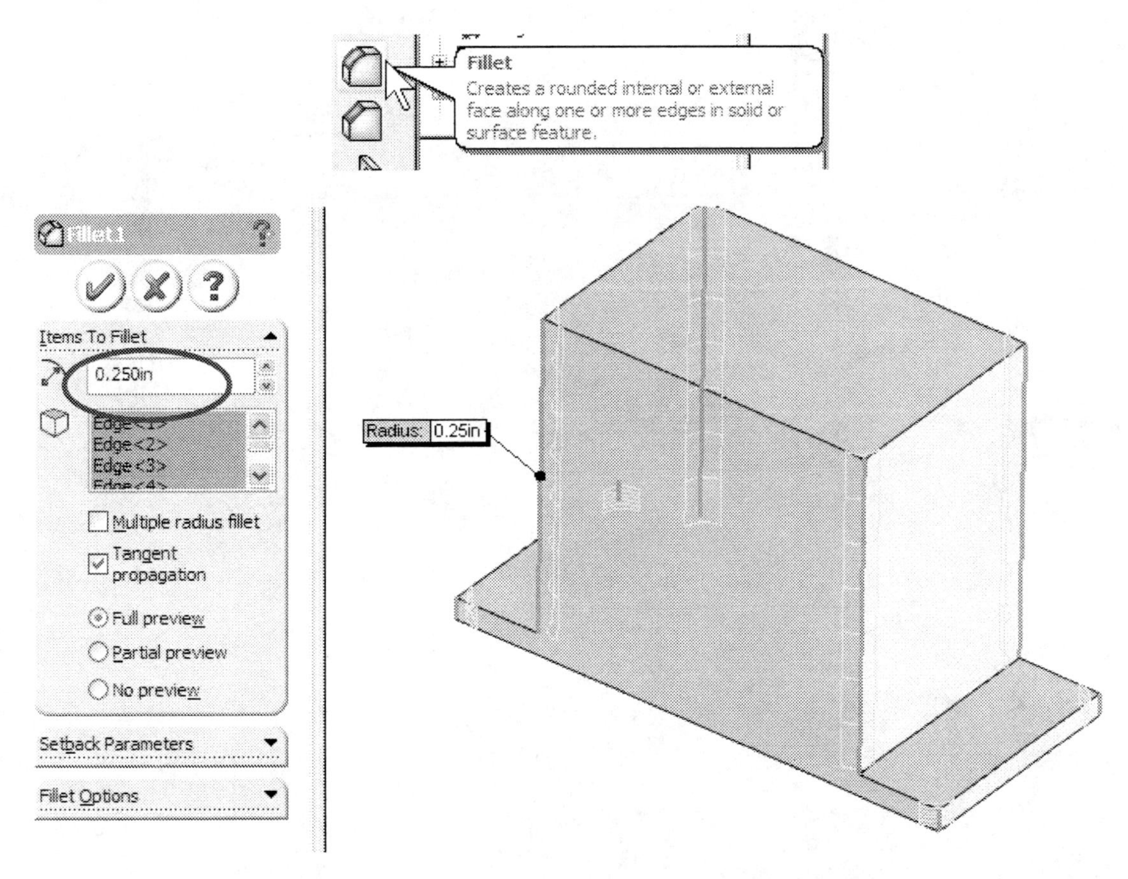

23. - Add a 0.125" fillet at the base of the Housing in the edges indicated in the next image.

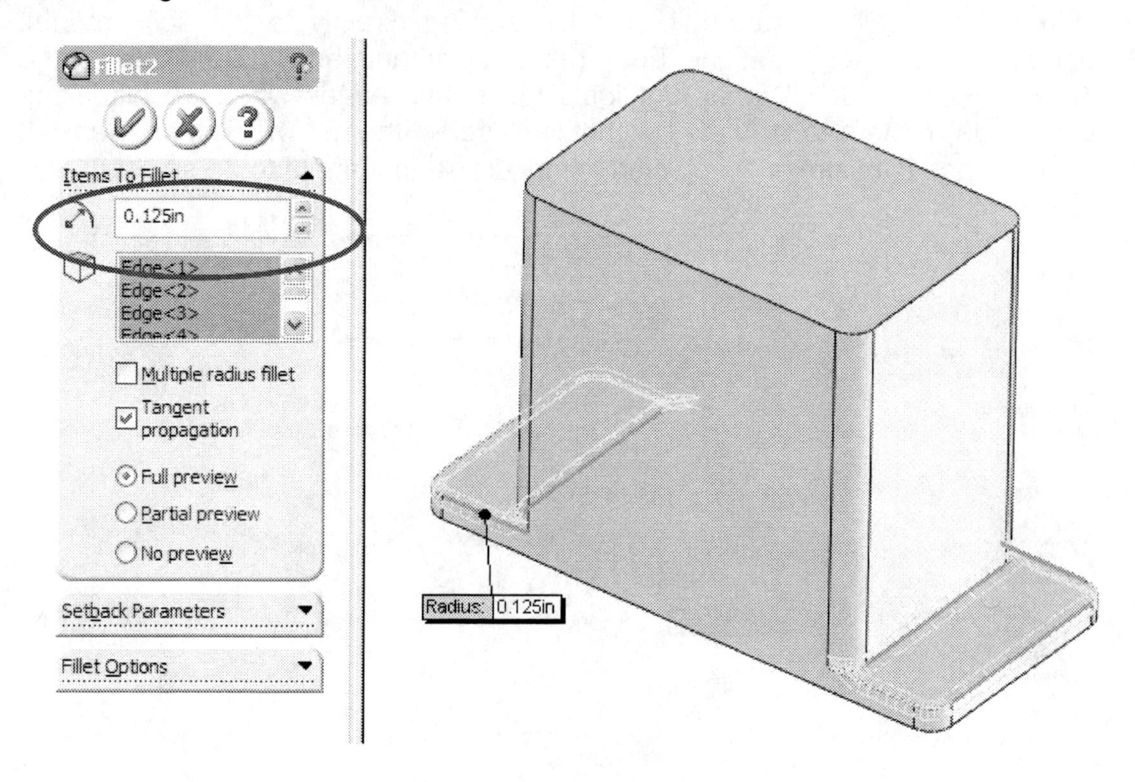

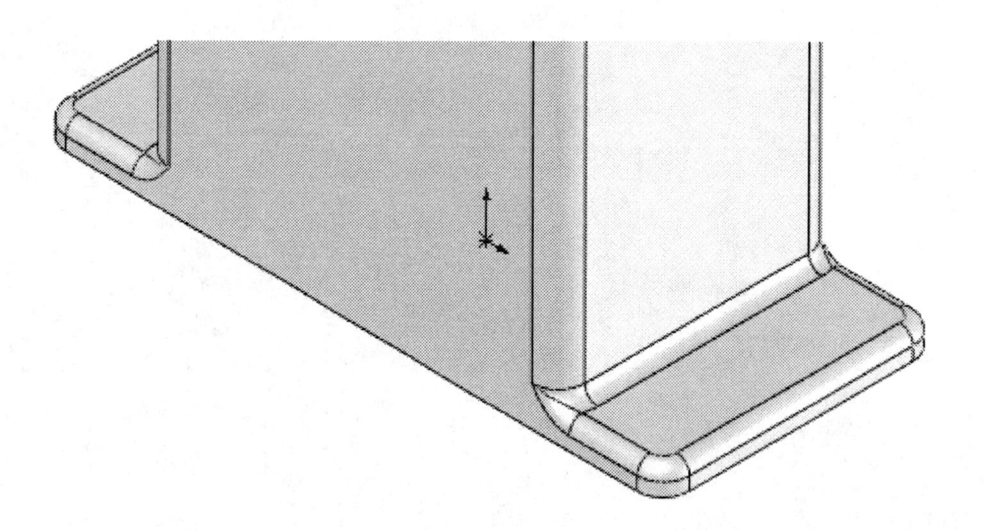

Finished fillets at the base.

24. - We will now remove material from the model using the "Cut Extrude" feature. Switch to a **Top View** and select the top face to create a new sketch as shown; this sketch will be the area to cut. Notice that we can dimension sketch geometry to existing model edges.

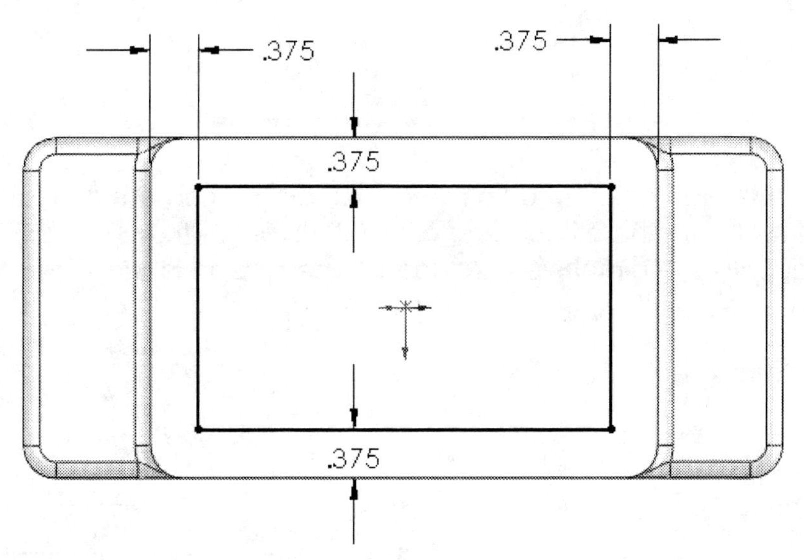

25. - If needed, switch to a "Hidden Lines Removed" mode to view the model without shading.

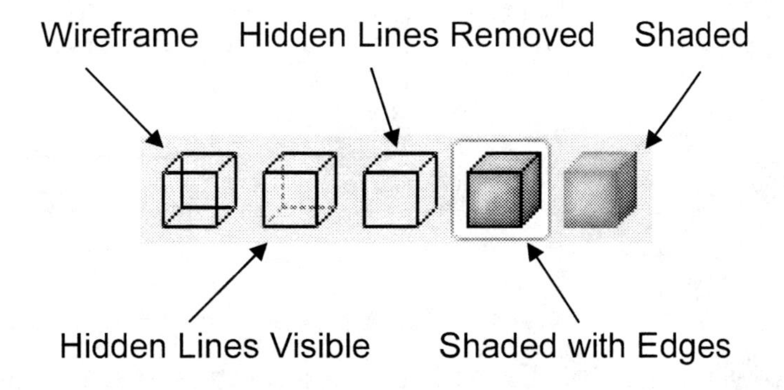

26. - In this feature, we will round the corners in the sketch using a **Sketch Fillet**. We can add the fillets as features, but in this step we chose to show you how to do it in the Sketch. Select the "Sketch Fillet" icon.

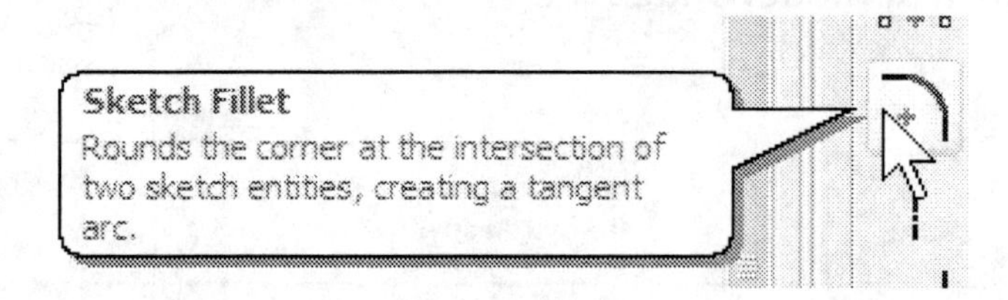

Set the fillet radius to 0.15", and click on the corners as indicated to round them. After clicking on all corners, click OK to finish the Sketch Fillet command. Notice that only one dimension is added. The reason is that SolidWorks adds an equal relation from all fillets to the one dimensioned.

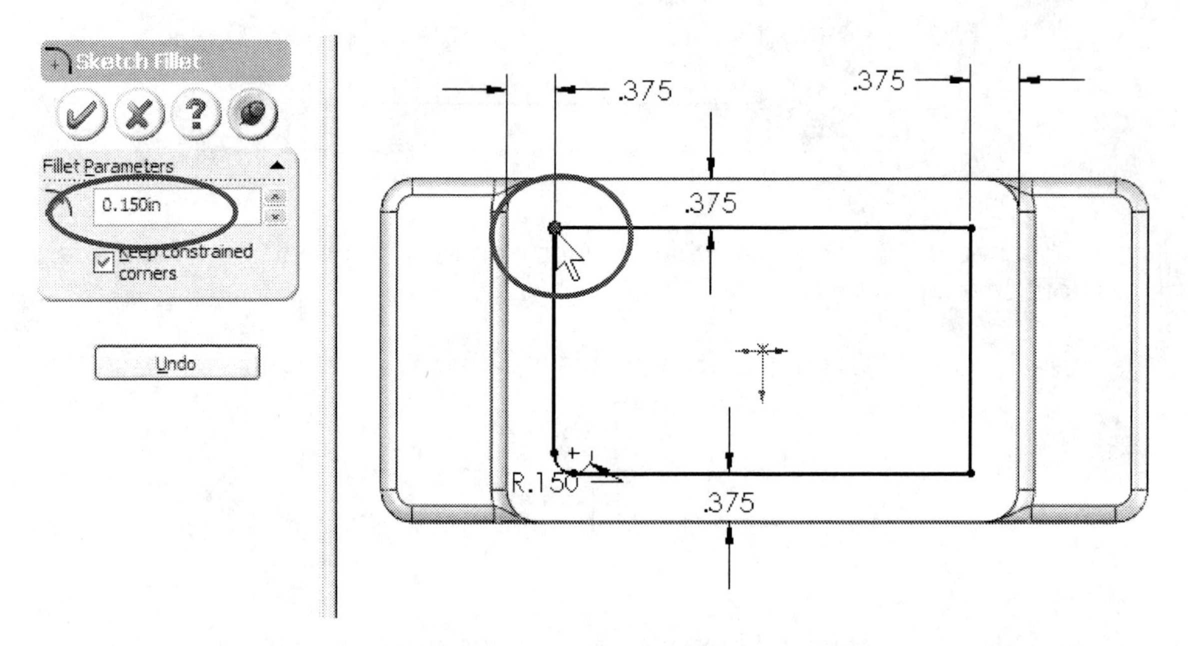

27. - Now we will select the **Extruded Cut** icon to remove material. Opposite to the Boss Extrude feature that adds material, the Cut feature, as it name implies, removes material.

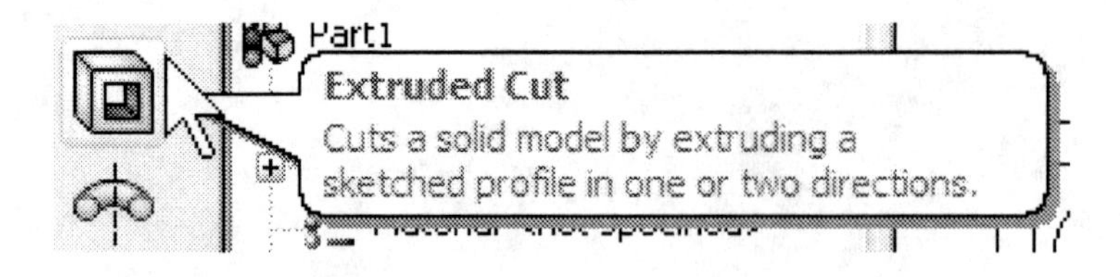

Select the options indicated and make the cut 3.5" deep.

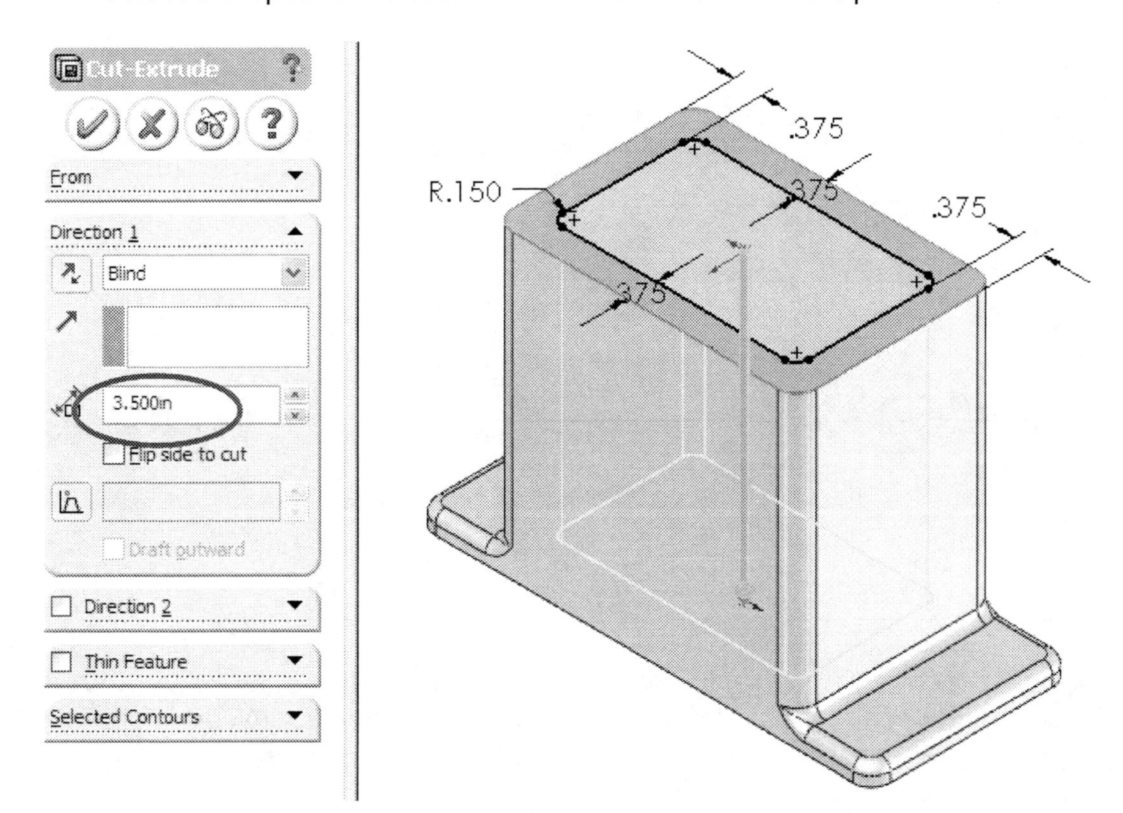

28. - In the next step we will add a simple round boss to the front of the Housing. Switch to a **Front View**, and make a sketch on the front face (we already know how to do this). Select the "Sketch" icon, and then click in the front most face. If you do it in this order, it will work as expected.

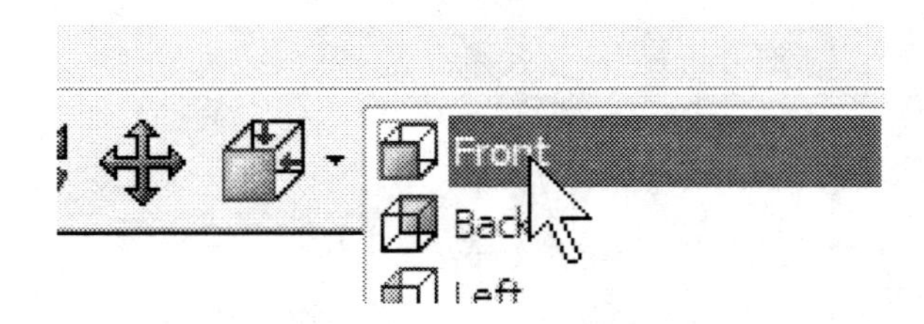

29. - Select the "**Circle**" sketch tool …

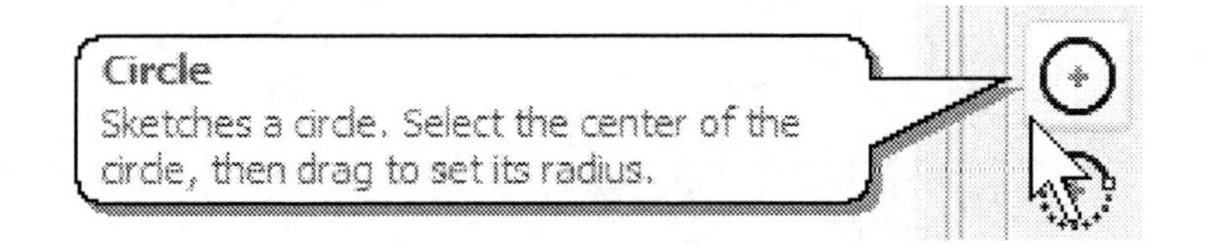

… and draw and dimension a circle as shown next.

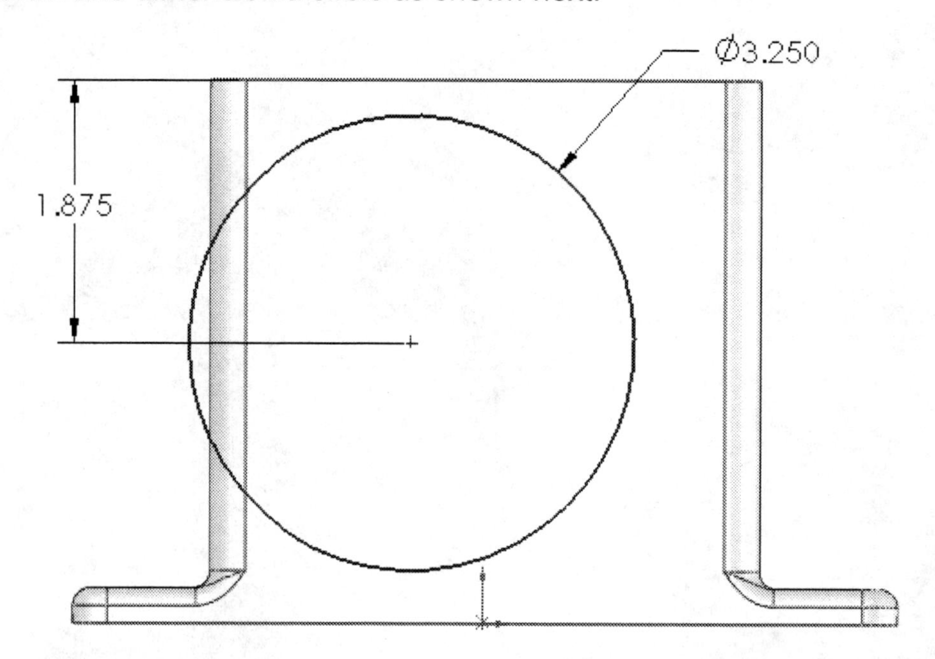

30. - To locate the circle in the center of the part, we will add a **Vertical Relation** between the center of the circle and the origin. SolidWorks allows us to align sketch elements to each other or to existing model geometry (edges, faces, vertices, etc.).

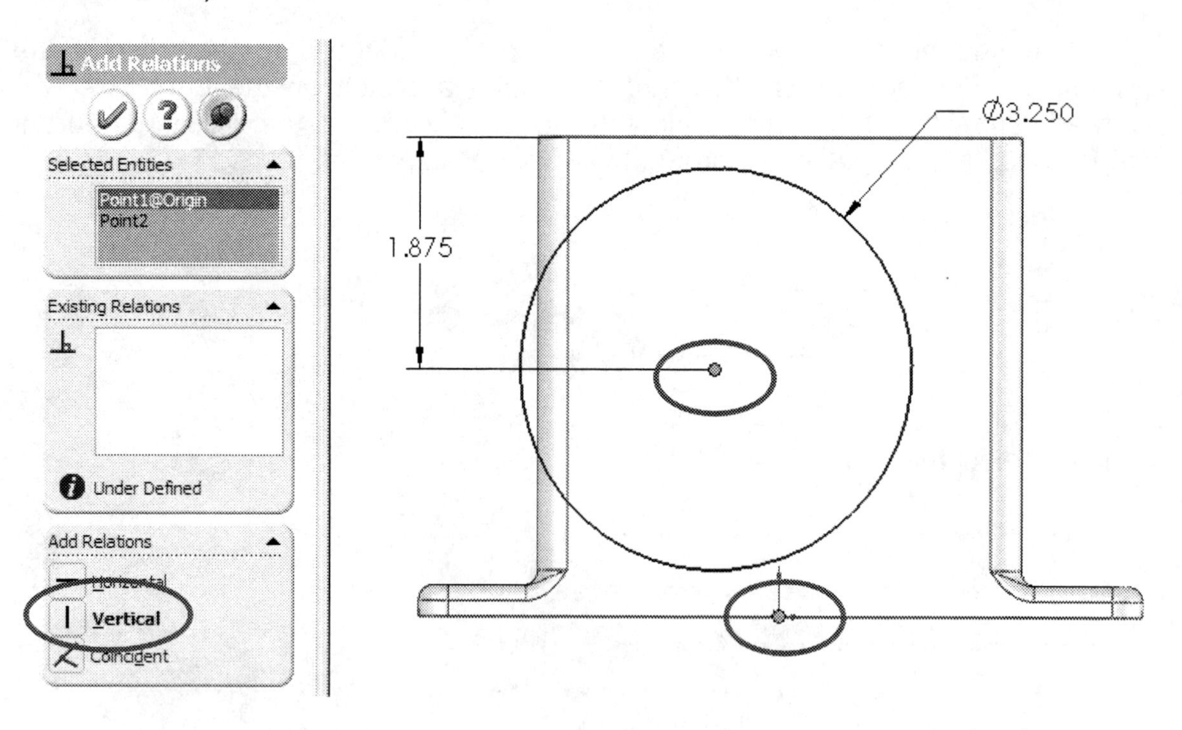

31. - When finished, extrude the Sketch 0.250". Notice the preview in the graphics area.

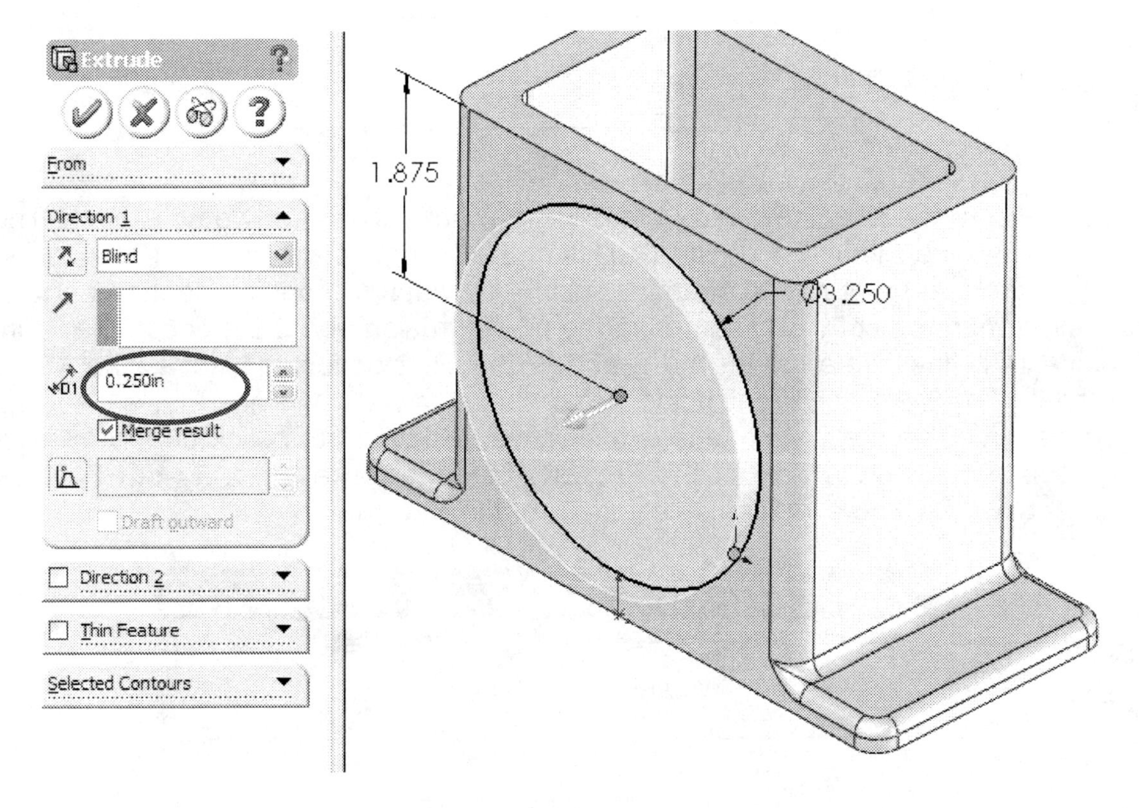

32. - Now we need to have an identical extrusion on the other side of the Housing, and to make it we will use the "**Mirror**" command to copy the one we just created. Switch to an Isometric view to help us visualize the Mirror preview and make sure we are getting what we expect.

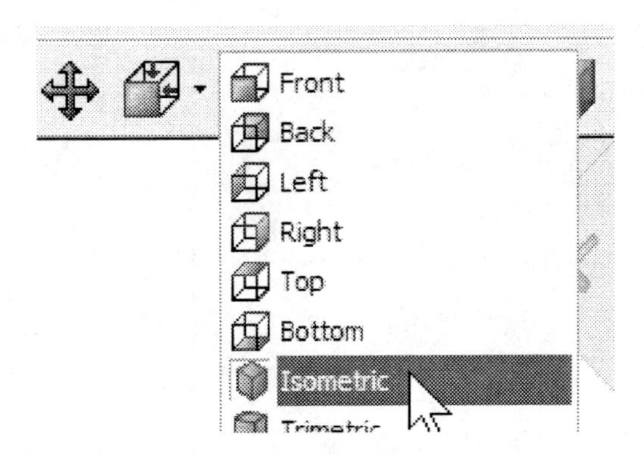

33. - Select the Mirror command from the Features Toolbar.

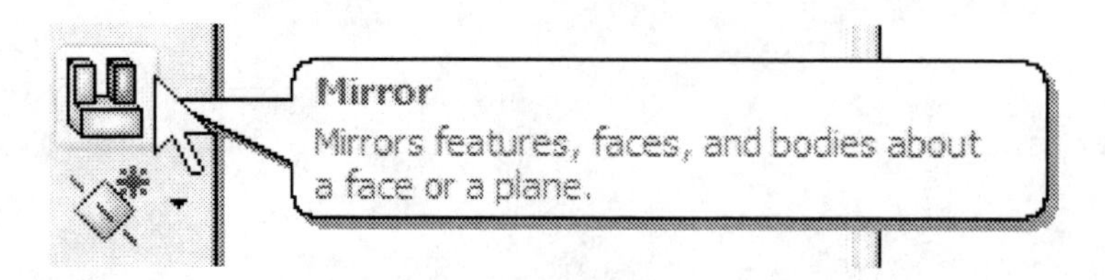

34. - From the Mirror Property Manager, we have to make two selections. The first one is the Mirror Face/Plane. This is the face or plane that will be used to mirror the feature. It has to be in the middle between the original feature and the desired mirrored copy. By centering the first extrusion about the origin, the Front plane is in the center of the part, and is the best option to use as the Mirror Plane. To select the Front plane, (make sure the Mirror Face/Plane selection box is pink; this means that this is the active selection box) click on the "+" sign next to "*Part1*" to the right of the Property Manager to reveal a **fly-out Feature Manager** from where we can select the Front Plane.

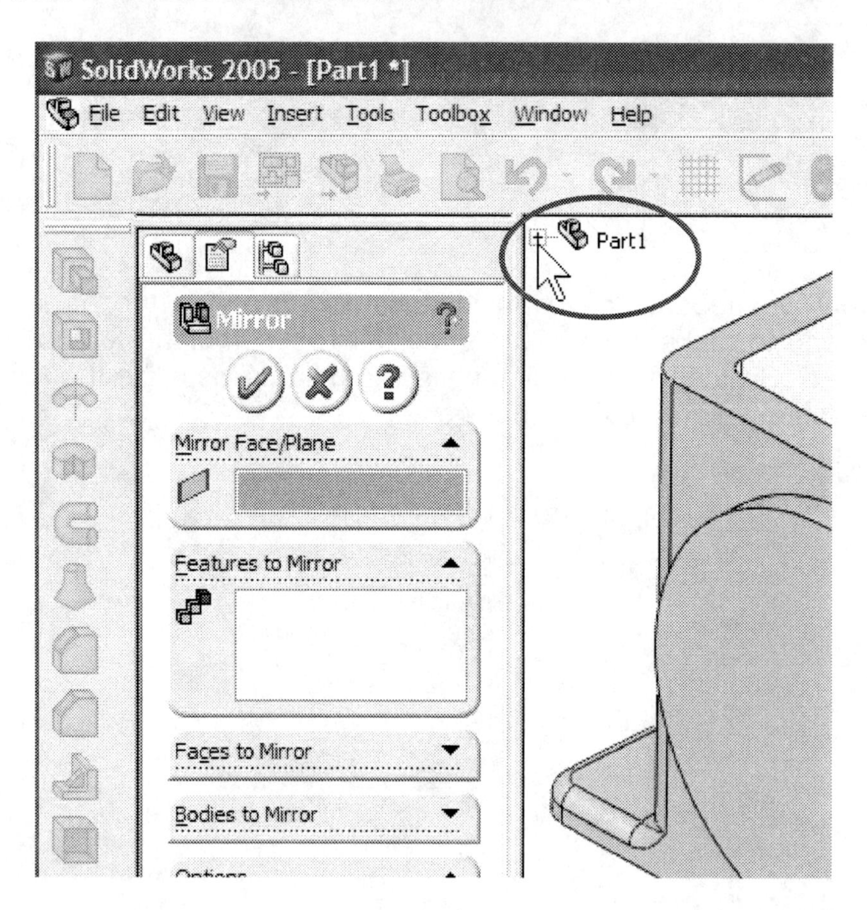

35. - After selecting the Front Plane from the fly-out Feature Manager, SolidWorks automatically activates the "Features to Mirror" selection box (now in pink), and if not already selected, select the last extrusion to mirror it. Notice the preview and click OK.

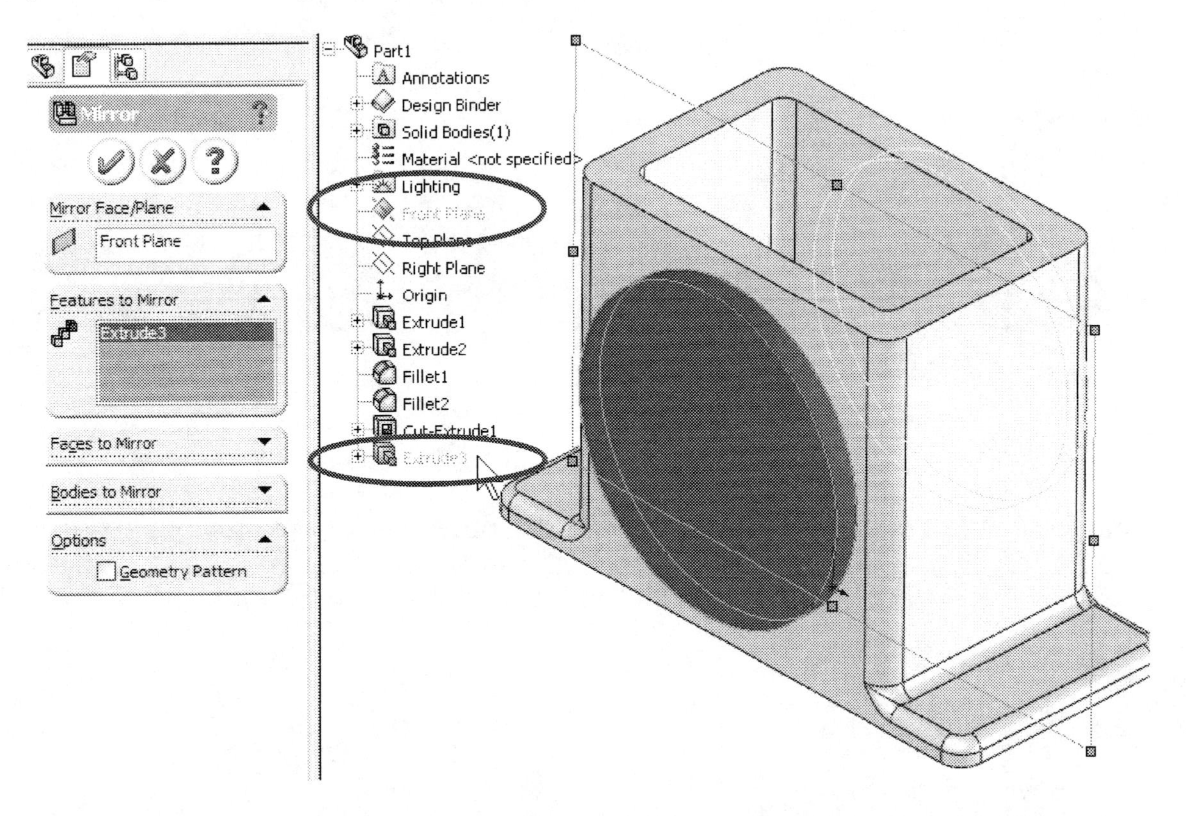

36. - In the next step we'll add the small boss at the right side of the Housing. Switch to a **Right View** and insert a sketch in the rightmost face as shown. Remember to add a Vertical Relation between the center of the circle and the origin as we did in step 30.

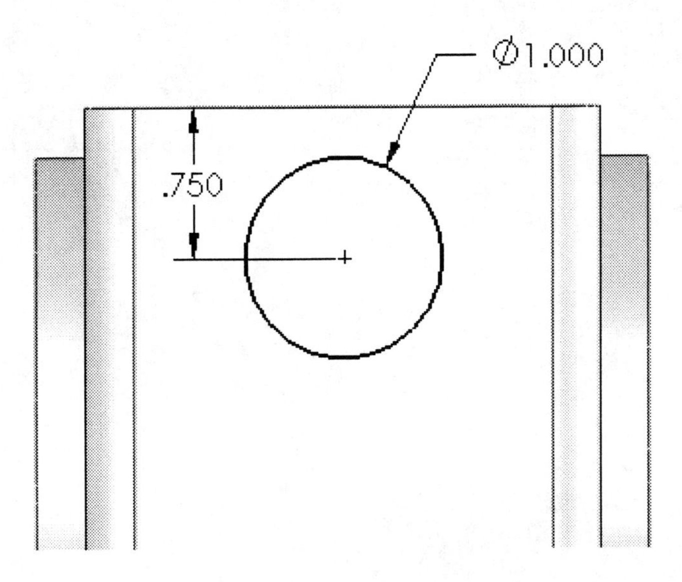

37. - Now we are ready to extrude the sketch 0.5" to make the side boss.

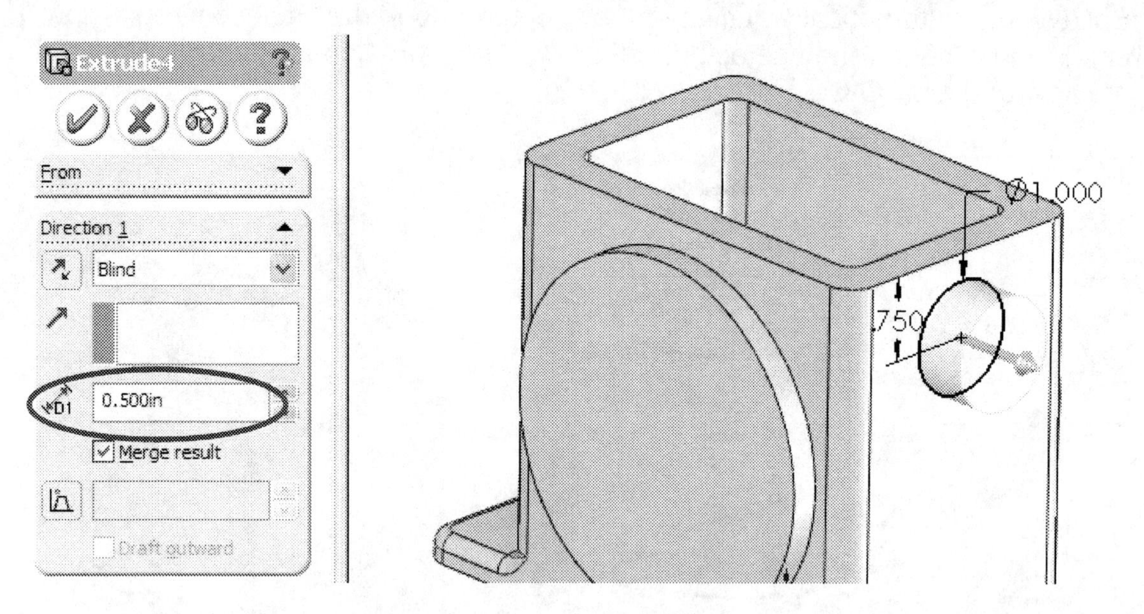

38. - Just like we did with the front circular boss, Mirror the previous extrusion about the Right Plane, which is also in the middle of the part. Use the fly-out Feature Manager to select the Right Plane and the previous Extrusion.

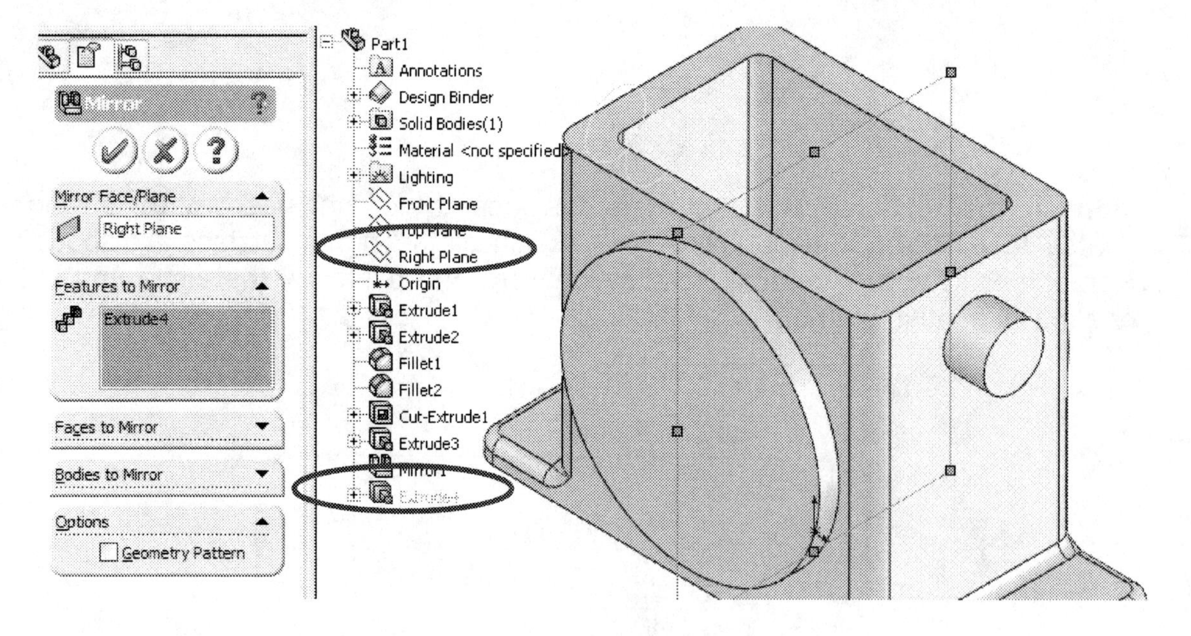

39. - We'll now make the cut in the front of the Housing. Switch to a Front View and create a sketch on the frontmost face. Draw a circle using the Circle tool and dimension it as shown. To center the circle about the circular extrusion, add a "**Concentric Relation**" selecting the circle and the round edge. Click OK to finish the "Add Relations" command.

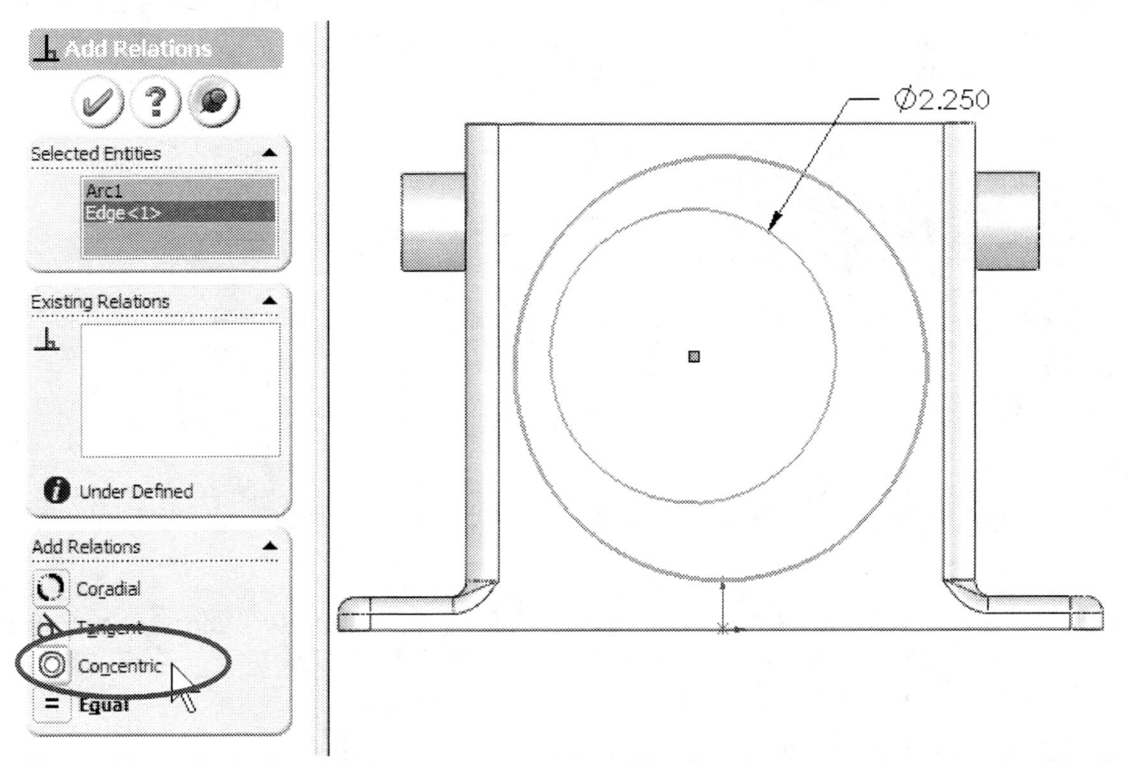

40. - Now that the circle is concentric with the boss, make a cut with the "**Through All**" option; this will make the cut go through the entire part regardless of its size. This way, if a parameter changes the width of the housing, the cut will still go all the way across the part.

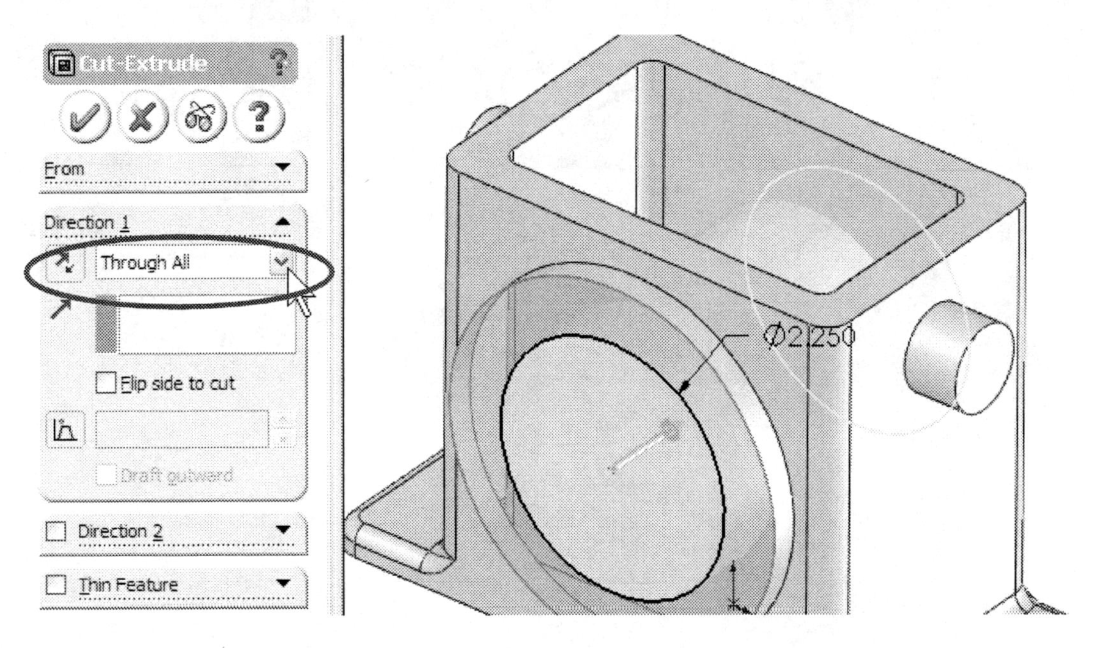

41. - We will now make a hole in the boss added in step 37 for a shaft. Switch to a right view and insert a sketch on the circular boss' face. We want a hole to be concentric with the boss. To do this we can add the circle and a concentric relation as we just did; however, this is a two step process. Instead, we will do it in one step: Select the "**Circle**" tool icon, and before drawing the circle, move the cursor and rest it on top of the circular edge as shown until the center of the circular edge is revealed. DO NOT CLICK ON THE EDGE. This highlight works only if you have a drawing tool active like Line, Circle, Arc, etc.

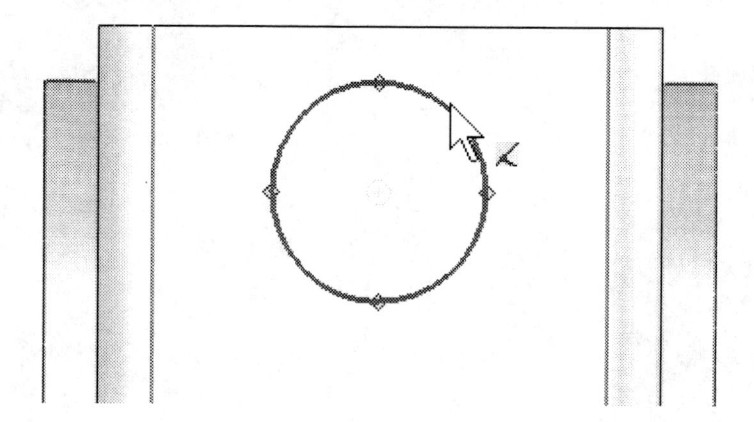

42. - Start drawing the circle at the center of the boss to automatically capture a concentric relation with the boss and dimension it 0.625" diameter.

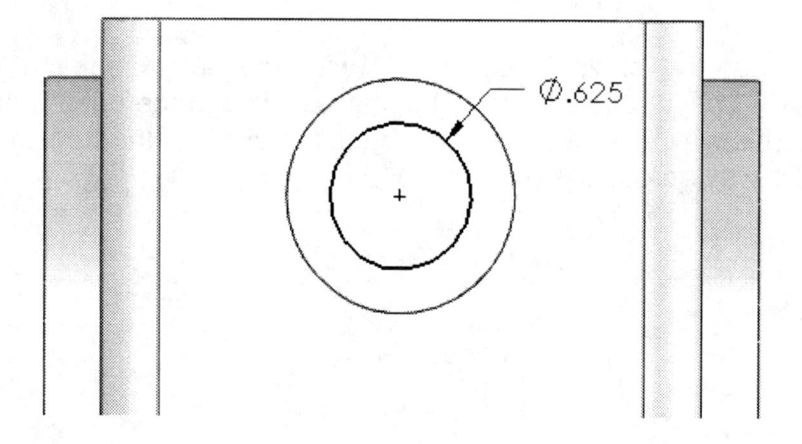

43. - Since this hole will be used for a shaft, we'll add a Bilateral **tolerance** to the dimension. Select the 0.625" dimension in the graphics area, and from the dimension's Property Manager, under "Tolerance/Precision" select "Bilateral". Now we can add the tolerances. Notice that the dimension changes immediately in the graphics area. This tolerance will be transferred to the Housing's drawing later on.

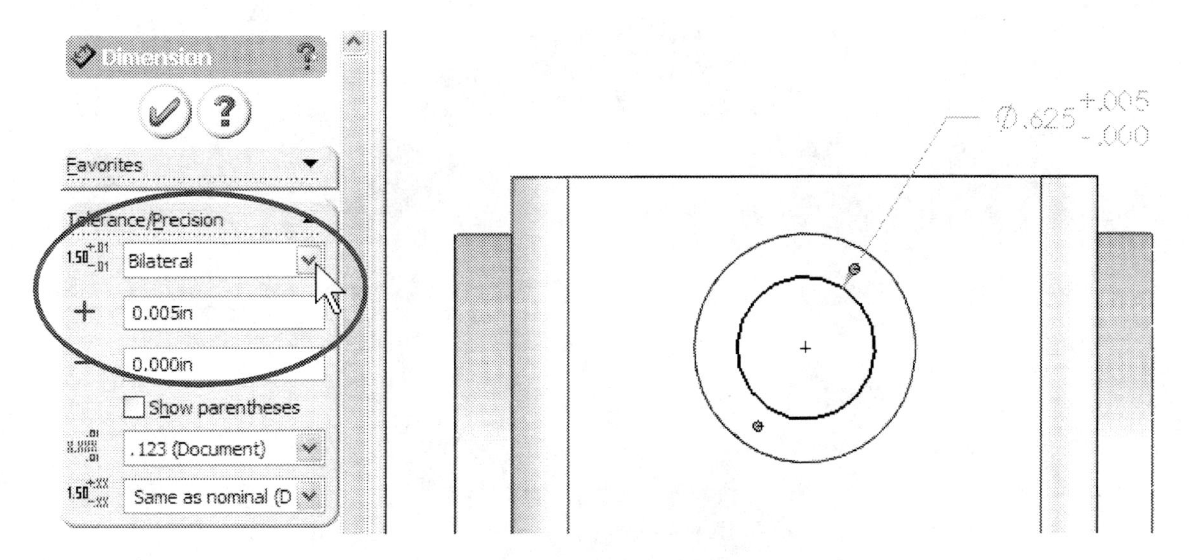

44. - Now make a Cut with the "Through All" option to complete the feature.

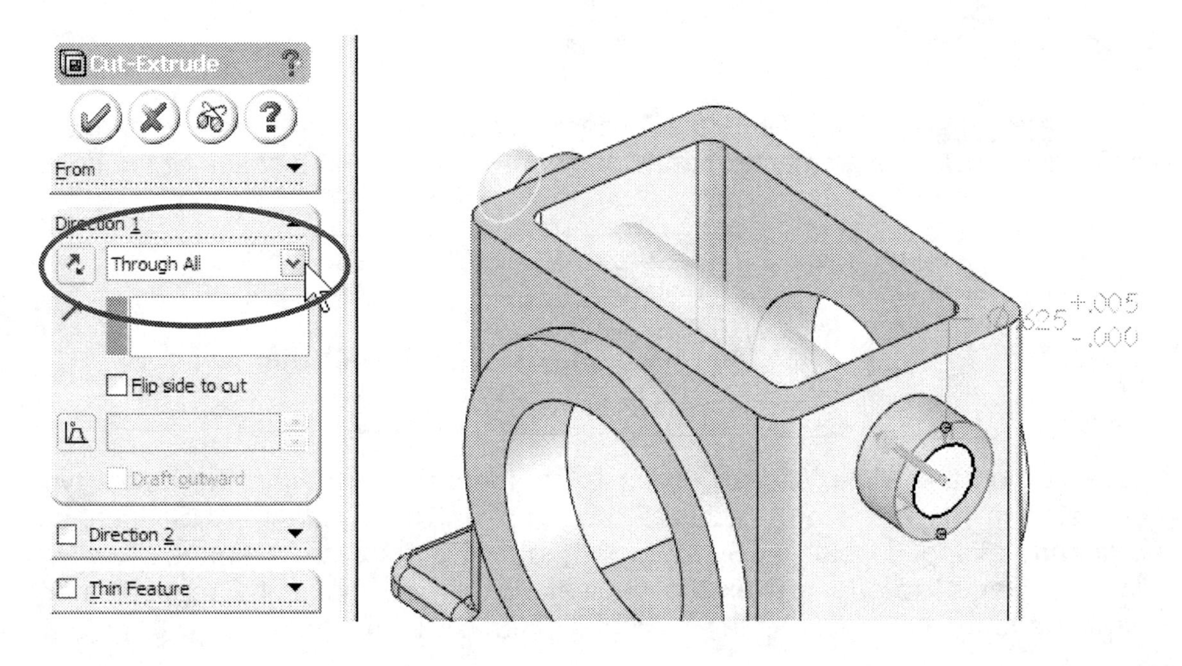

45. - For the next feature we'll make a ¼"-20 tapped hole in the front face. SolidWorks provides us with a tool to automate the creation of simple, Countersunk and Counterbore holes, tap and Pipe tap by selecting a fastener size, depth and location. The "**Hole Wizard**" command is a two step process: in the first step we define the hole type and size, and in the second step we define the location of the hole(s). To add the tapped hole, switch to a Front View.

First, select the frontmost face (where we want to put the tapped hole),

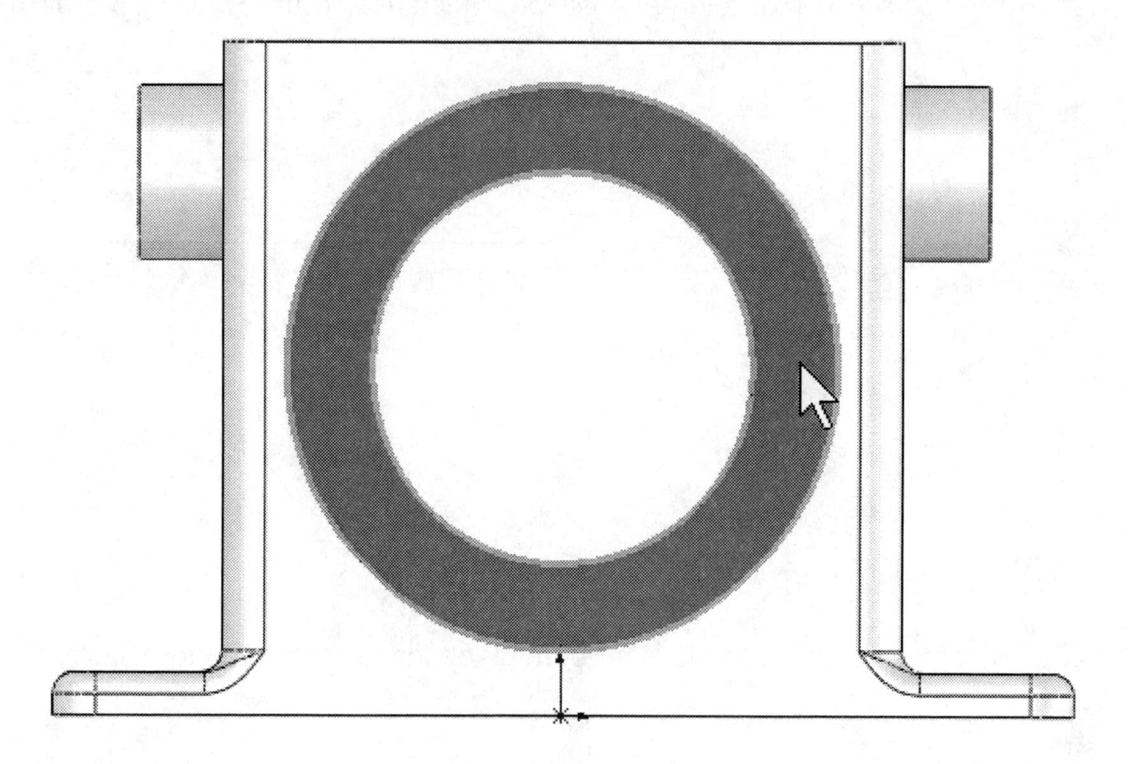

and then select the "Hole Wizard" icon. The order is very important; otherwise we'll have a different behavior when we define the location of the hole.

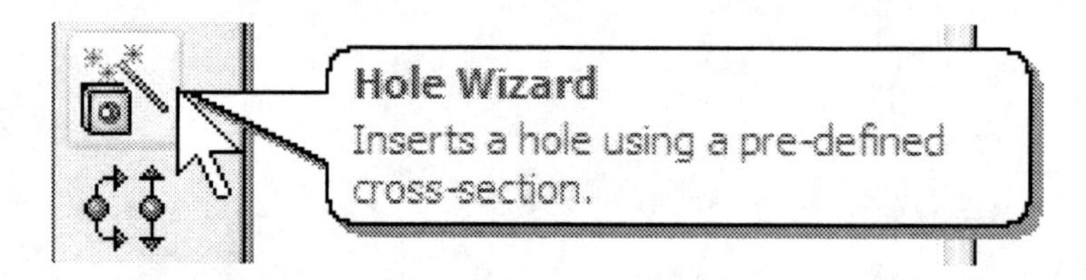

46. - When the "Hole Wizard" dialog is presented, we'll define the hole's type and size in the first step. Activate the "**Tap**" tab and select "ANSI Inch" for Standard, "Tapped Hole" for Screw type and "¼-20" for Size. Change "Tap Drill Type & Depth" to "Up to Next" as indicated, this will make the tapped hole's depth up to the next face the tap finds behind it. Click on "Next".

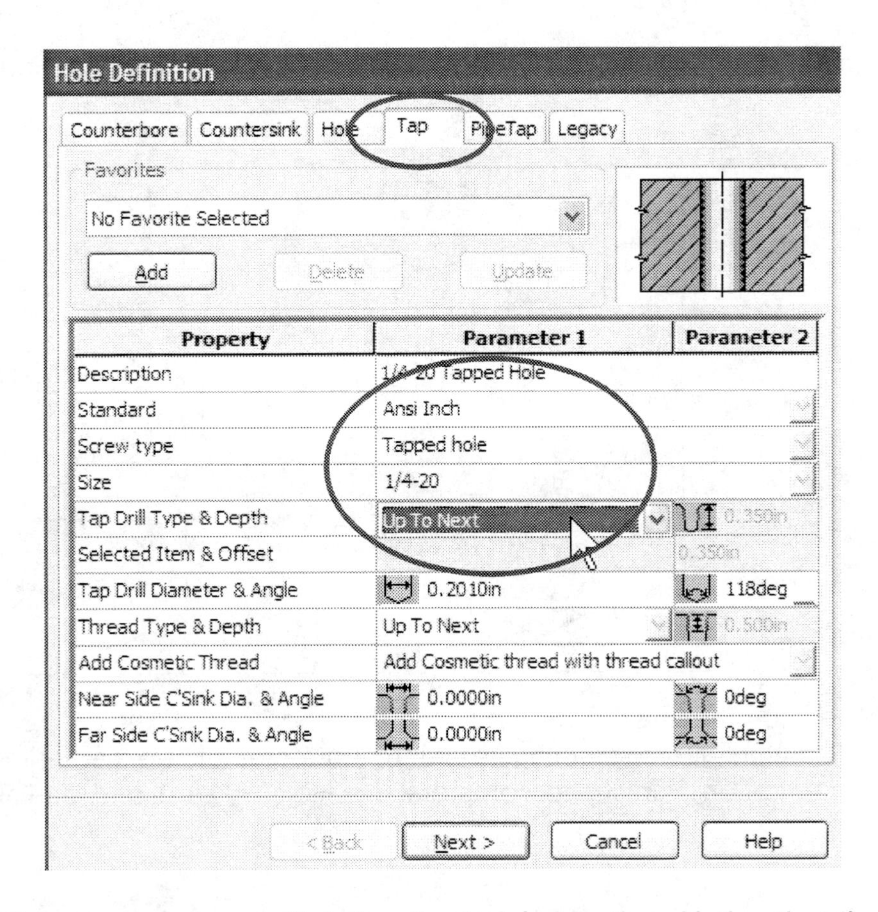

47. - In the second step we'll define the hole's location. Notice that the "**Sketch Point**" tool is active by default. By pre-selecting the face, SolidWorks automatically adds a point to the locating sketch; draw a centerline starting at the center of the circular extrusion…

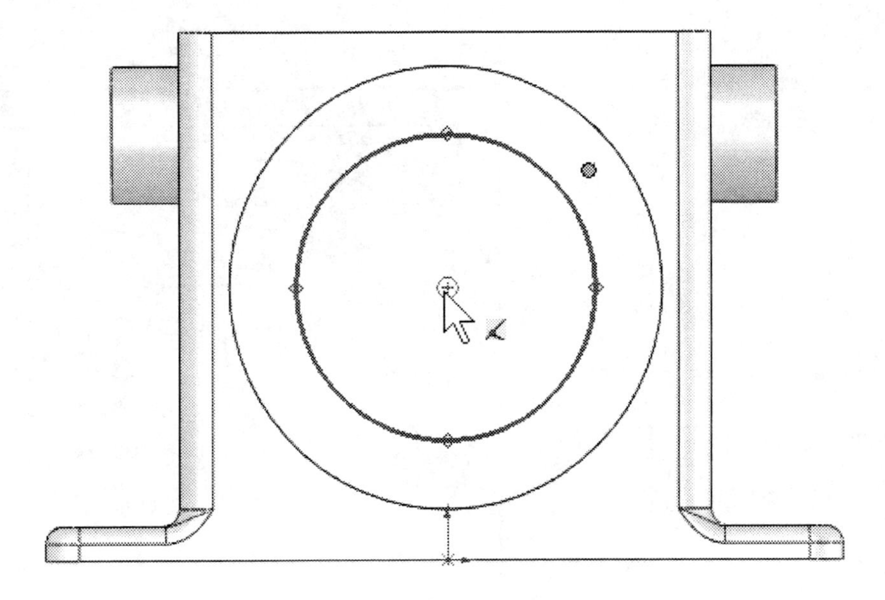

… and draw it to the right as shown. We'll use this centerline to define the hole's location.

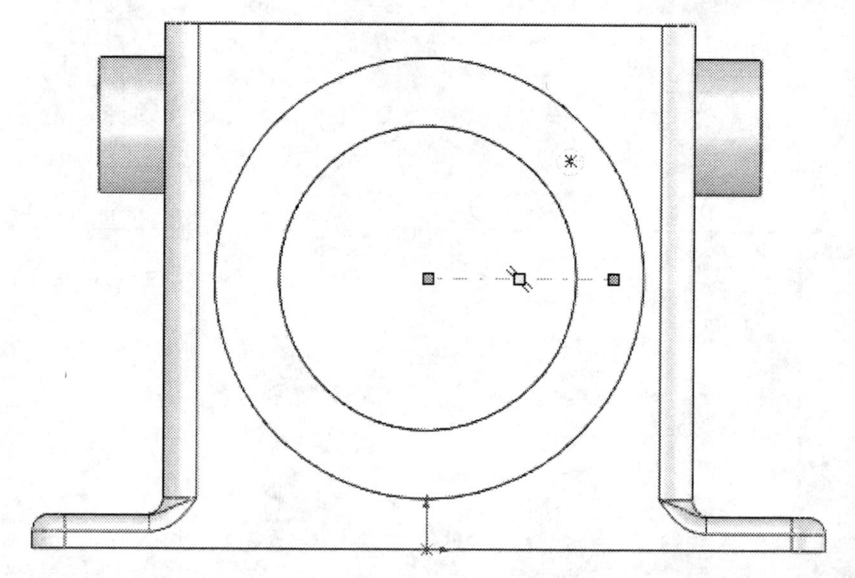

Dimension the line 1.375" long, and then add a "Coincident" relation using the "Add Relation" icon selecting the right endpoint of the centerline and the Point. Then click on "Finish" to complete the Hole Wizard and add the tapped hole.

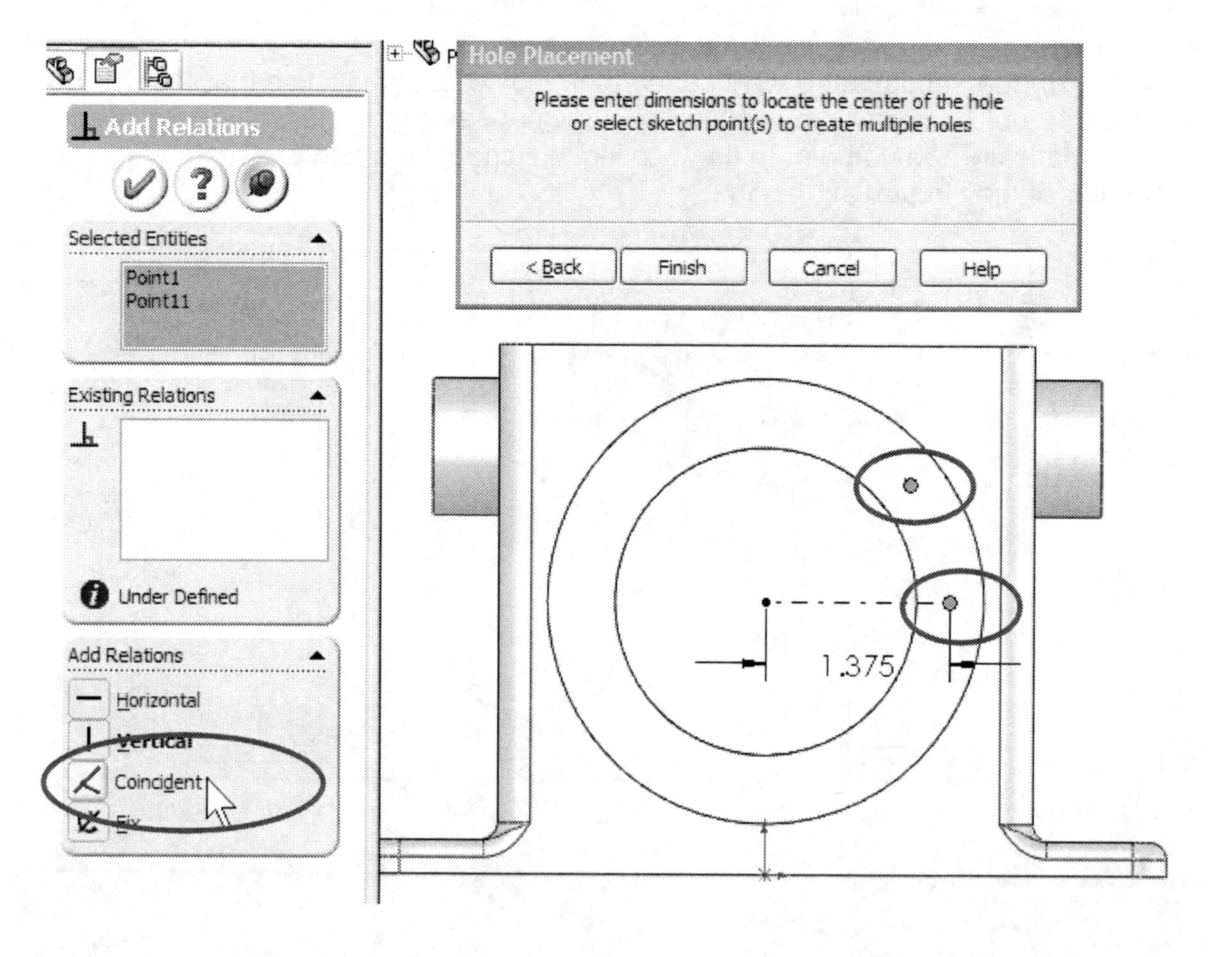

This is the finished "Hole Wizard" for the ¼"-20 Tapped Hole.

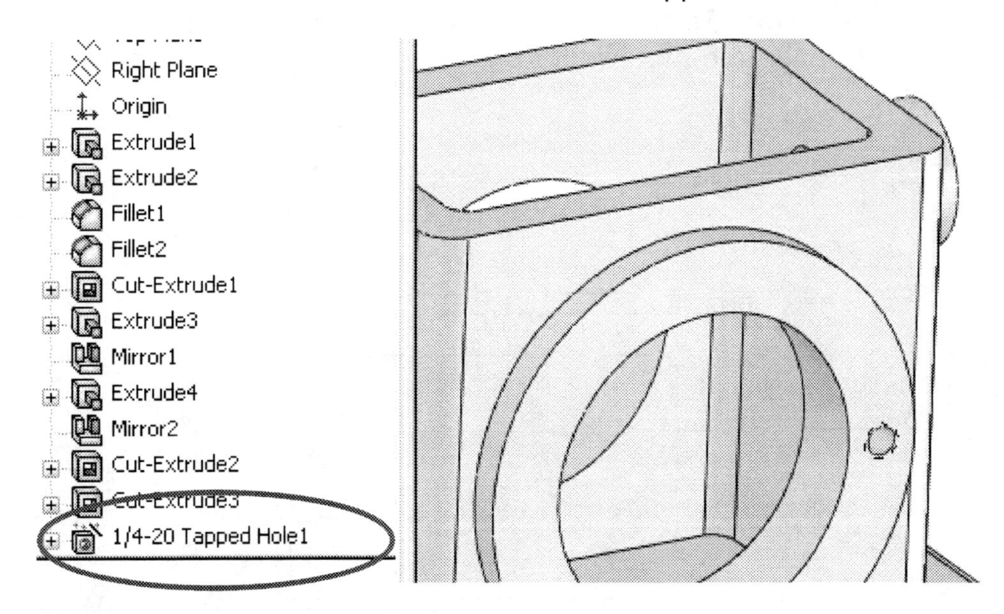

48. - After making the Tapped hole we decide that we want to make the walls of the Housing thinner, and need to make a change to our design. In order to do this, we find the feature that we need to modify in the Feature Manager, and make a click on it with the Right Mouse Button. From the pop-up menu, select "**Edit Sketch**".

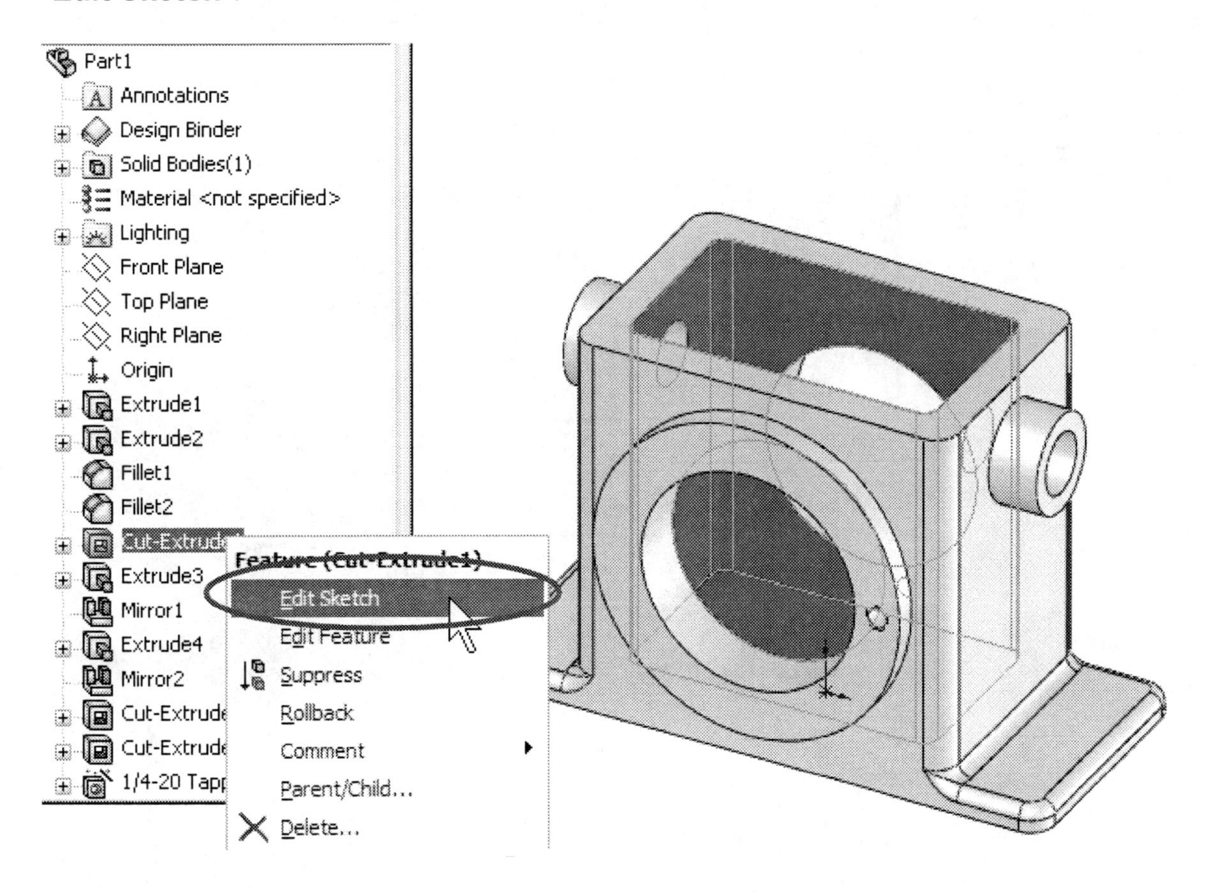

49. - We are now editing a previously made Sketch. Switch to a top view if needed for visualization. To change a dimension, double click on it to display the "Modify" box. Change the two dimensions indicated from 0.375" dimensions to 0.25" as shown.

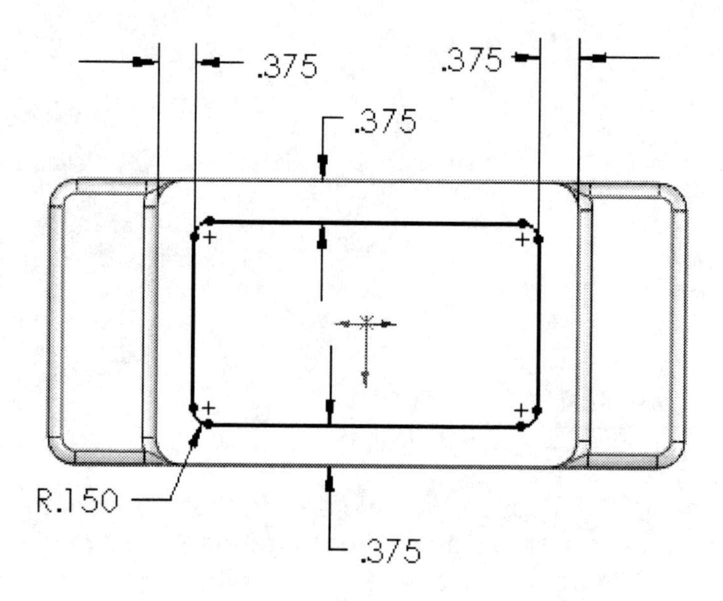

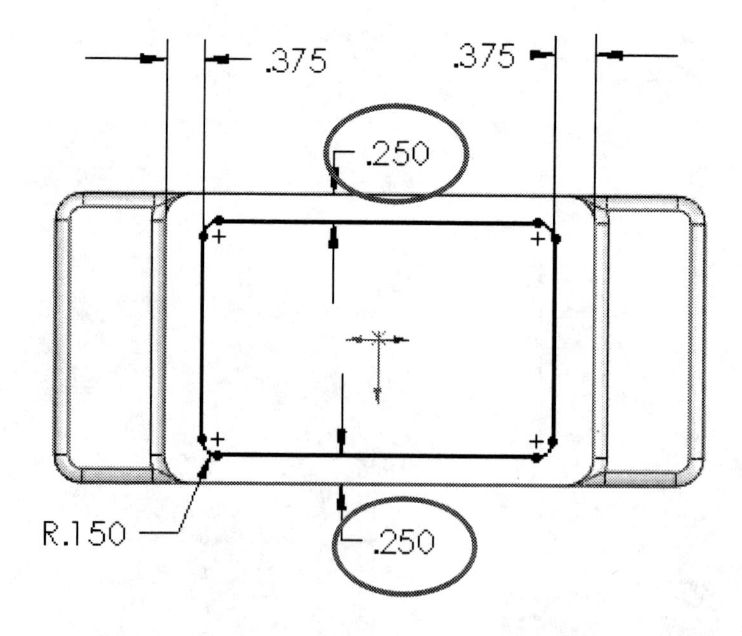

50. - After changing the sketch dimensions we cannot make a Cut Extrude, because we had already made a cut with that sketch; what we have to do is to "**Exit Sketch**" or "**Rebuild**" to update the model with the new dimension values. This step is only meant to show the reader how to go back to an existing feature's sketch and make changes to it if needed.

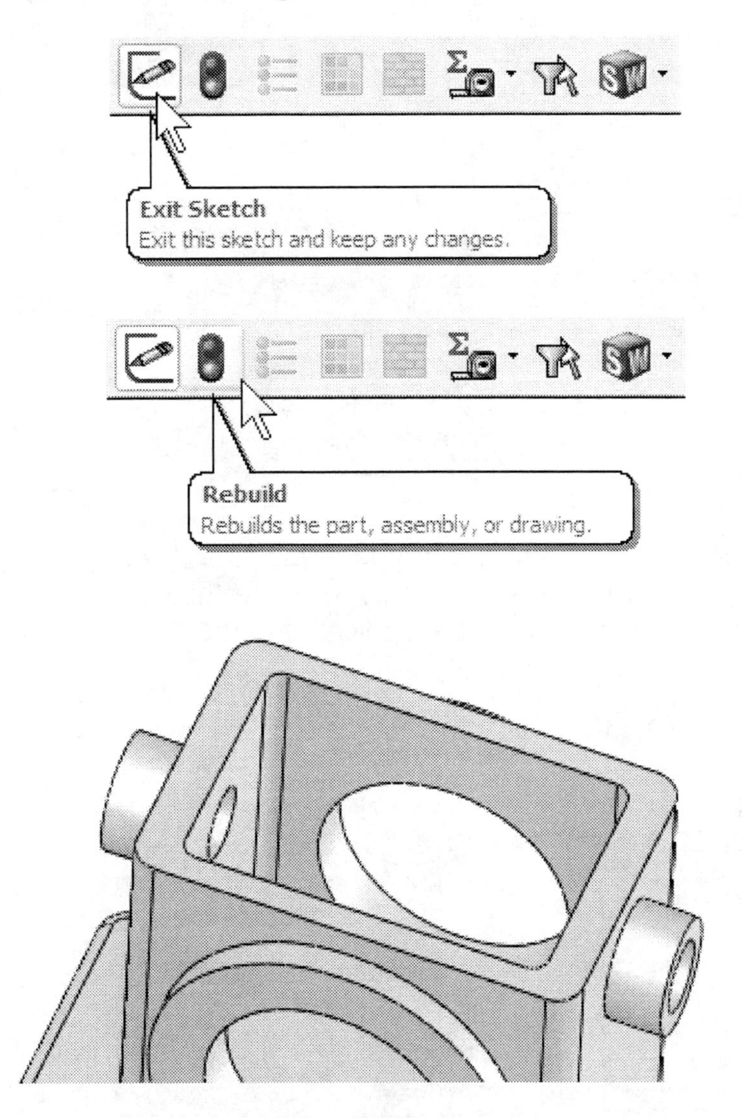

51. - Once the first tapped hole is made, we need to add five more to complete the flange mounting holes. To do this we will make a **Circular Pattern** to copy the first hole in a circular pattern. Select the Circular Pattern icon from the Features Toolbar.

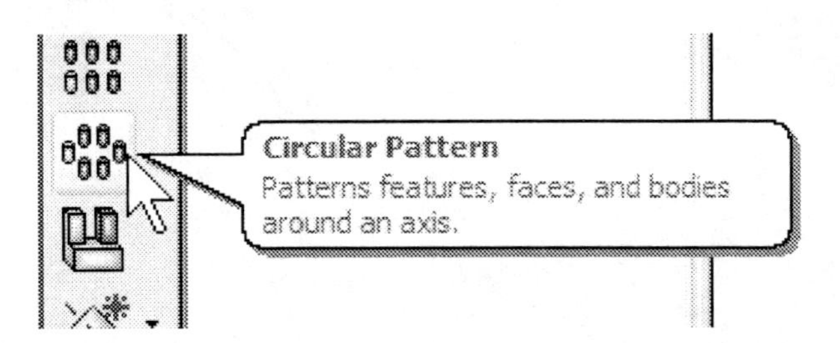

52. - The Circular Pattern requires a model edge or axis to make copies about it. SolidWorks automatically adds an axis to every cylindrical face of the part, which means that we already have axes that we can use. To make these axes visible, go to the "View" menu and select "**Temporary Axes**". (We can also control the visibility of other reference geometry from this menu).

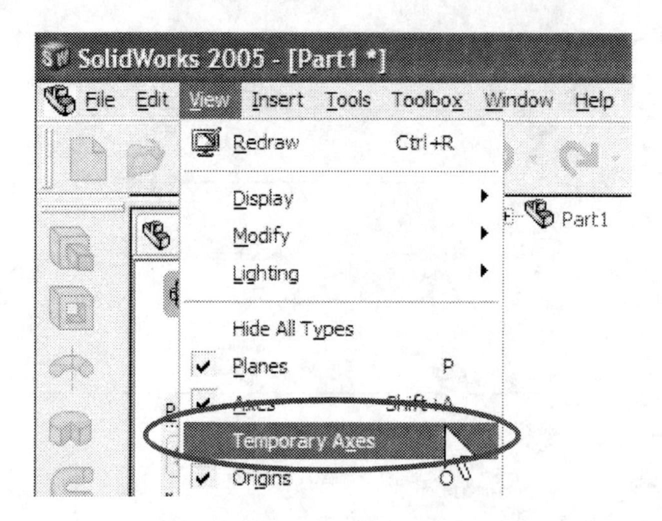

53. - In the Circular Pattern's Property Manager the "Parameters" selection box is active; this is where we select the axis to make the copies about. Select the axis in the center of the circular cut as indicated in the next image.

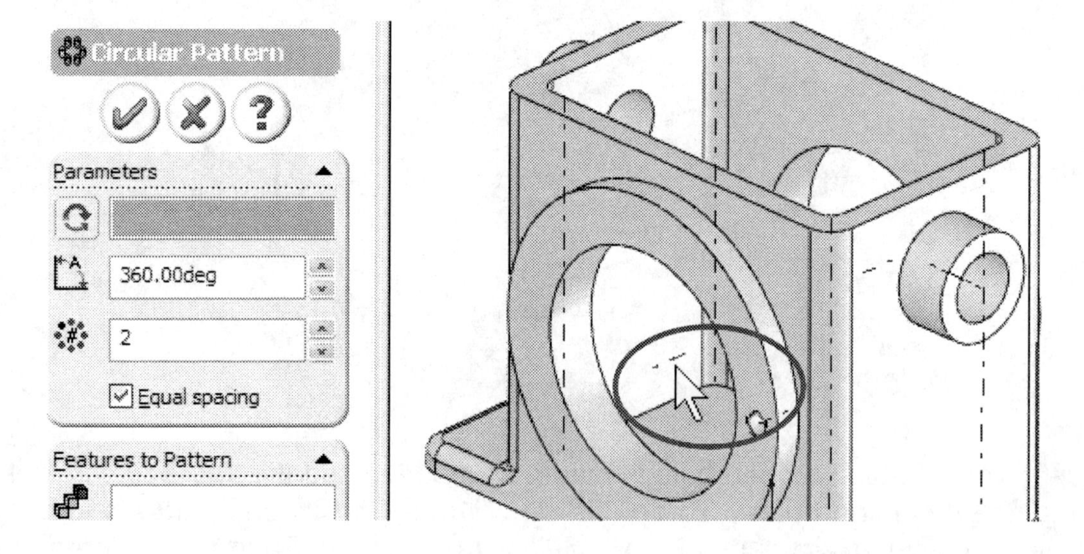

54. - Now click inside the "Features to Pattern" selection box to activate it. Select the "¼-20 Tapped Hole1" feature from the fly-out Feature Manager, change the number of copies to six (this count includes the original!) and make sure the "Equal spacing" option is selected to equally space the copies in 360 degrees. Notice the preview in the graphics area and click OK to finish the command. Now we can turn off the Temporary Axes the same way we turned it on.

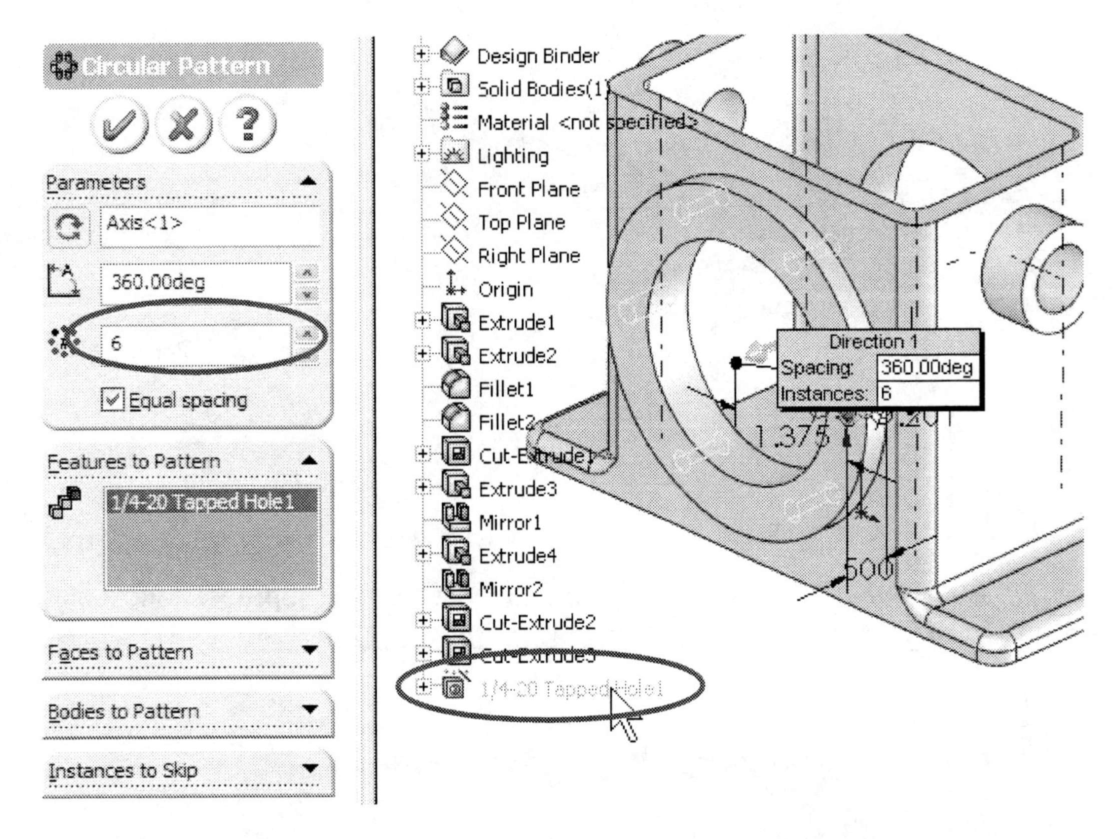

55. - Since we need to have the same six holes in the other side of the Housing, we will use the "Mirror" command to copy the Circular Pattern about the Front Plane to add the same holes on the other side of the housing. Review the Mirror command from steps 33 and 38 if needed. Make the mirror about the Front Plane selecting the "CircPattern1" feature created in the previous step. After the mirror, your part should look like this:

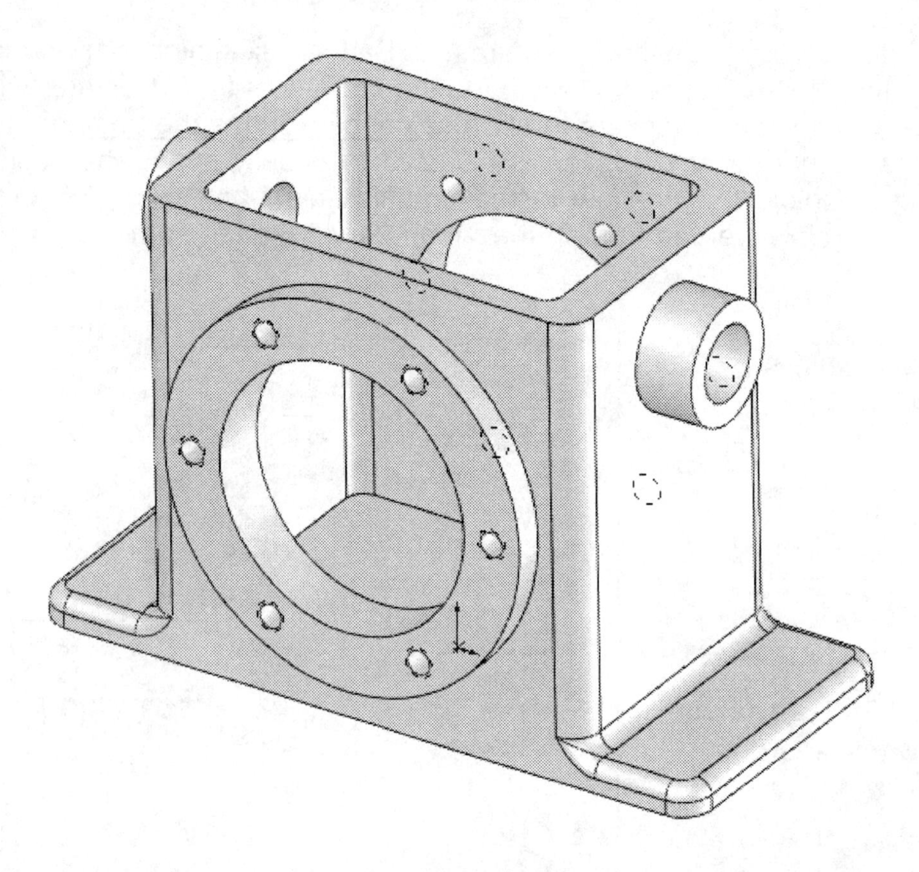

56. - We will now add four #6-32 tapped holes to the topmost face using the **Hole Wizard**. Switch to a Top View, select the top face,

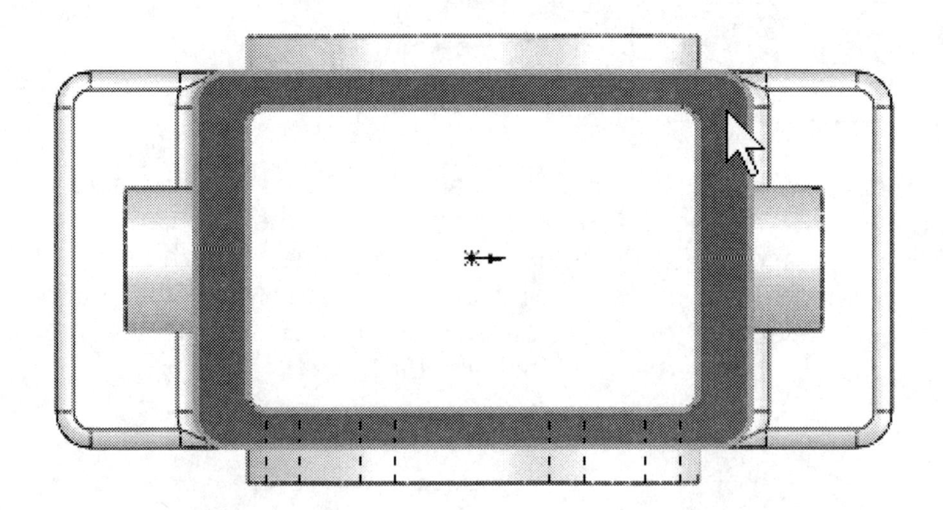

and then select the Hole Wizard icon.

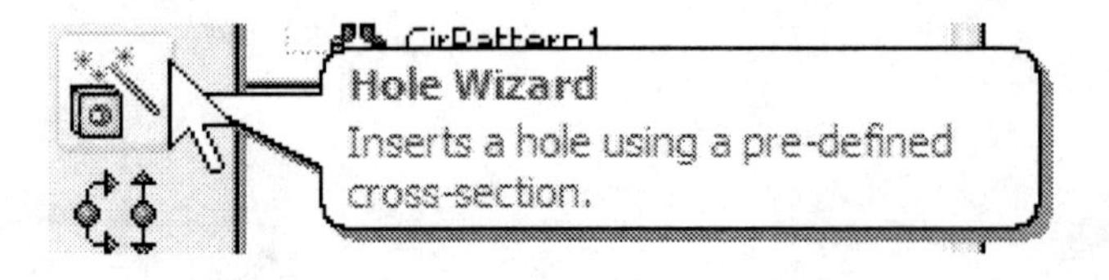

57. - Remember that the first step in the Hole Wizard is to define the hole's type and size. Activate the "**Tap**" tab and the #6-32 Tapped Hole options as shown; change the tapped hole's depth to 0.75". The "Blind" condition tells SolidWorks to make the hole a certain depth. Click "Next" to go to the second step to locate the holes.

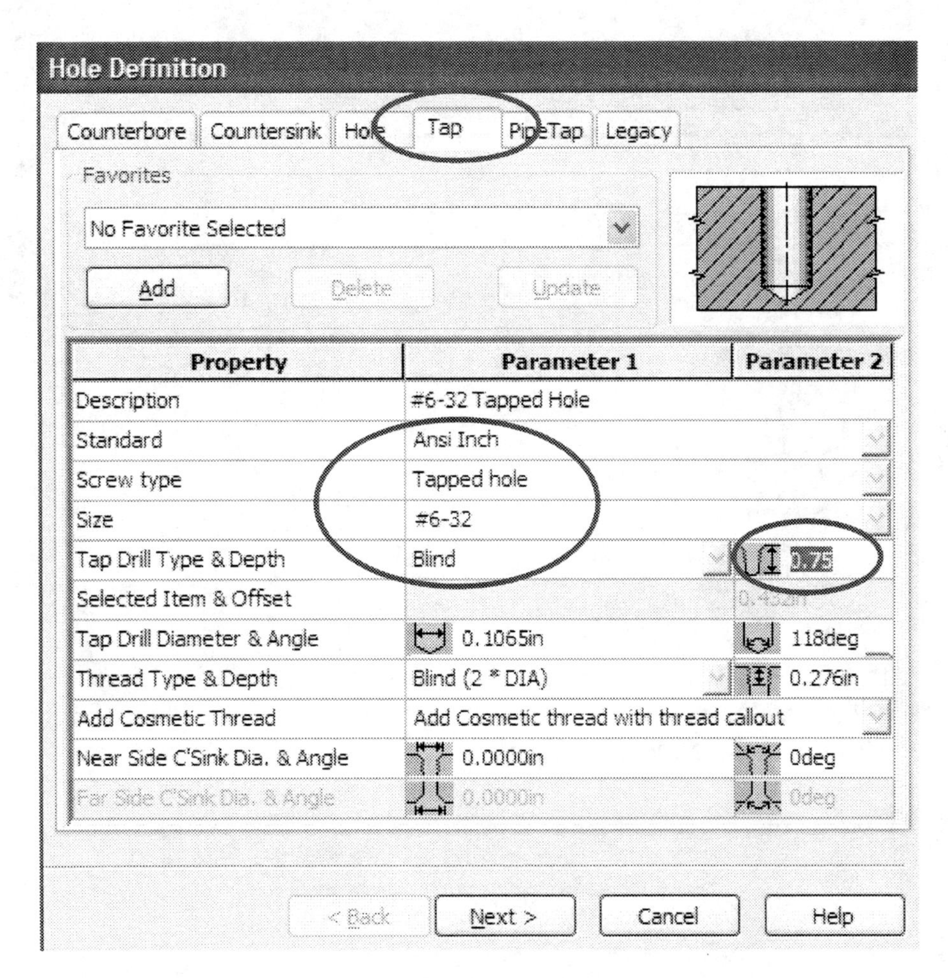

58. - In the second step we will define the hole's location. Notice that the **Sketch Point** tool is active and a Point has already been added where the face had been pre-selected.

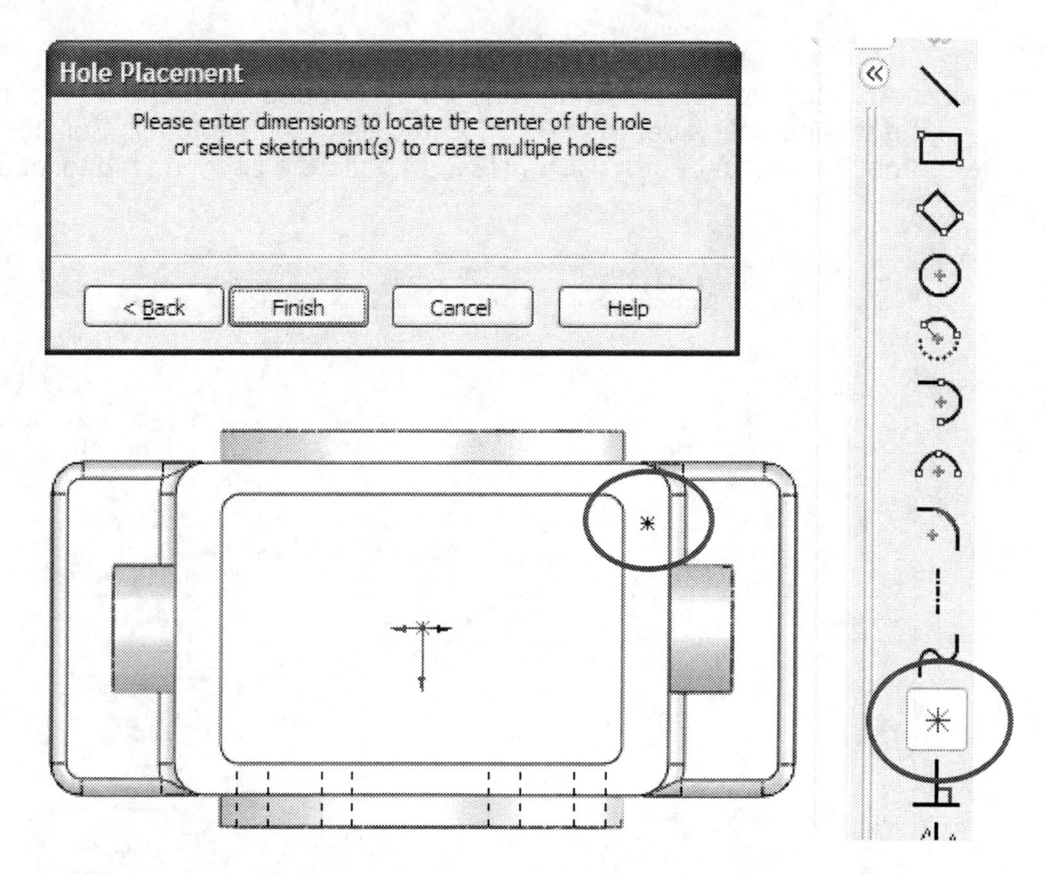

59. - With the "**Point**" tool active, click on the top face to add three more points, and add relations to make them concentric to the corner fillets. Click "Finish" when ready to complete the operation.

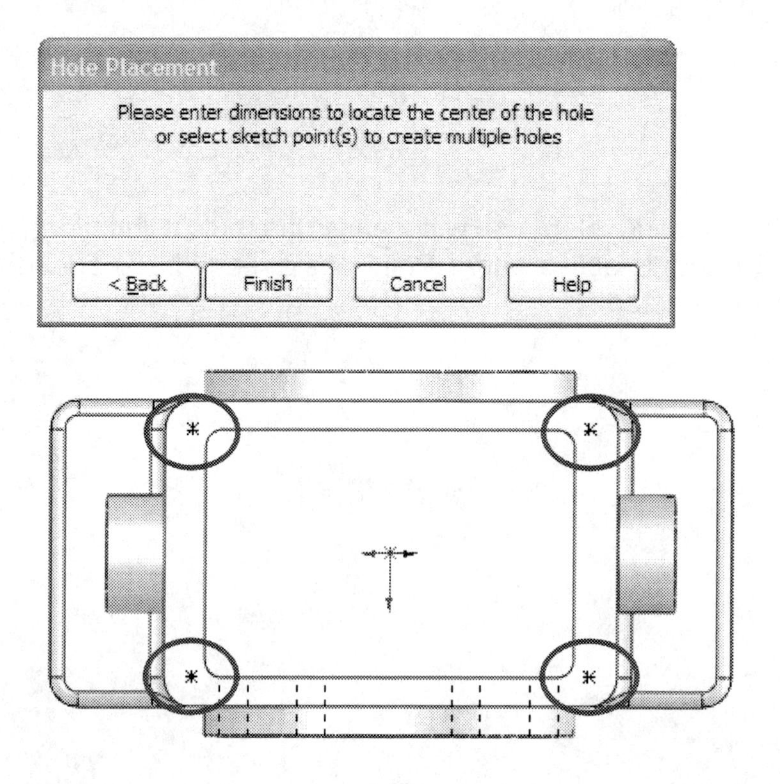

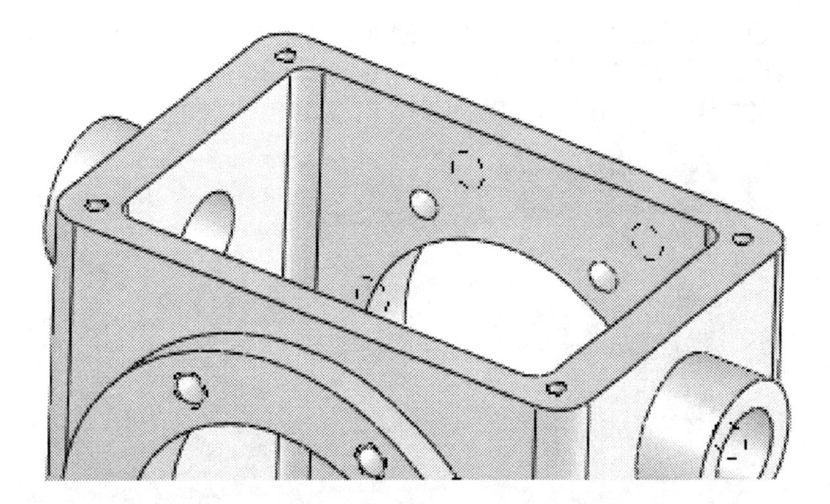

60. - We are now ready to make the holes at the base of the housing. Go to a top view and make a new Sketch in the face selected. Draw a circle and dimension it, then make a Cut using the "Through All" option.

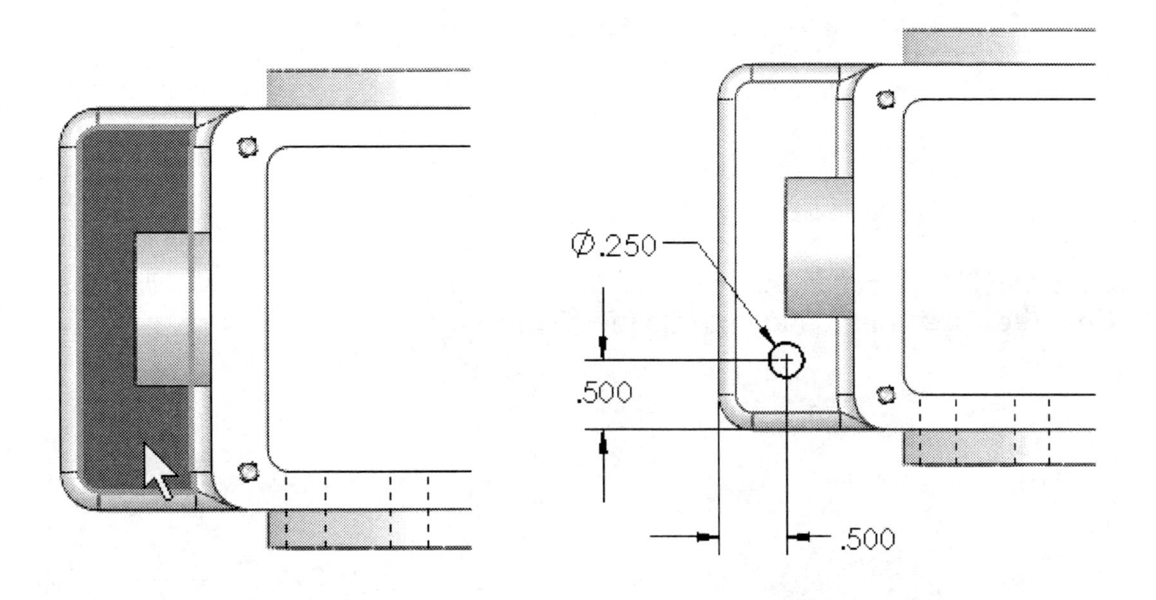

61. - In this step we will create a **Linear Pattern** of the previously made hole. A linear pattern allows us to make copies of one or more features along one or two directions (usually along a model edge). Select the Linear Pattern icon from the Features Toolbar.

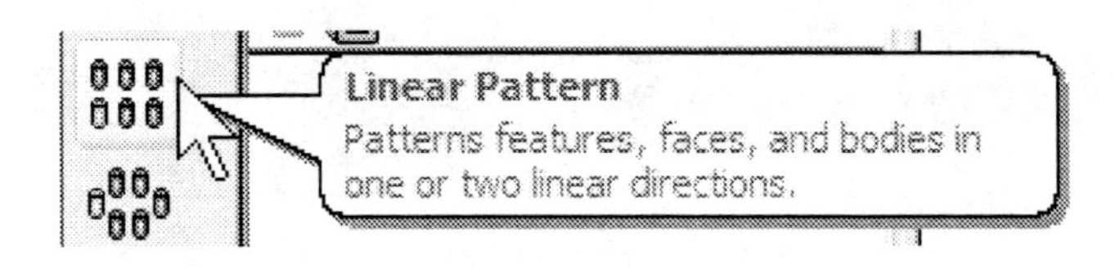

62. - In the Linear Pattern's Property Manager, the "Direction 1" selection box is active; select the edge indicated for the direction of the copies. This is the direction the copies will follow. Any linear edge can be used for direction.

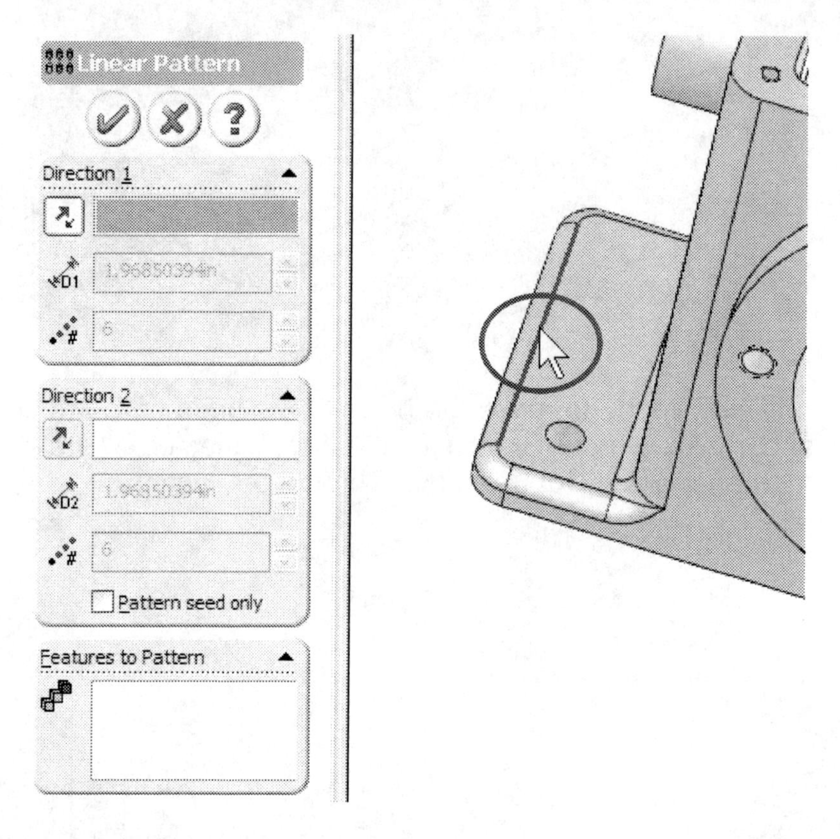

NOTE: If the Direction Arrow is pointing in the wrong direction, click on the "**Reverse Direction**" button next to the "Direction 1" selection box.

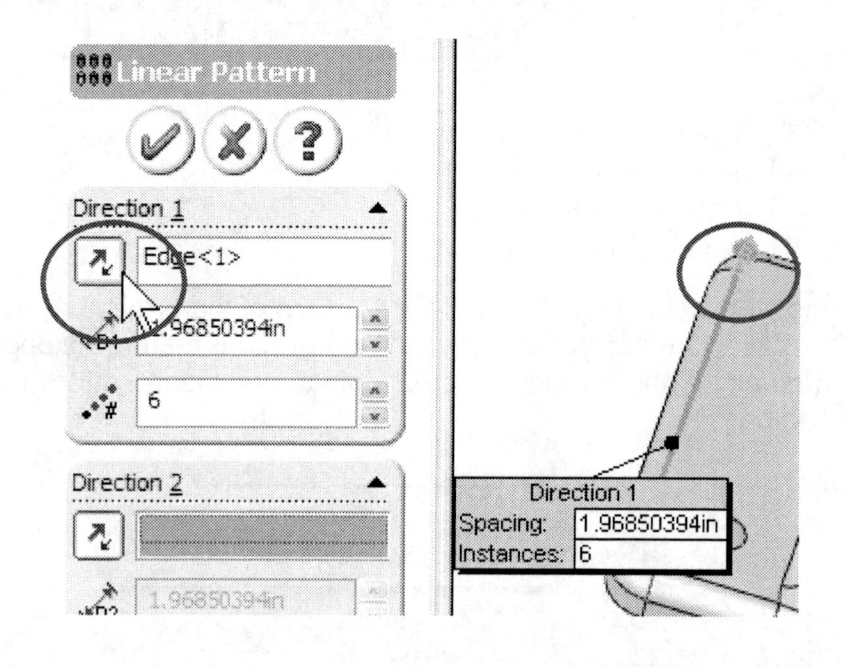

63. - Now click in the "Features to Pattern" selection box to activate it and select the previous cut operation from the fly-out Feature Manager. Change the spacing between the copies to 0.75" and total copies to 3 (this value includes the original). Click OK to finish the command.

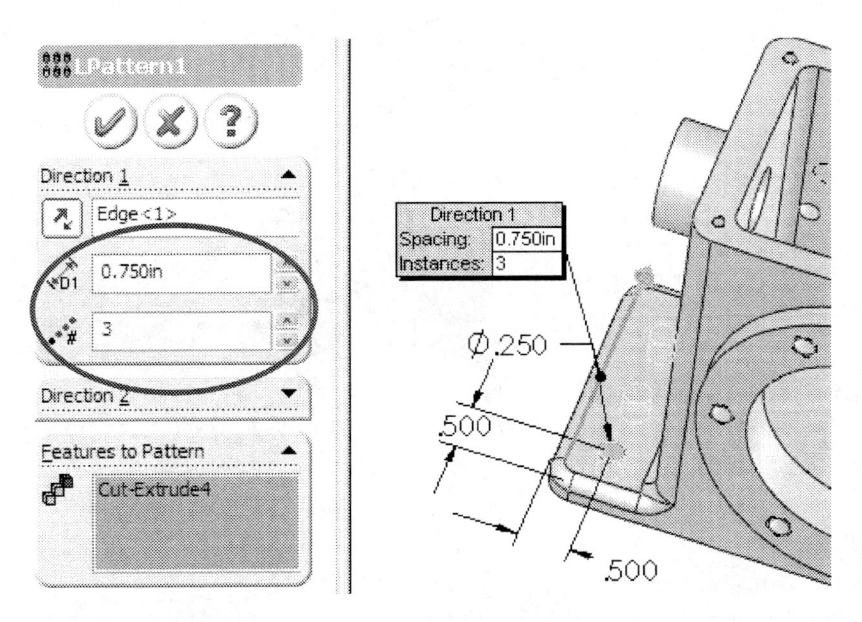

64. - To copy the previous linear pattern to the other side of the Housing, select the "Mirror" icon to mirror the Linear Pattern about the Right Plane to copy the holes.

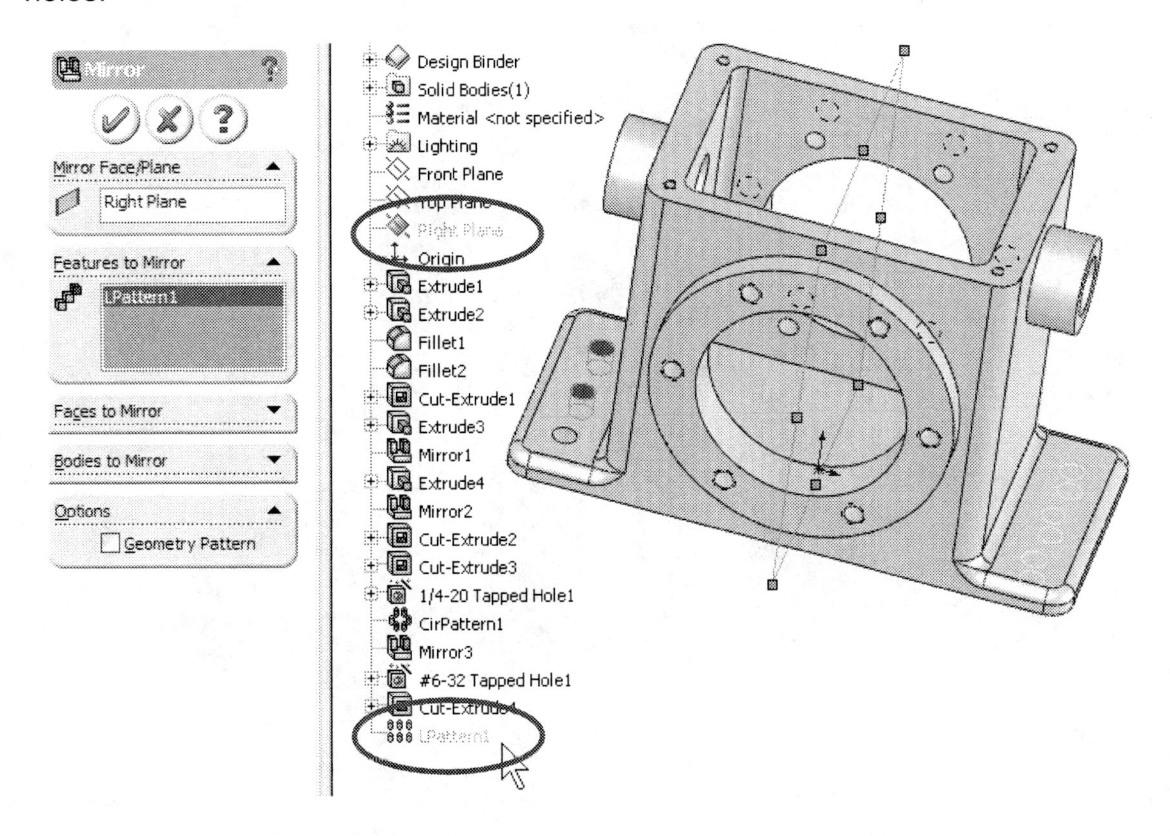

65. - Add a 0.125" fillet to the indicated edges to finish the Housing as a finishing touch. Save the finished part as "Housing" and close the file.

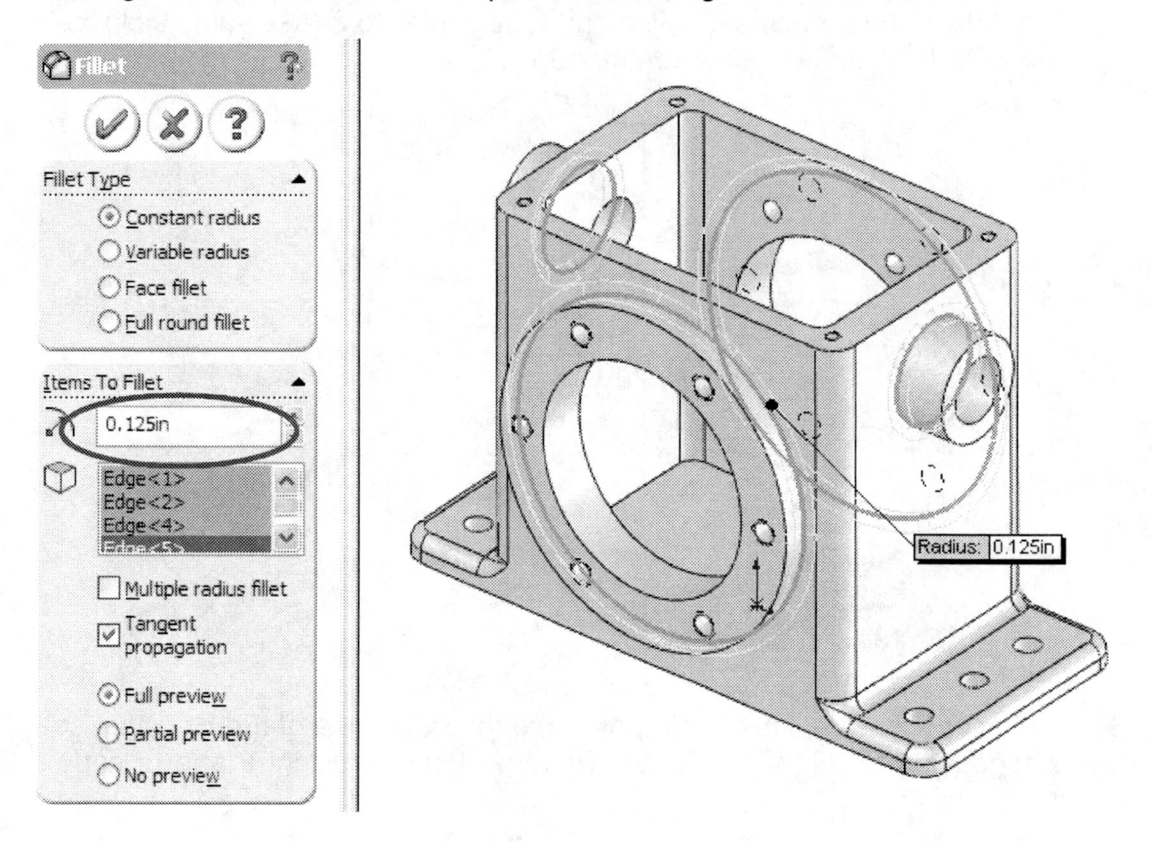

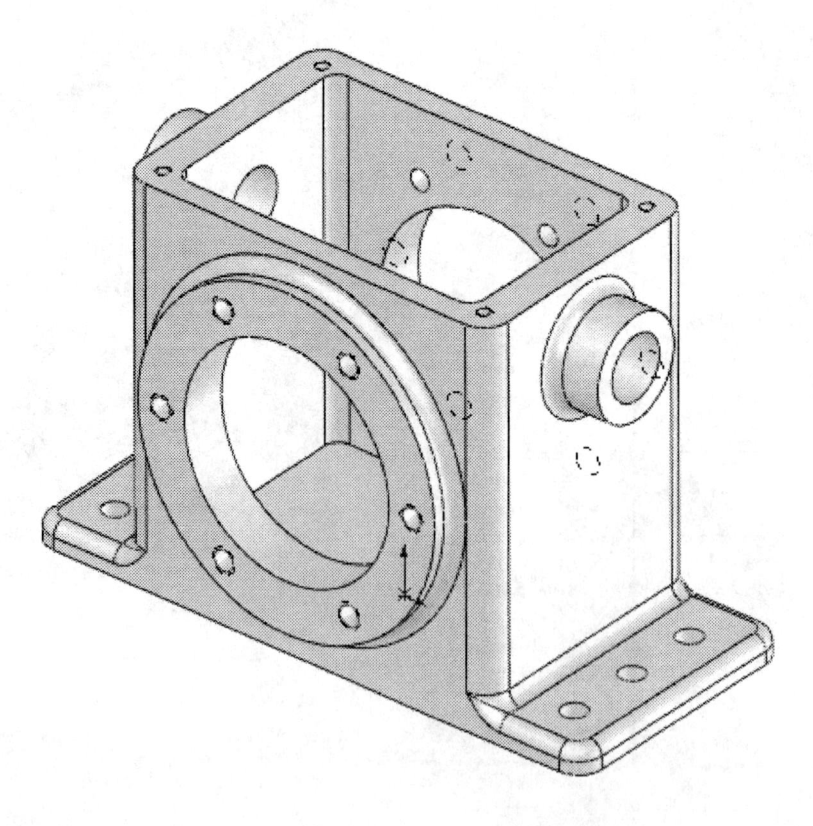

The Side Cover

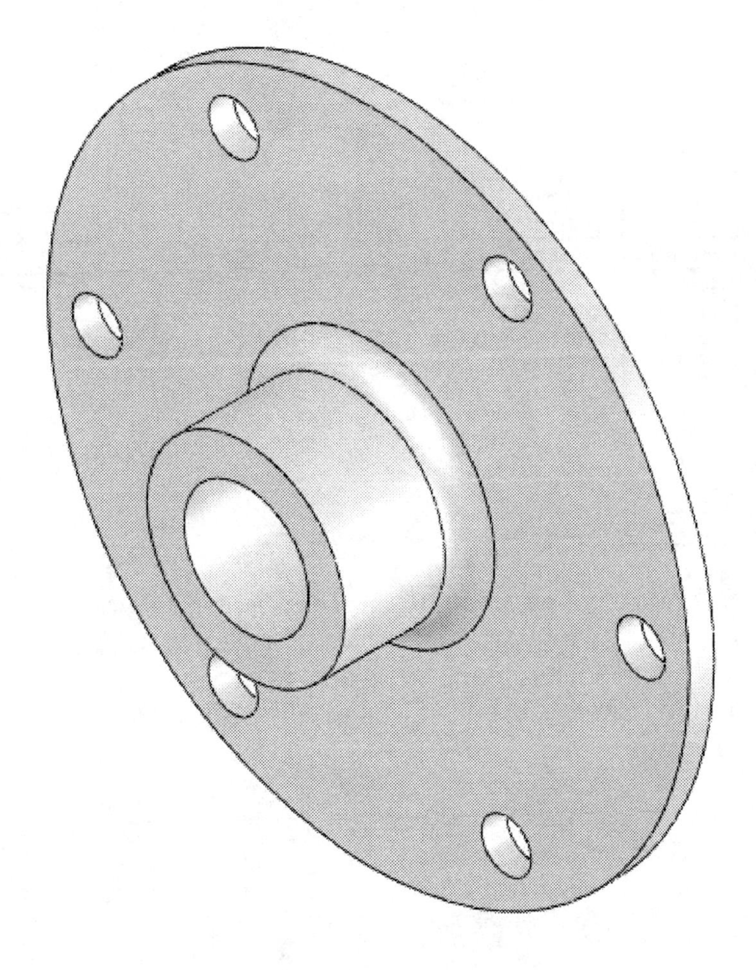

In making the Side Cover part we will learn: Revolved Feature, Sketch Trim and Extend, and construction geometry. The sequence of features we will follow for the Side Cover is:

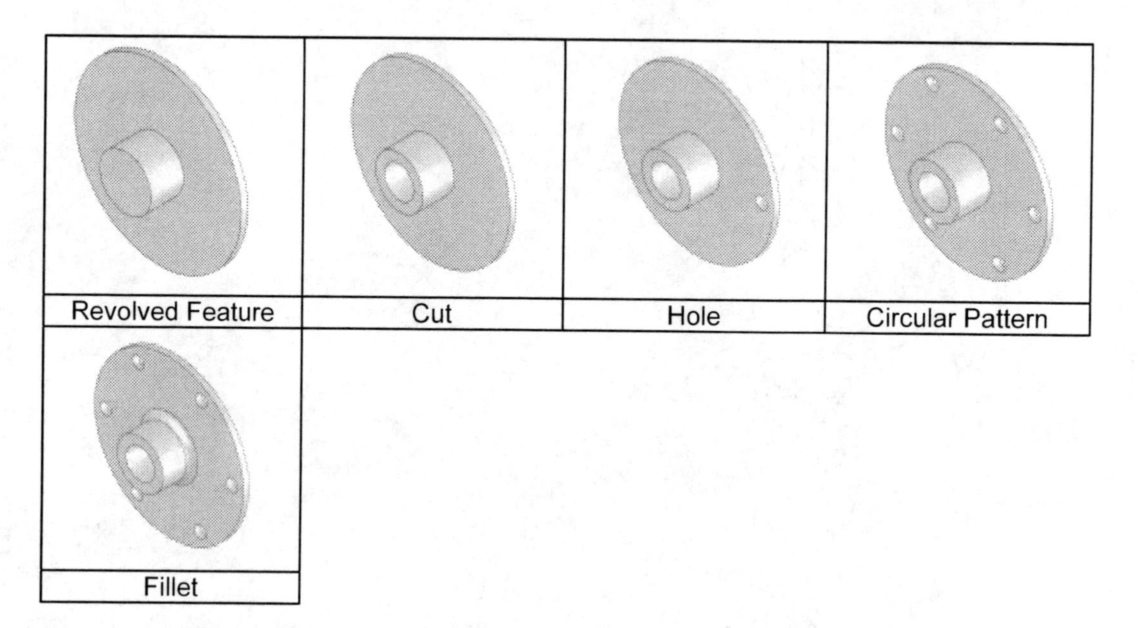

66. - The first feature we will create is a **Revolved Feature**. A revolved feature, as its name implies, is created by revolving a sketch or profile about an axis. Make a new part and in this case, the first sketch will be located in the Right Plane. Click on the "Sketch" icon and select the Right Plane from the screen as indicated in the image.

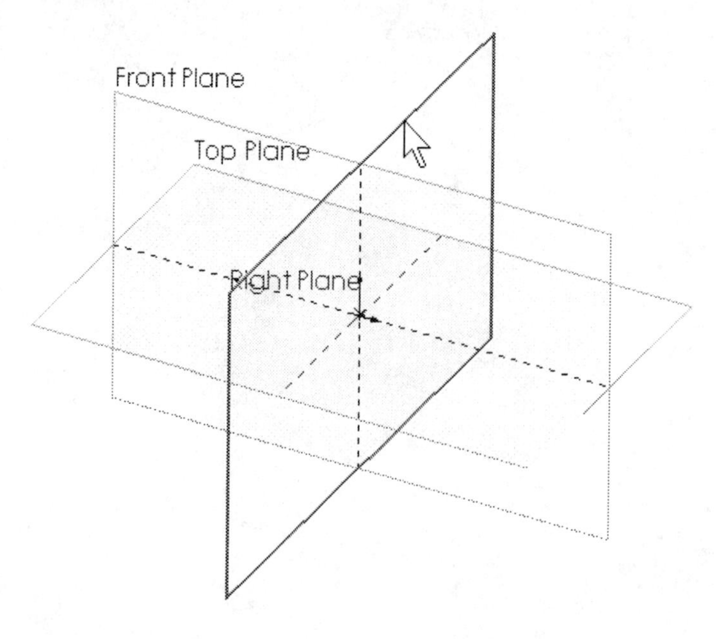

67. - Select the Rectangle icon from the Sketch Toolbar and draw the following sketch using two rectangles, starting at the Origin and to the left (there will be two lines overlapped in the middle).

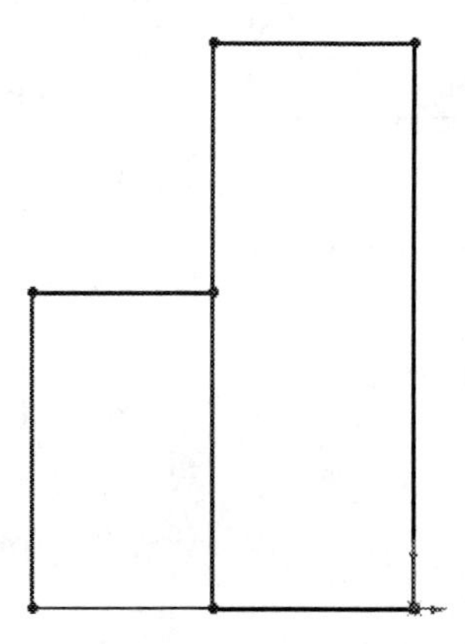

68. - It is good practice to have single, non-intersecting profiles, and no more than 2 lines sharing a single endpoint. It is possible to use sketches with intersecting lines; this is a very powerful technique. However it will not be covered in this book. For beginners it is a good idea to start with a single contour sketch, and advance to more powerful techniques later on.

To clean up the sketch, we will use the "**Trim**" command.

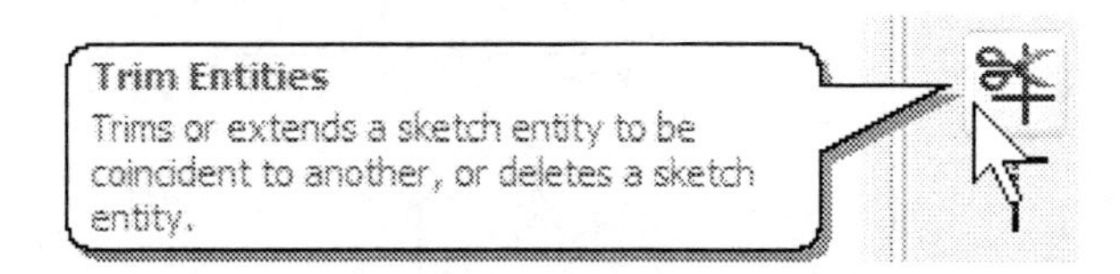

69. - The Trim tool allows us to cut sketch entities against other geometric elements. After selecting the "Trim Entities" icon, select the "**Power Trim**" option. The Power Trim allows us to click and drag across the entities that we want to trim. Click and drag the cursor starting HERE crossing the two lines; as you cross them they will be trimmed.

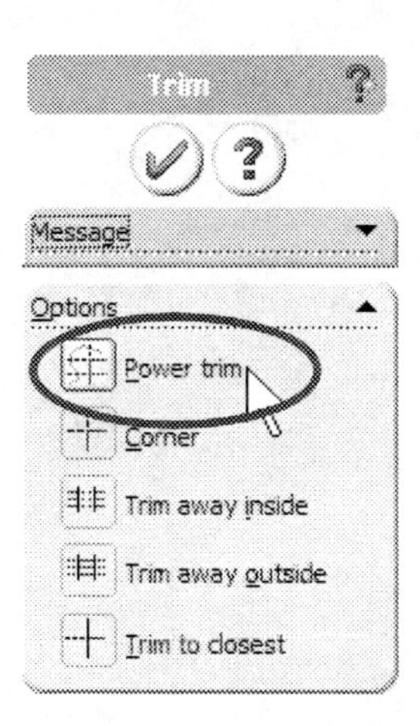

70. - The next step is to extend the short line to the Origin to close the sketch and have a single profile. Select the "**Extend Entities**" icon and click on the short line indicated; a preview will show you how the line will be extended. If the extension does not cross a line, you will not see a preview.

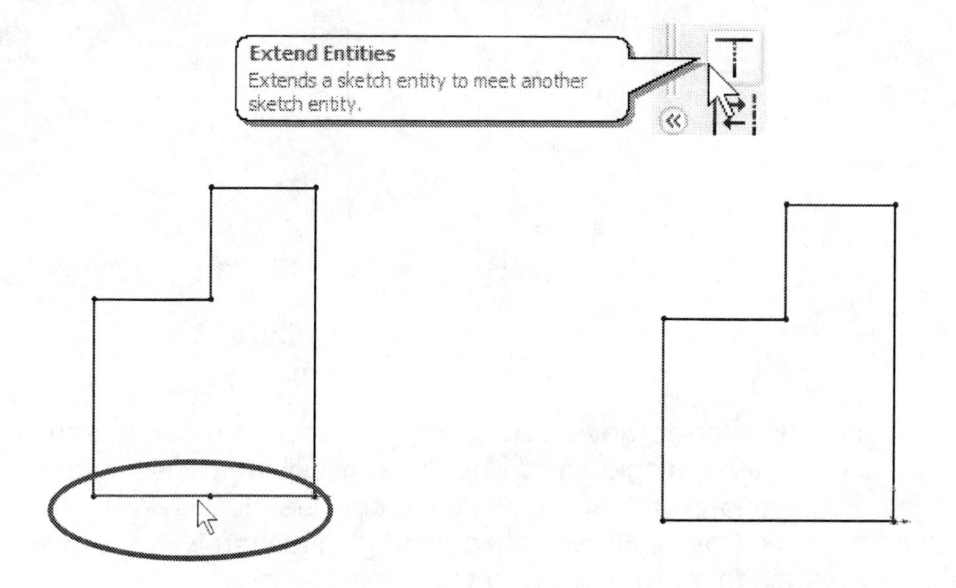

71. - Add the following dimensions to the sketch.

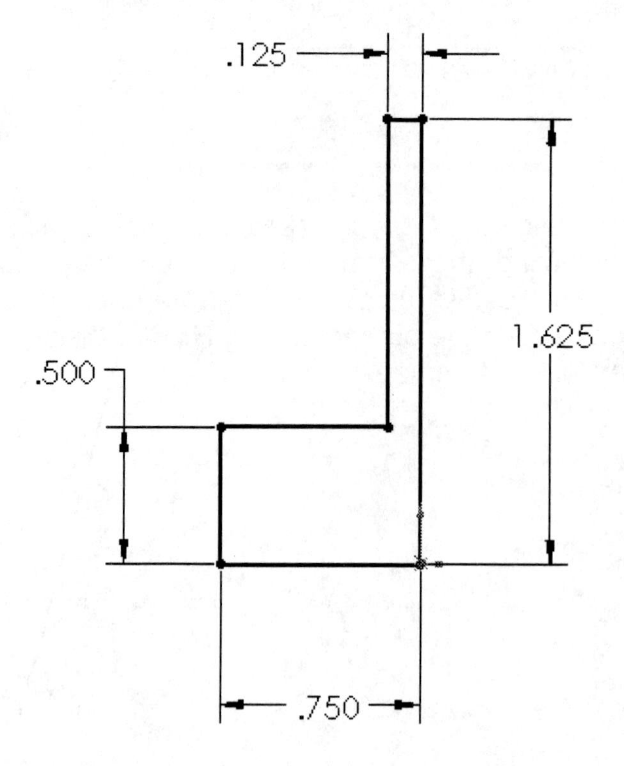

72. - Now that the Sketch is fully defined we'll create the **Revolved Boss/Base**. Select the Revolved Boss/Base icon from the toolbar.

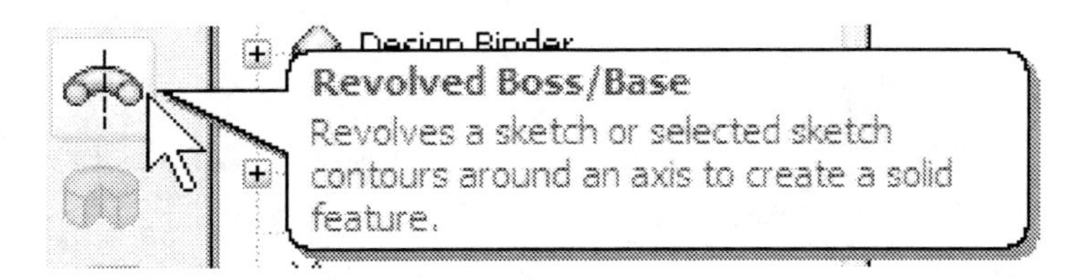

73. - The Revolve Property Manager is presented and ready for us to select an axis to make the revolved feature about; if the sketch has one centerline, it is automatically selected as the default axis of rotation. Select the line indicated as the axis of revolution to make the revolved base.

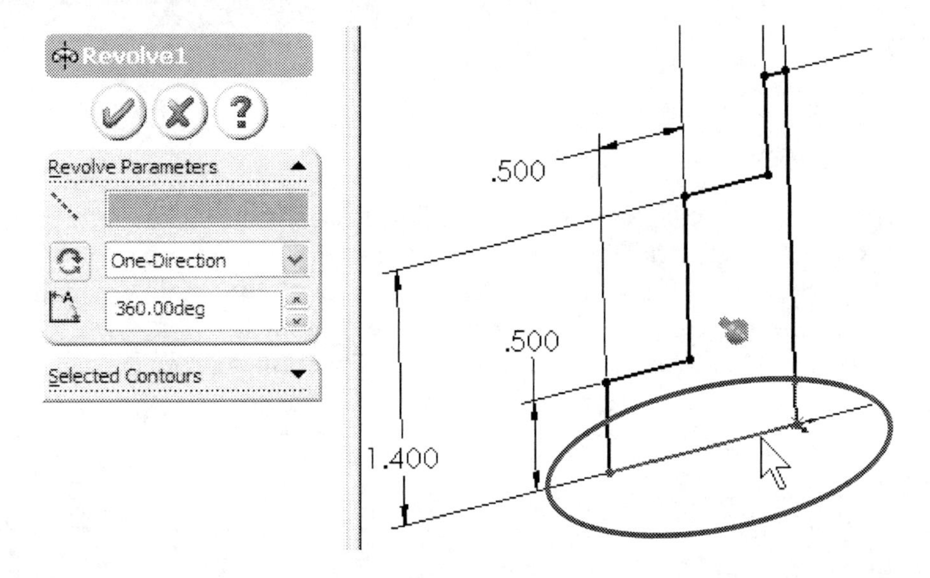

74. - Notice the preview when the line is selected. The default setting for a revolved feature is for 360°. Click OK to complete the revolved feature.

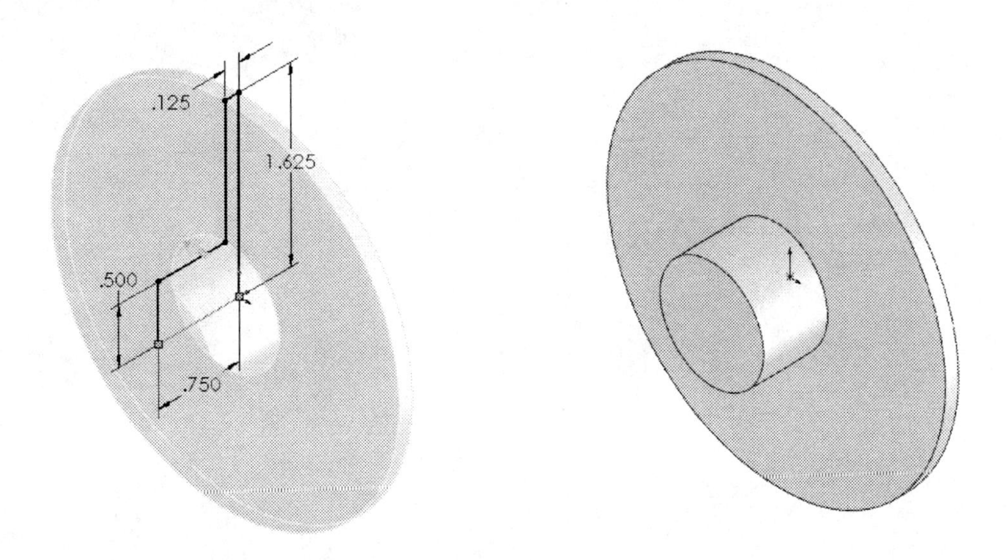

75. - Now switch to a Front view. We'll make the hole in the center of the cover. Select the frontmost face of the cover and make a sketch as indicated. Make a cut using the "Through All" option.

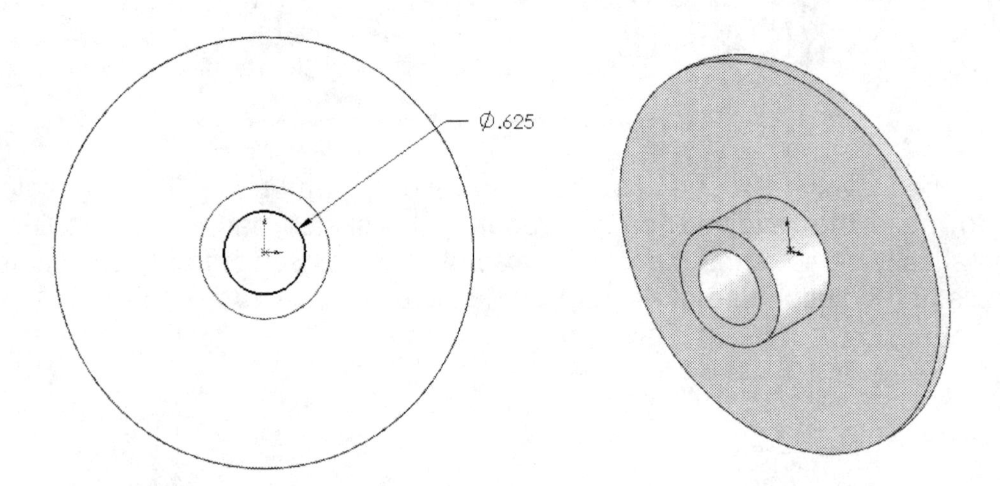

76. - The next step will be to make the first hole for the screws to pass through. We'll make one hole, and then use a Circular Pattern to make the rest. In this case we'll use the Cut Extrude feature instead of the Hole Wizard to show a different approach. Select the face where we want to make the first hole, and draw the next sketch. Make a Cut using the "Through All" option.

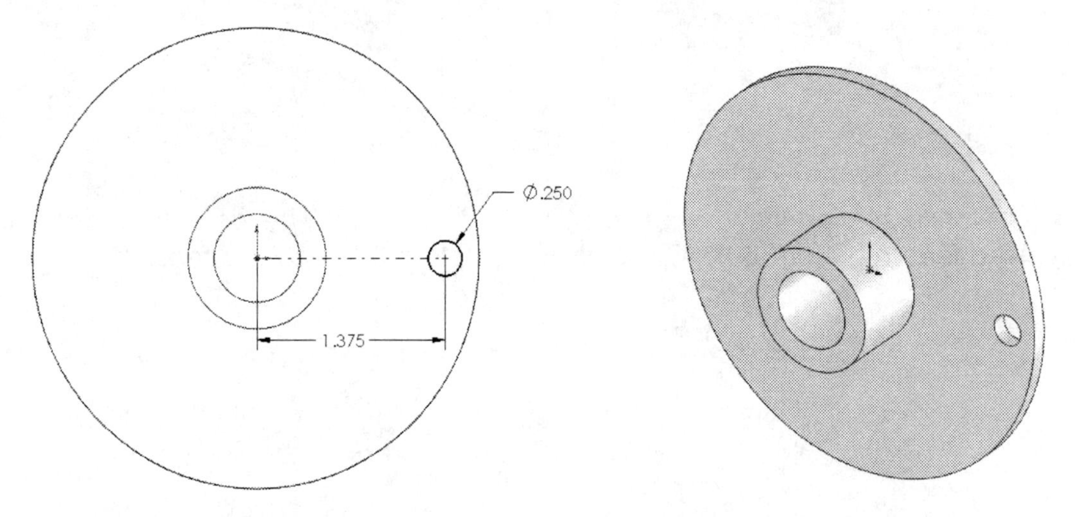

77. - A different way of doing this feature is to make the sketch by drawing a circle and then converting it to construction geometry. This way you can dimension the circle's diameter, and locate the Hole's center coincident to the construction circle and horizontal (or vertical) to the Origin. To convert any element to construction geometry, simply select it in the graphics area and activate the "For construction" check box in the element's Property Manager or click the "**Construction Geometry**" icon in the Sketch Toolbar.

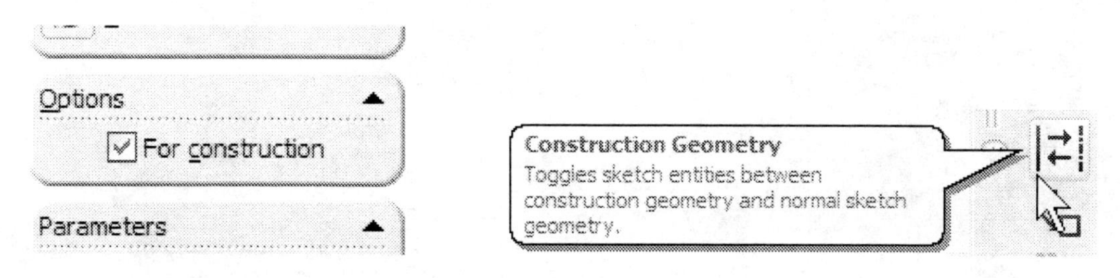

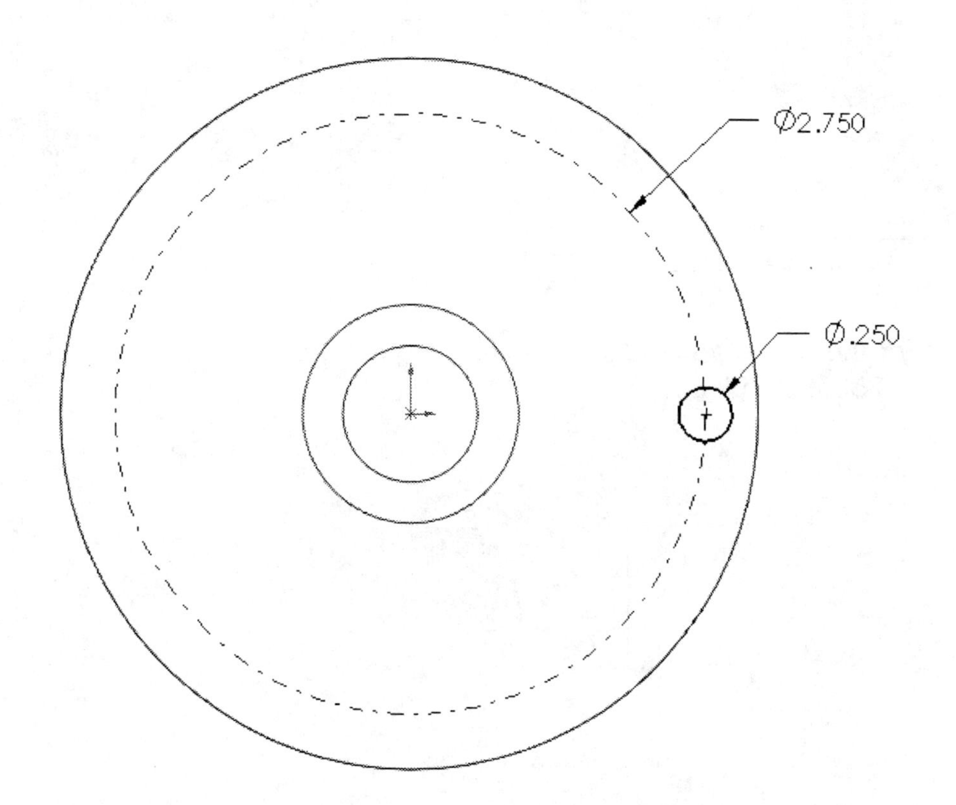

The biggest advantage of making the sketch using this technique is that you can add a diameter dimension for the circle of holes instead of a radius.

78. - To complete the rest of the holes, we'll make a circular pattern. Select the "Circular Pattern" icon, and turn on the temporary axes ("View" menu, "Temporary Axes"). Select the axis at the center of the part to make the copies about it. Now click in the "Features to Pattern" selection box and select the last cut operation from the fly-out Feature Manager if it is not already selected. Change the number of copies to six; remember this count includes the original. Click OK to complete the command and turn off the Temporary Axes.

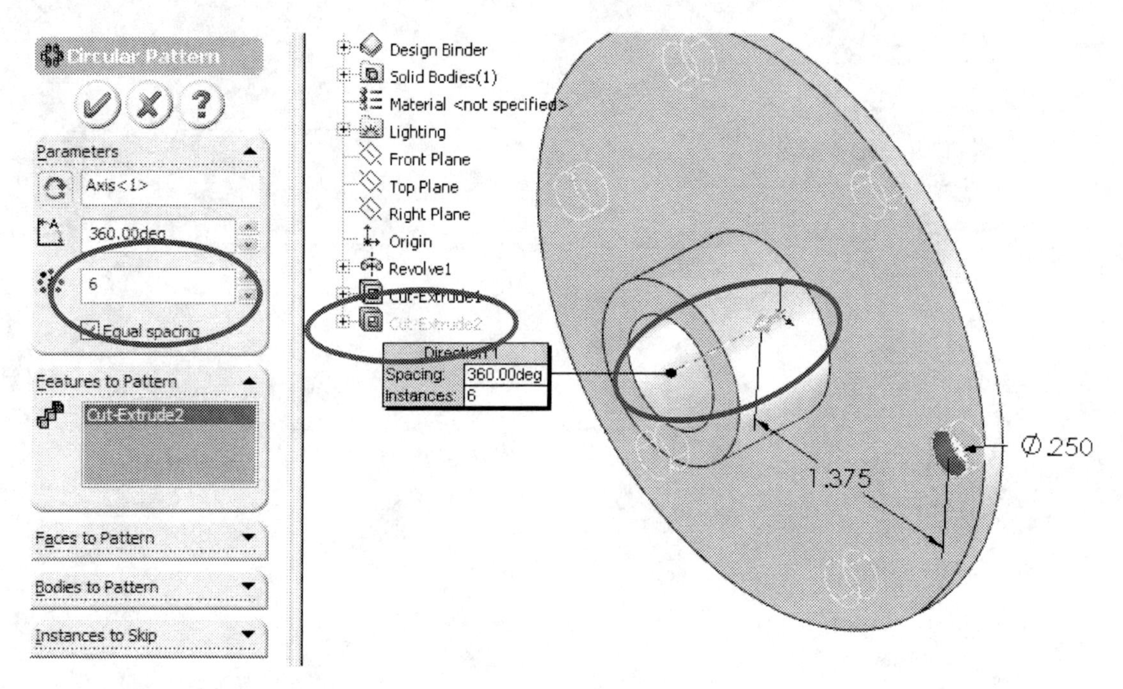

79. - For the final step, select the Fillet command and round the edge as shown with a 0.125" fillet radius.

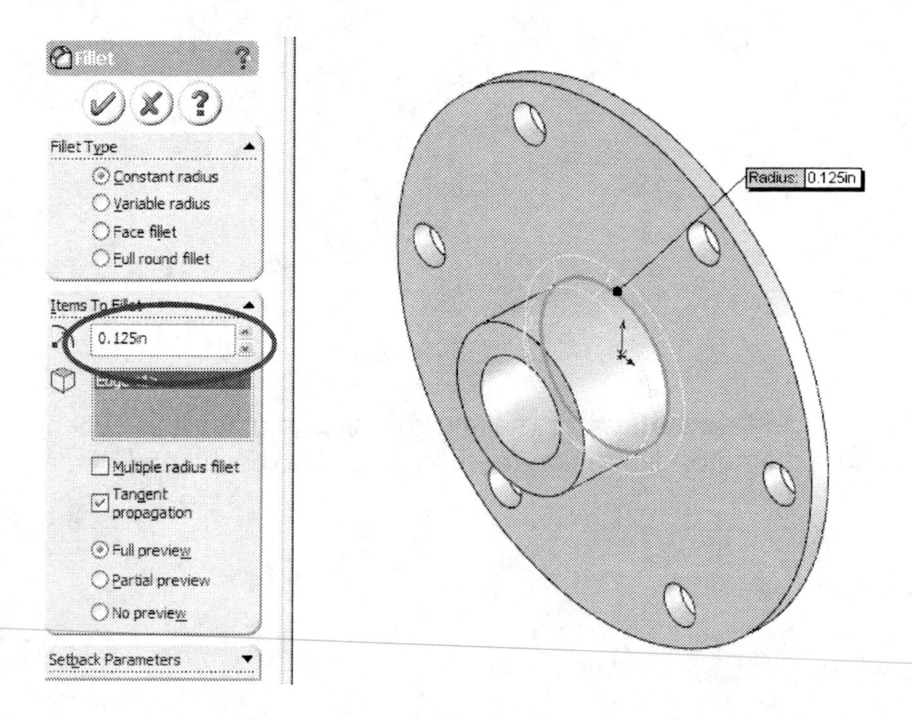

80. - Save the finished component as "Side Cover" and close the file.

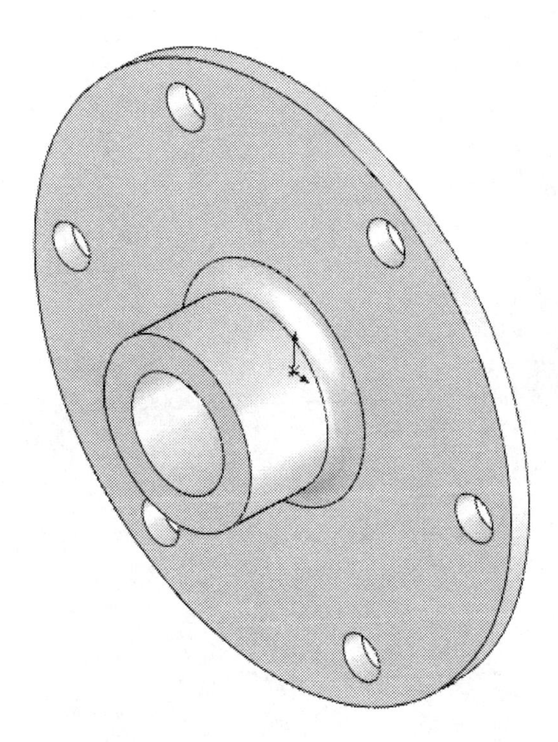

Notes:

The Top Cover

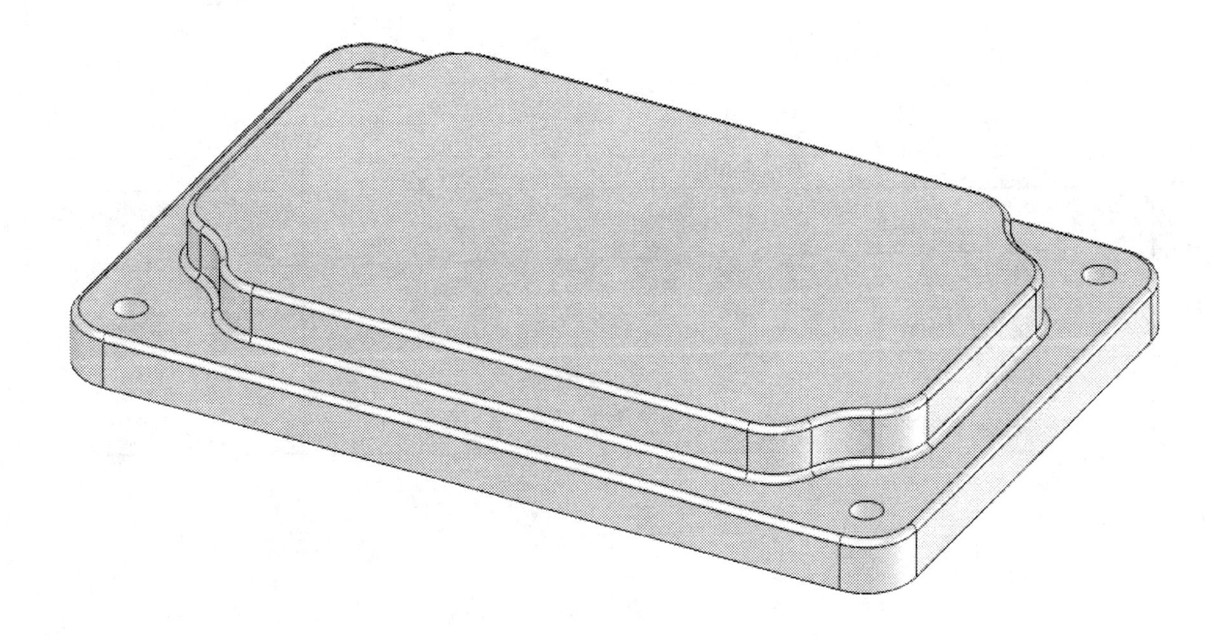

For the Top Cover part we will follow the next sequence of features. In this part we will learn a new feature called Shell, new options for Fillet and new extrude end conditions. Plus, we'll practice some of the previously learned features and options.

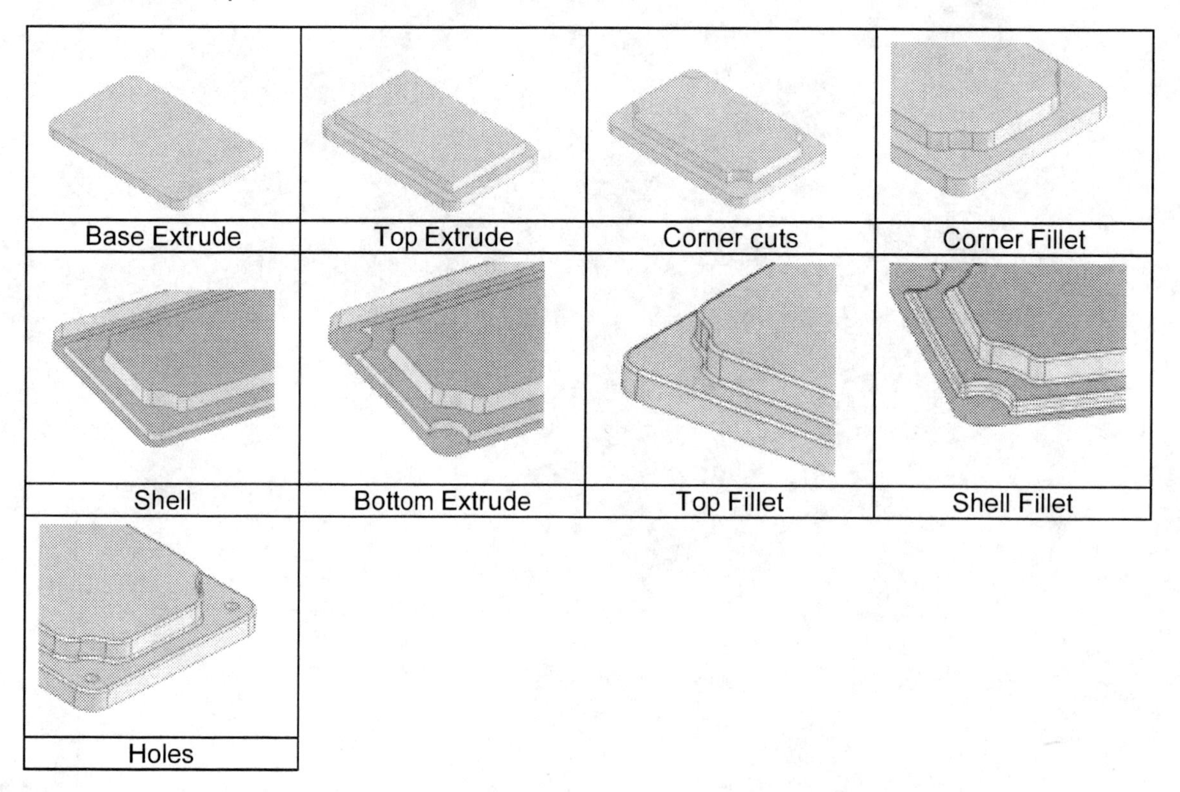

81. - For the Top Cover part, first create a new part, and just like the Housing, we'll start from the Top plane. Click on the Sketch icon and select the top plane to make a sketch as shown.

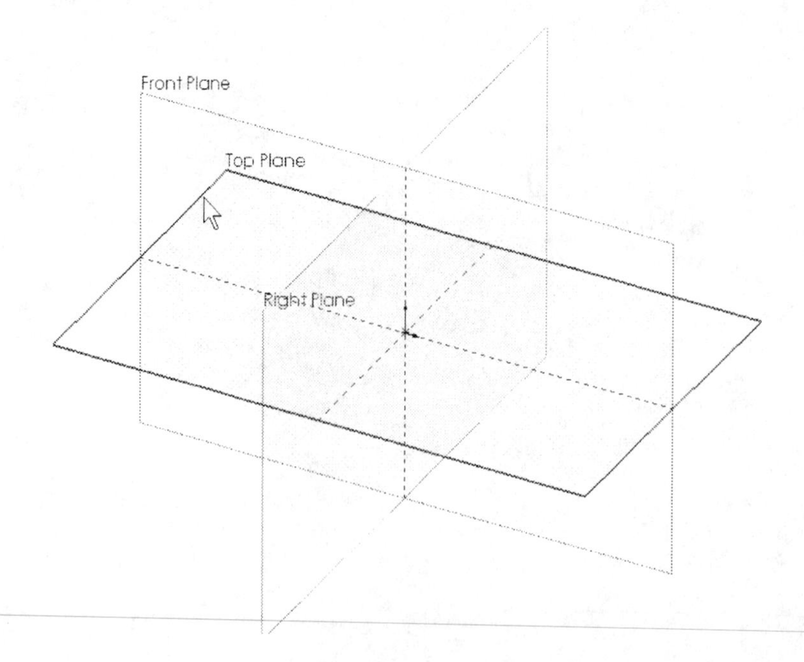

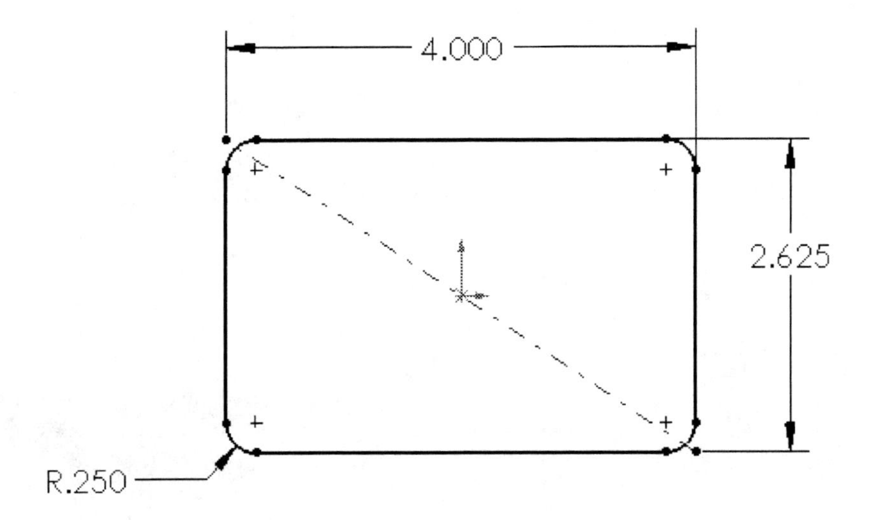

82. - And extrude it 0.25".

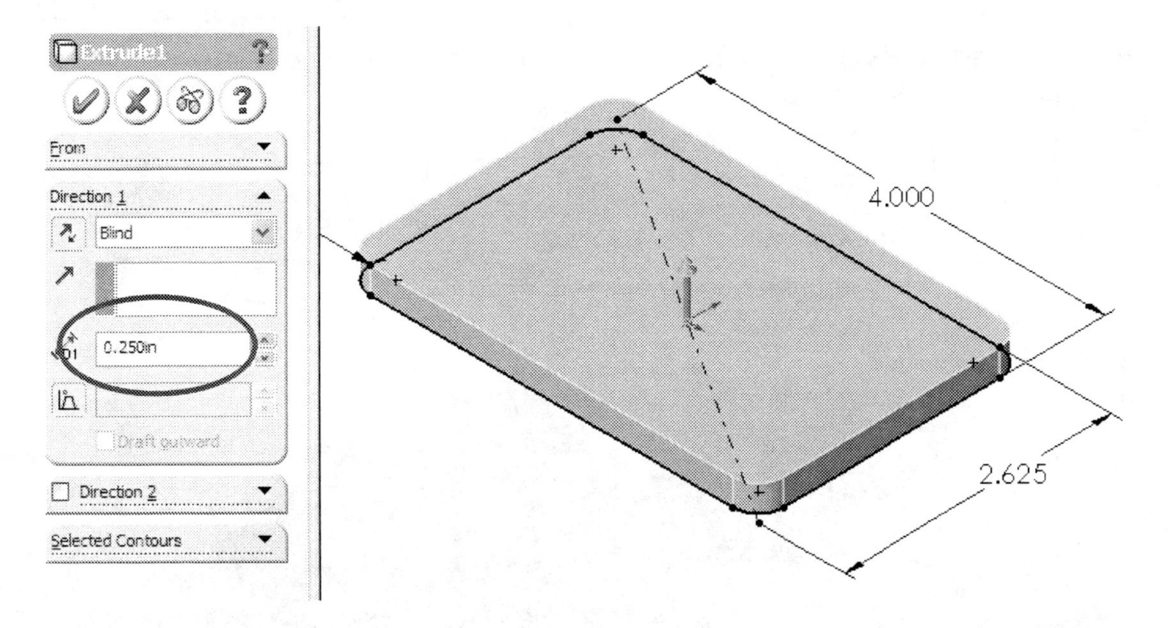

83. - For the second feature, we'll make an extrusion of similar shape to the first one, but of smaller size. For this feature we will use the Offset Entities function. Switch to a top view and insert a sketch on the top face.

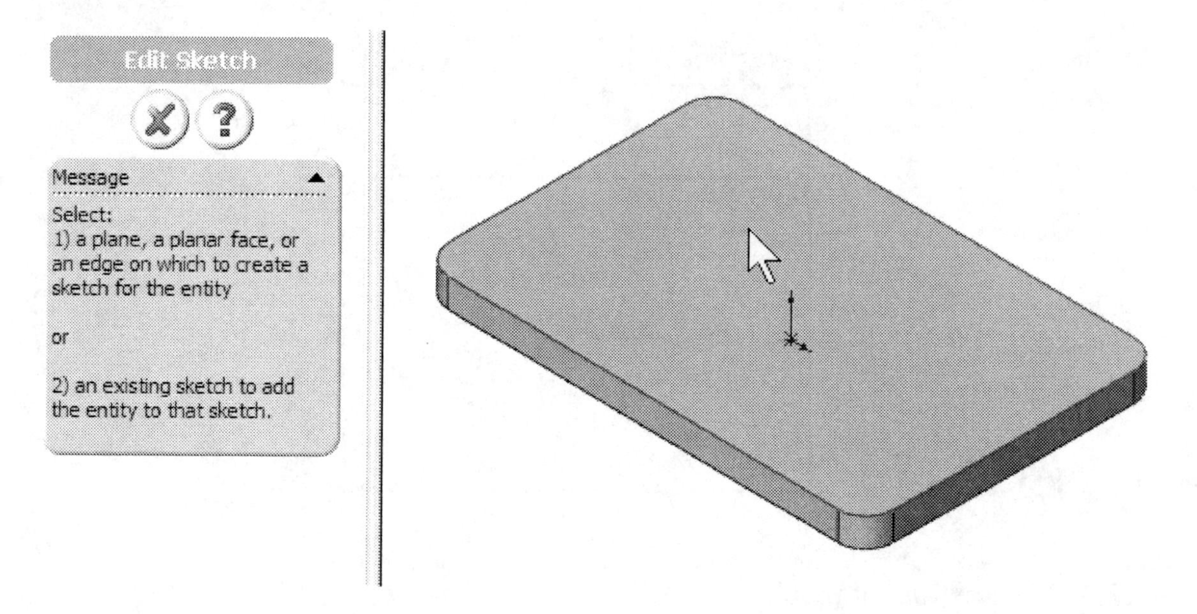

84. - After inserting the sketch, select the top face of the first feature and click on the "**Offset Entities**" icon from the Sketch Toolbar. You will see a preview of the offset of the face edges.

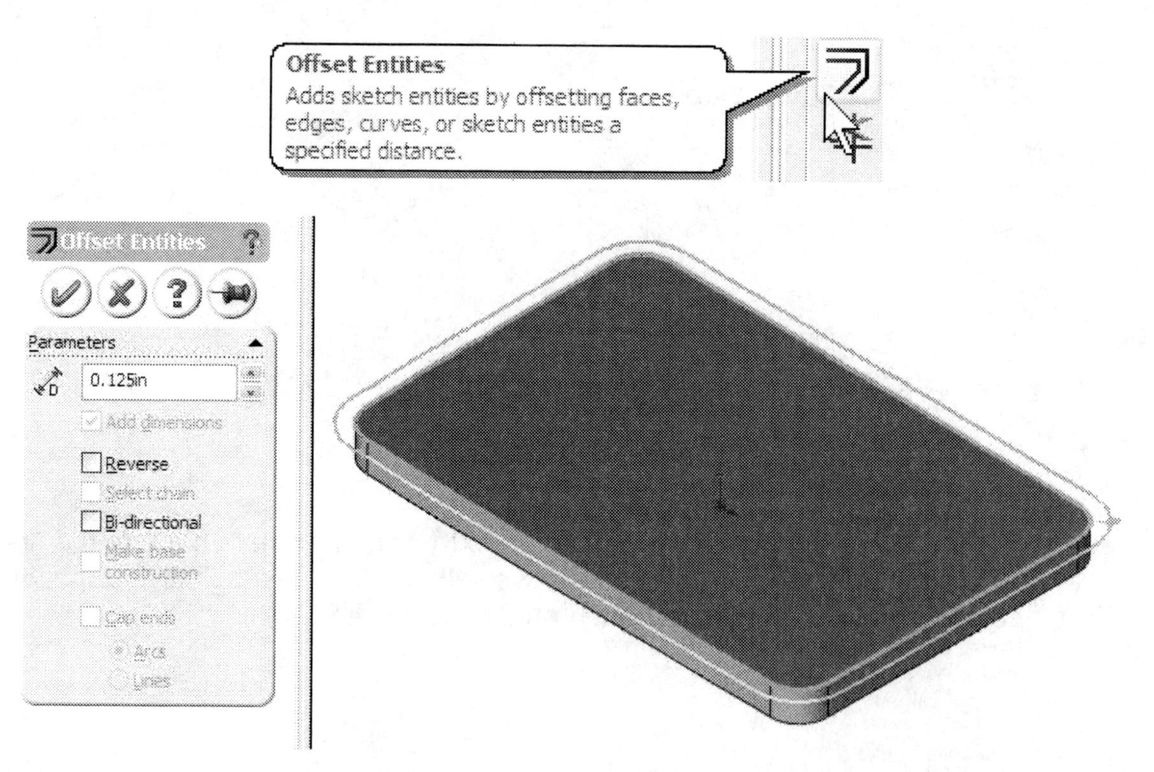

85. - Change the value to 0.375" and then click in the "Reverse" checkbox to make the offset inside, and not outside. Notice that when we change the direction the preview updates accordingly. Click OK when done.

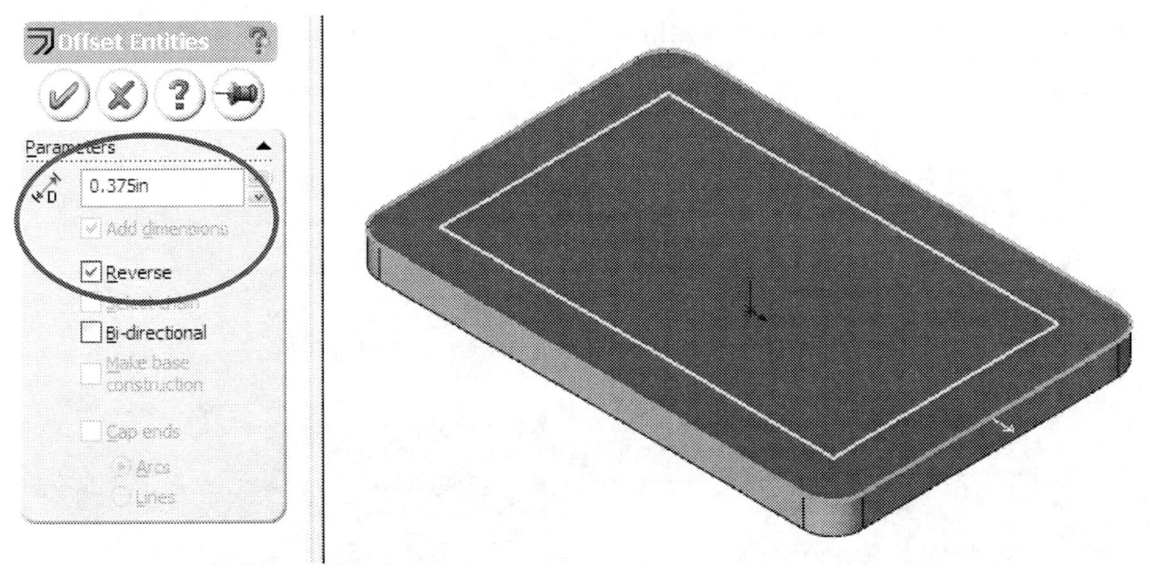

86. - The Sketch is now Fully Defined, this is because the sketch geometry is related to the edges of the face and only the offset dimension is added. The SolidWorks offset command is powerful enough to eliminate the rounds in the corners if needed.

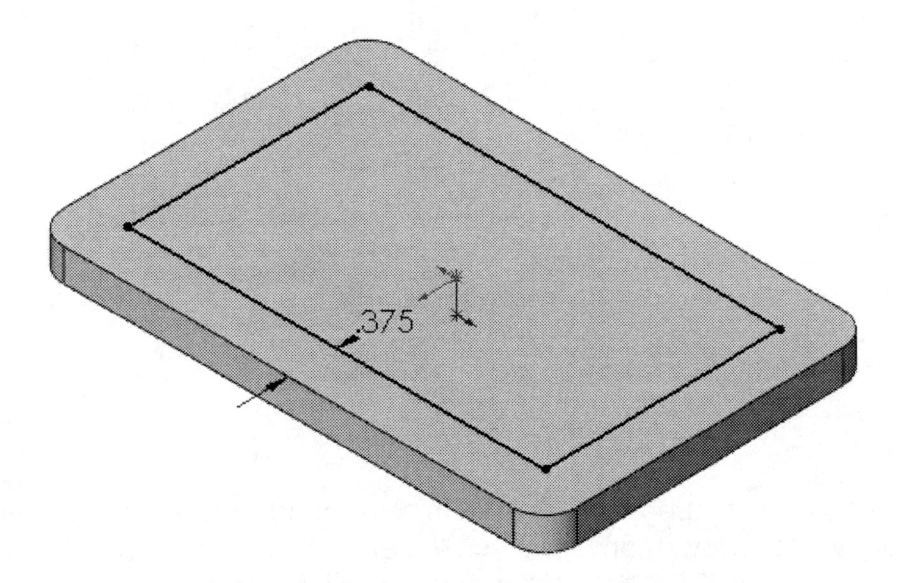

87. - To make the second feature, select the Extrude icon and extrude the sketch 0.25".

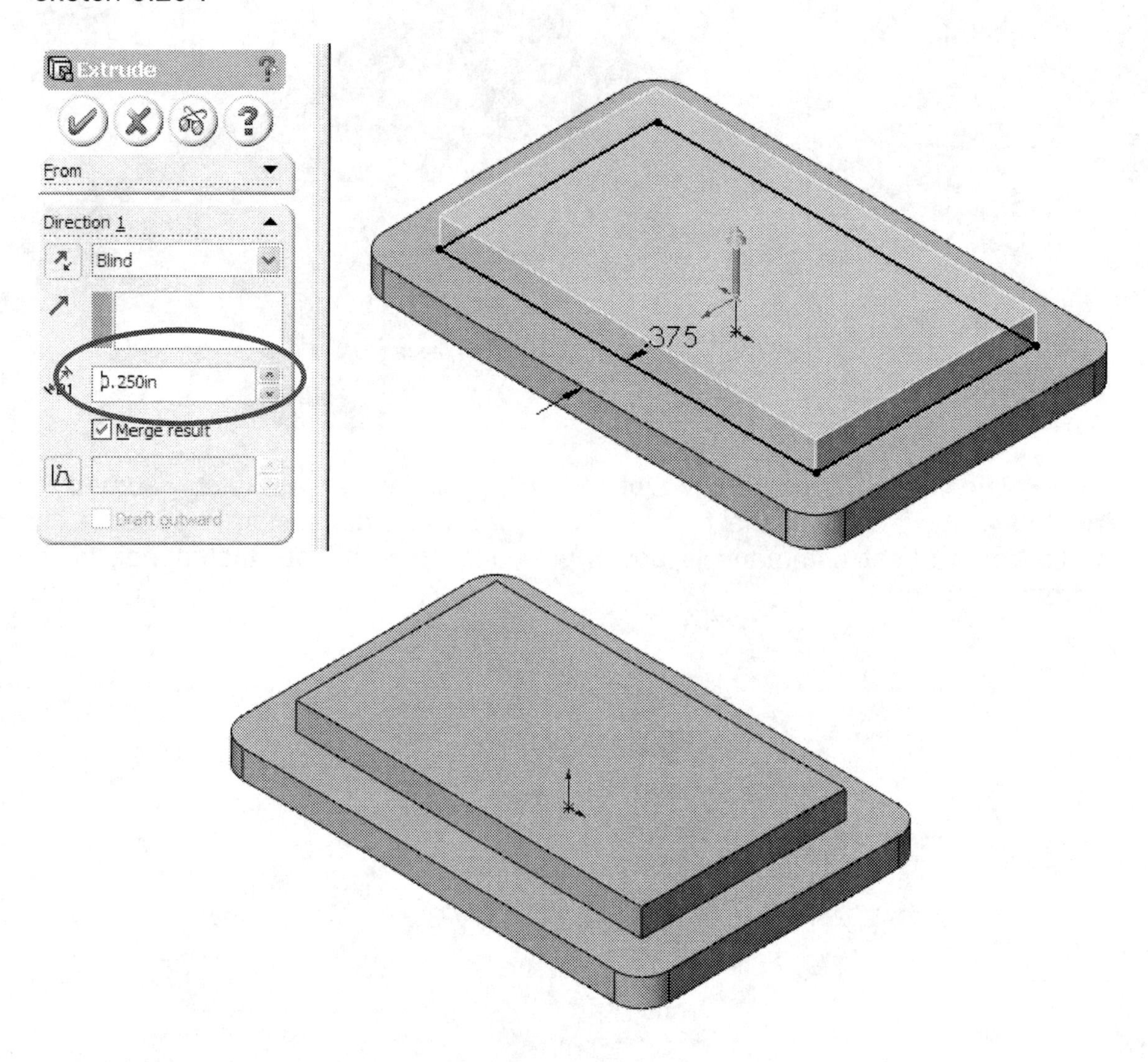

88. - For the next operation, we will make round cuts in the corners of the second feature to allow space for a screw head. Switch to a Top View and insert a sketch in the topmost face making sure the center of the circle is coincident to the corner. Add the two centerlines indicated; we'll need them in the next step.

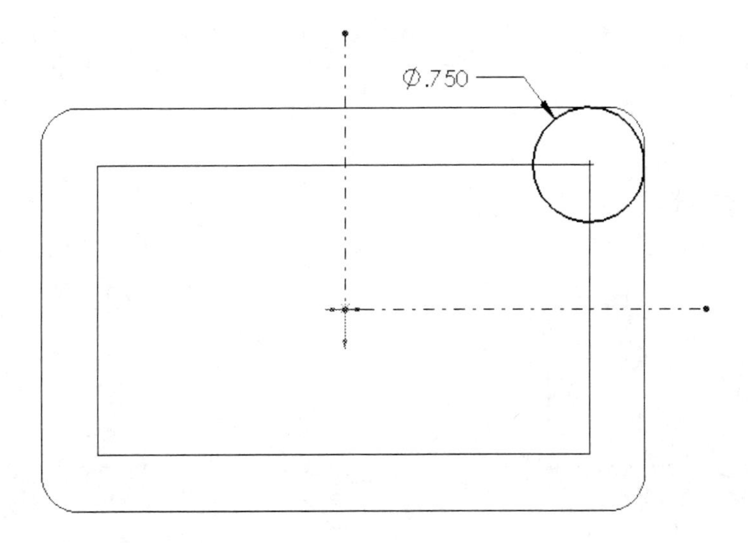

89. - We will now use the **Mirror Entities** tool to make an exact copy of the circle about the vertical centerline, and then both circles about the horizontal centerline to make a total of four equal circles. Select the Mirror Entities icon from the Sketch Toolbar.

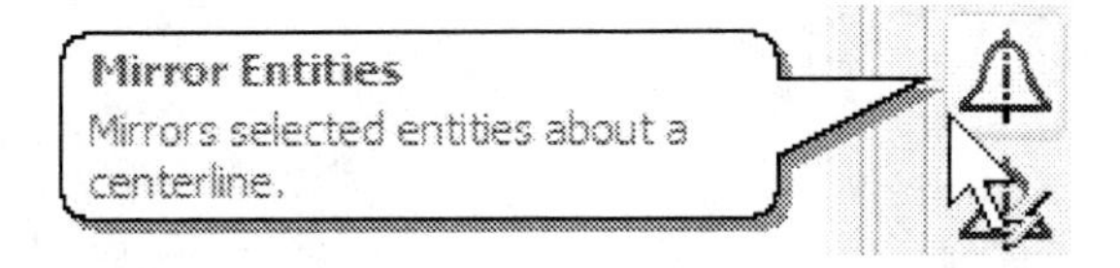

90. - In the "Entities to Mirror" selection box select the circle. Then click inside the "Mirror About" selection box to activate it (it will turn pink) and select the vertical centerline. Notice the preview of the mirrored circle. Click OK to complete the first sketch mirror.

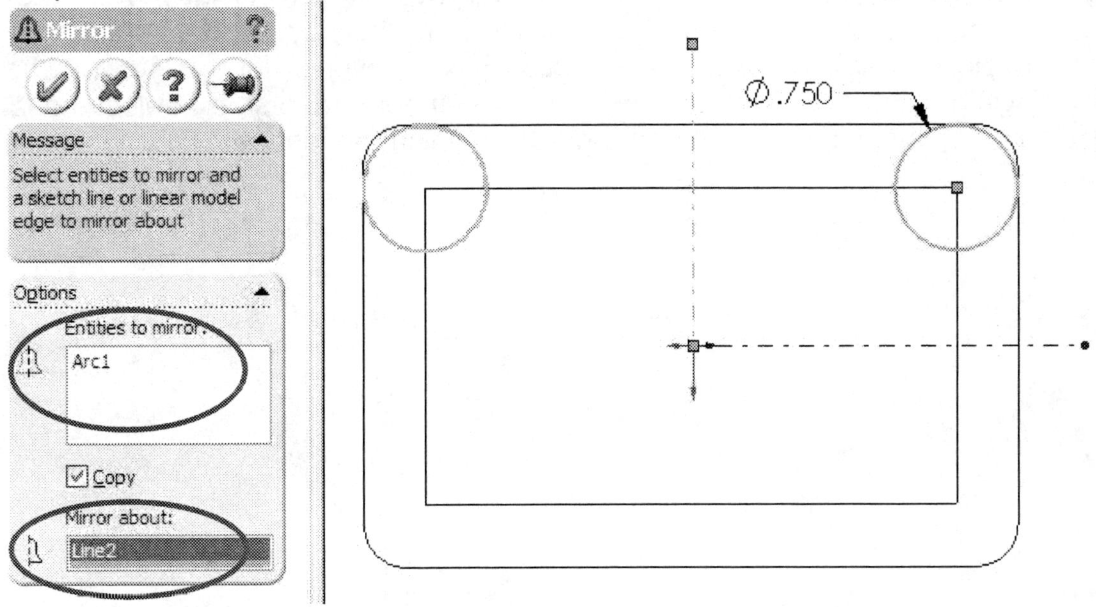

91. - We will now repeat the Mirror Entities command selecting both circles under the "Entities to Mirror" selection box, and the Horizontal centerline in the "Mirror About" selection box.

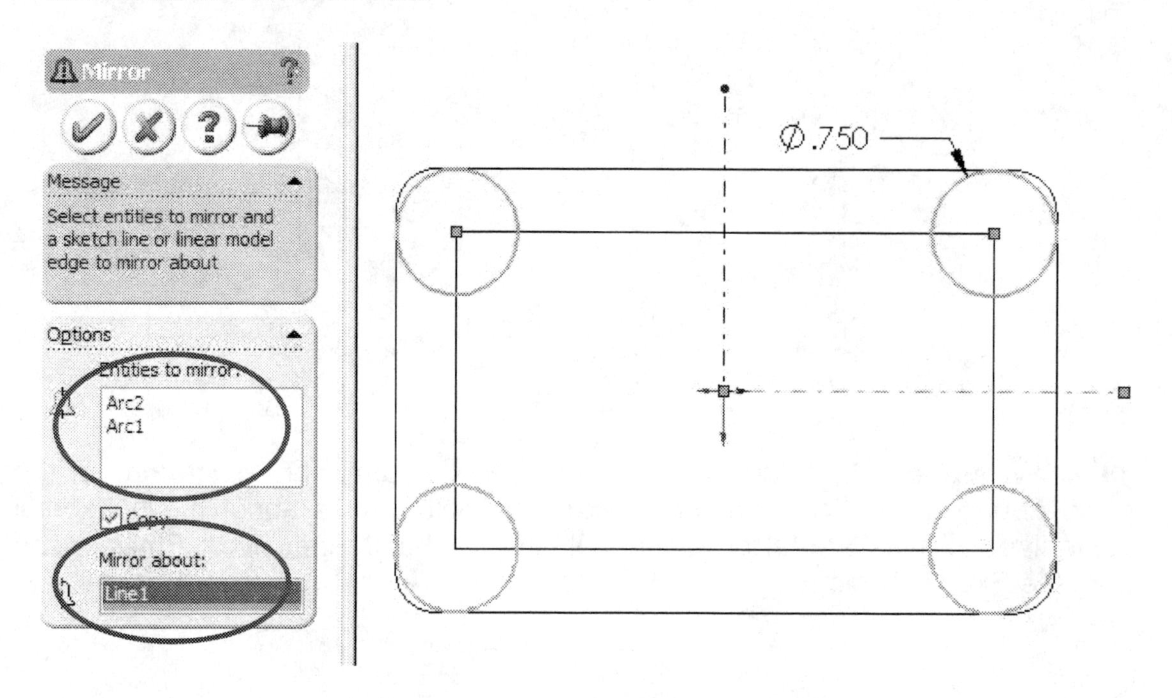

92. - Now we are ready to make the cut. In this step we'll cut all four corners at the same time. SolidWorks allows us to have multiple closed contours in a sketch for one operation, either Cuts or Extrusions as long as they don't intersect each other or touch at one point. To add intelligence to our model, we'll also use a special end condition in the cut feature called **Up to Surface**; with this end condition we can define the stopping face for the cut. Select the "Extruded Cut" icon, and select the "Up To Surface" option from the "Direction 1" options drop down selection box. Now a new selection box is revealed and is active, where we have to select the face where we want the cut feature to stop. Select the face indicated as the end condition and click OK to finish the feature.

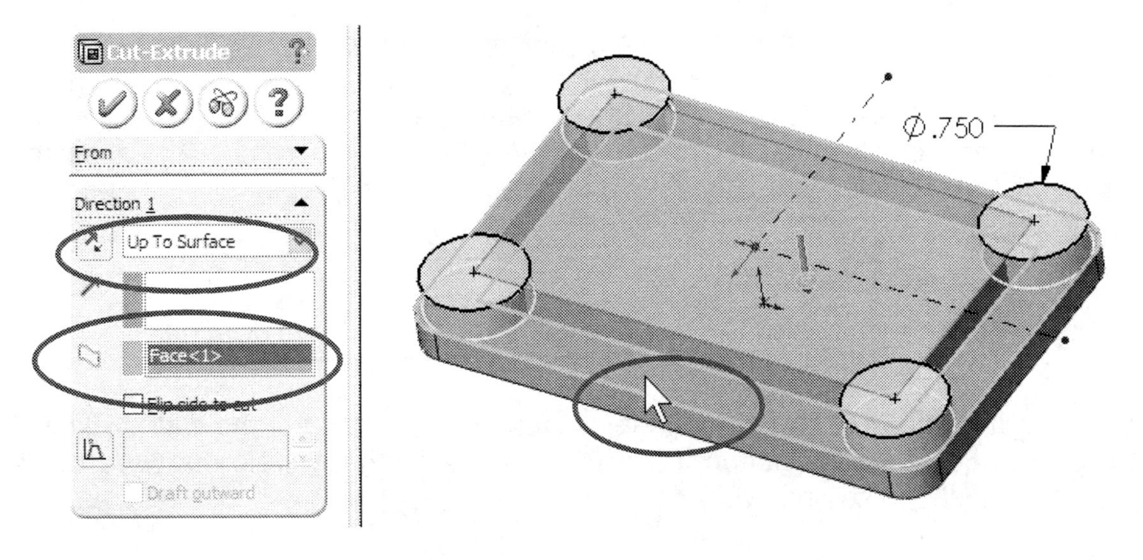

Our part should now look like this:

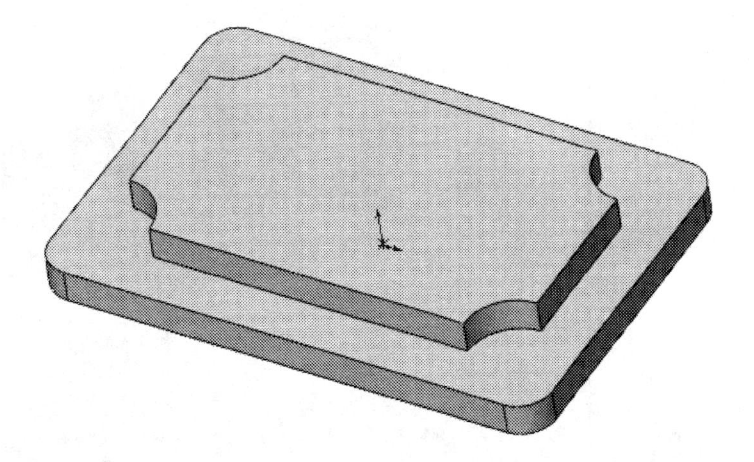

93. - Add a 0.25" Fillet to the edges indicated to round the corners.

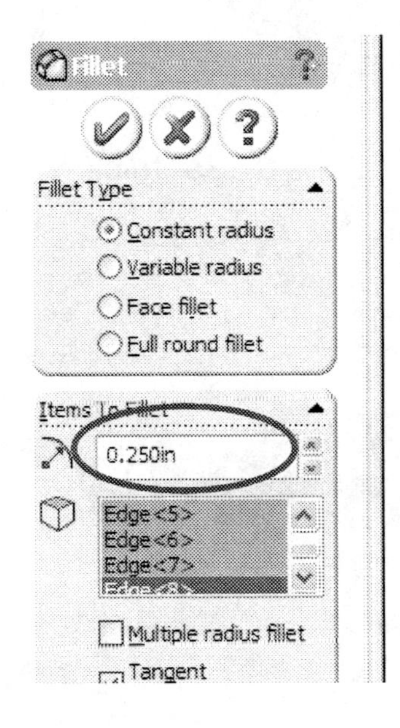

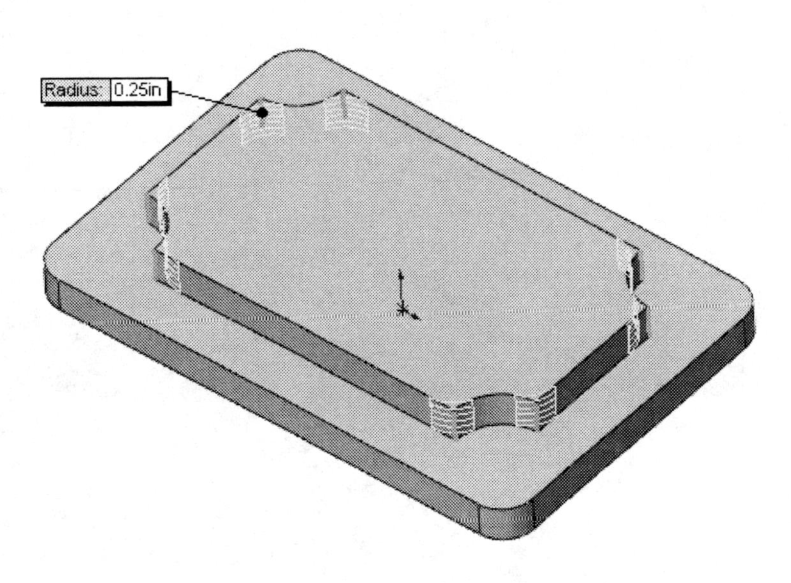

94. - Since this is going to be a cast part, we want to remove some material from the inside, and make its walls a constant thickness. The **Shell** feature is the correct tool for this purpose. The Shell creates a constant thickness part by removing one or more faces from the model making the remaining faces the same thickness. Select the "Shell" icon from the Features Toolbar;

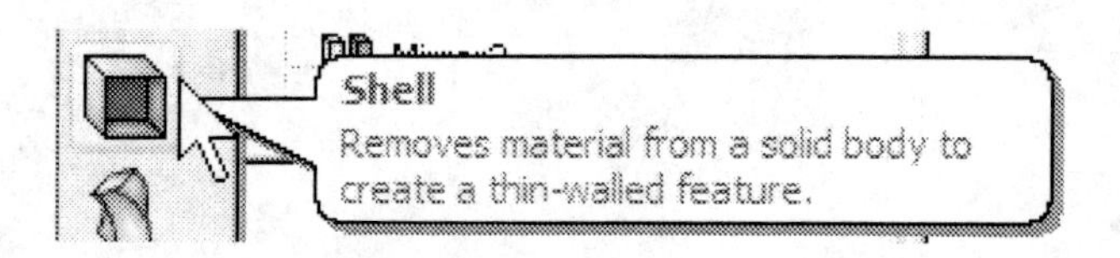

95. - In the Shell's Property Manager under "Parameters" set the wall thickness to 0.125" and select the bottom face, this is the face that will be removed. Click on OK to finish the command.

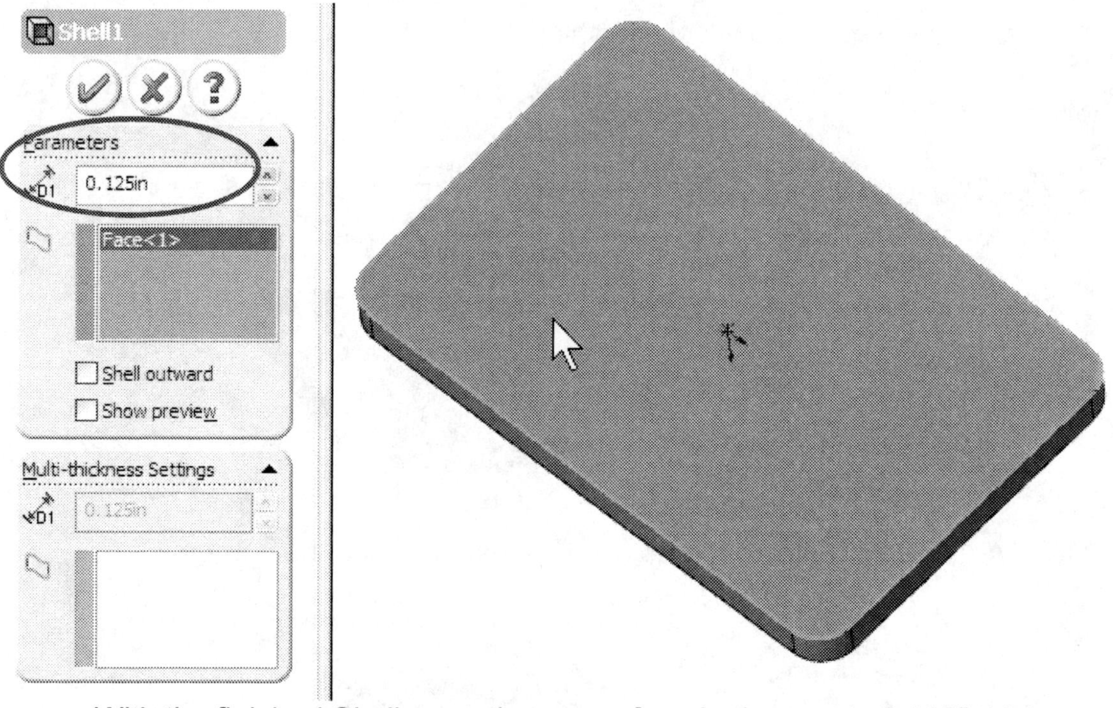

With the finished Shell operation every face in the part is 0.125" thick.

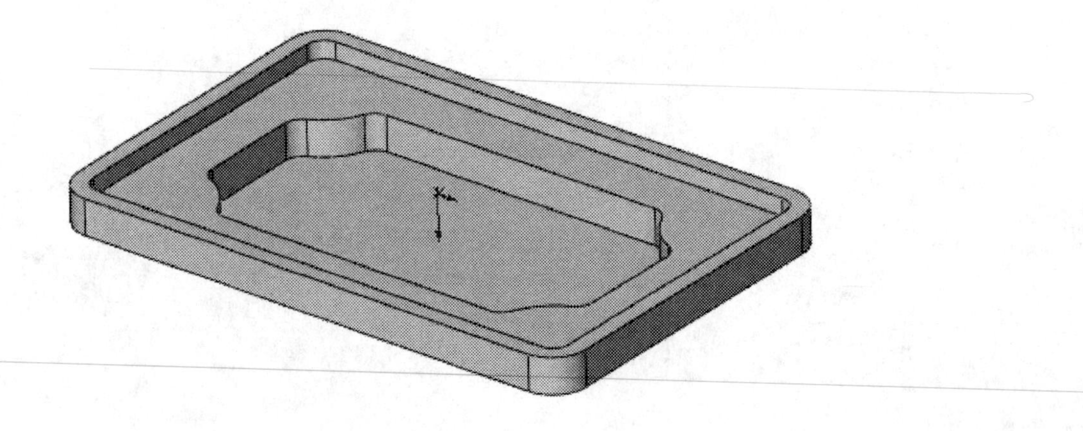

96. - The shell operation left the corners weak since it removed material from underneath them, and now we have to reinforce them by adding a simple extrusion to have enough support for a screw. Switch to a Bottom View and insert a sketch on the bottommost face as indicated, making sure to make the circles concentric to the fillet edges. Don't worry too much about the size of the other circles.

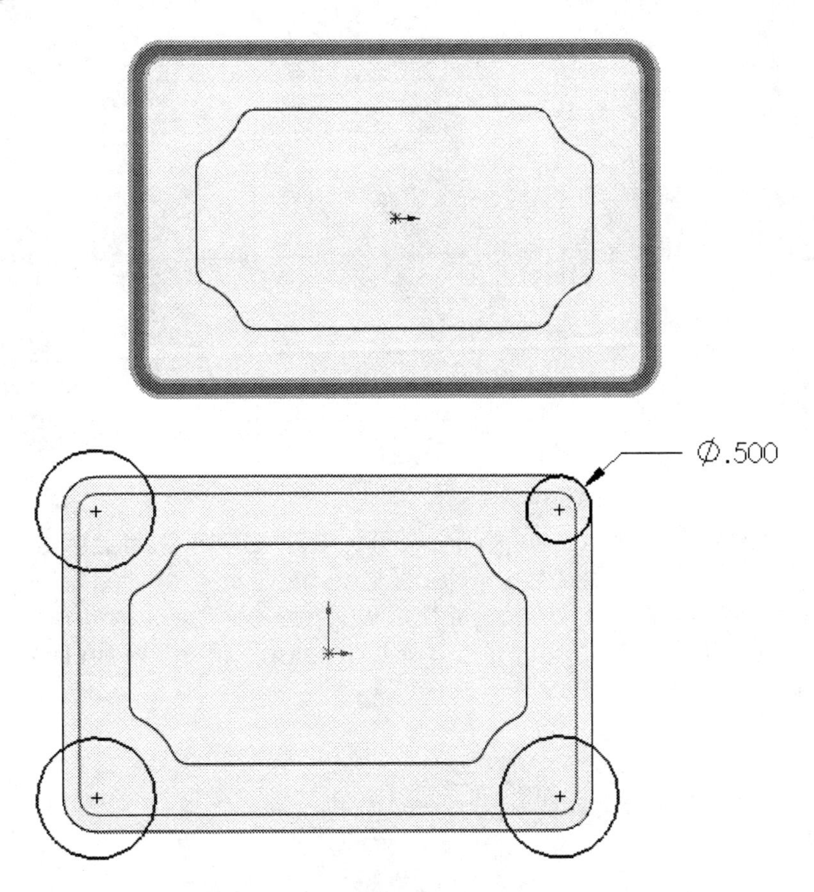

97. - In this case, instead of making the circles using the "Mirror Entities" command as we did before, we will use sketch relations to make all circles equal. This is just to show a different way to accomplish the same result. Select the Add Relation icon from the Sketch Toolbar, select all four circles and under "Add Relations", then select **Equal** to make all circles the same size. By dimensioning only one circle, the sketch is now fully defined.

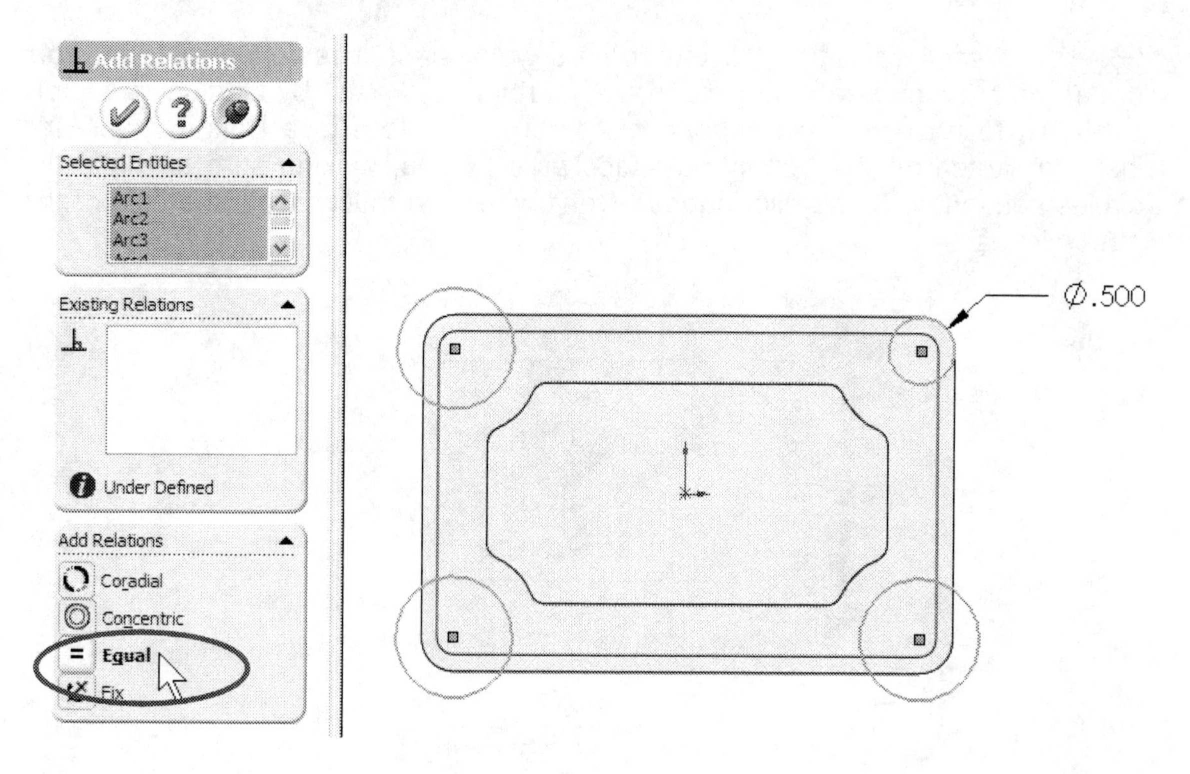

98. - We will now make the extrusion using the "Up to Surface" end condition as we did in the last cut. Select the face indicated in the next image as the end condition and select OK. By doing it this way we can be sure that if the extrusion on top changes, the corner cuts and this extrusion will maintain the design intent.

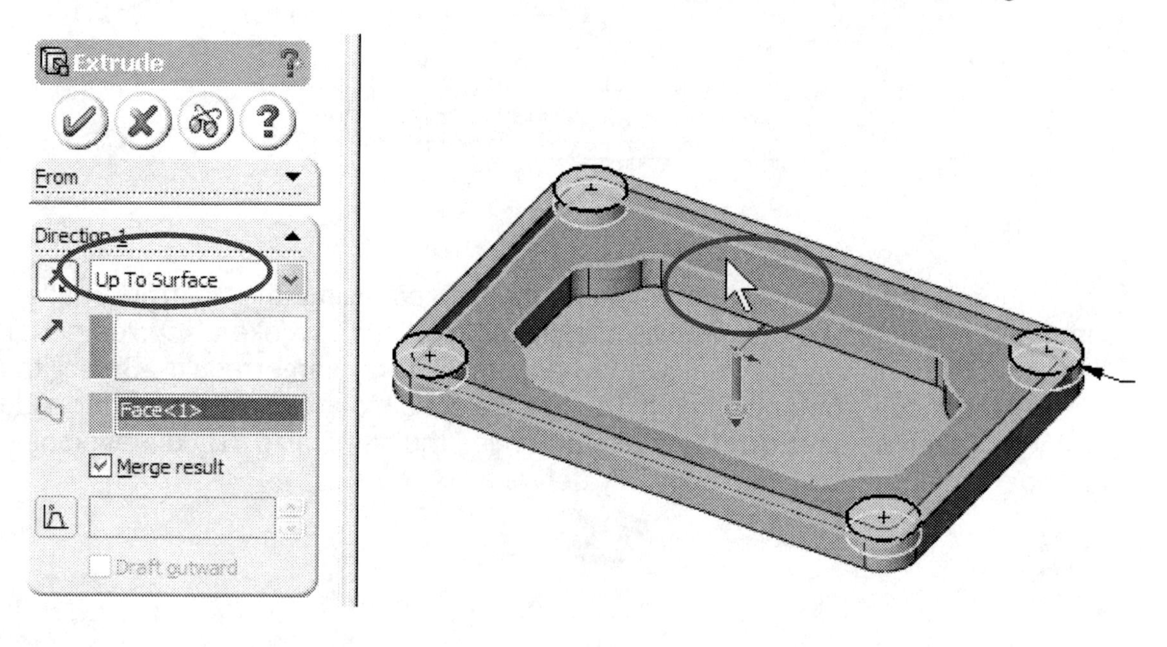

Our part should now look like this:

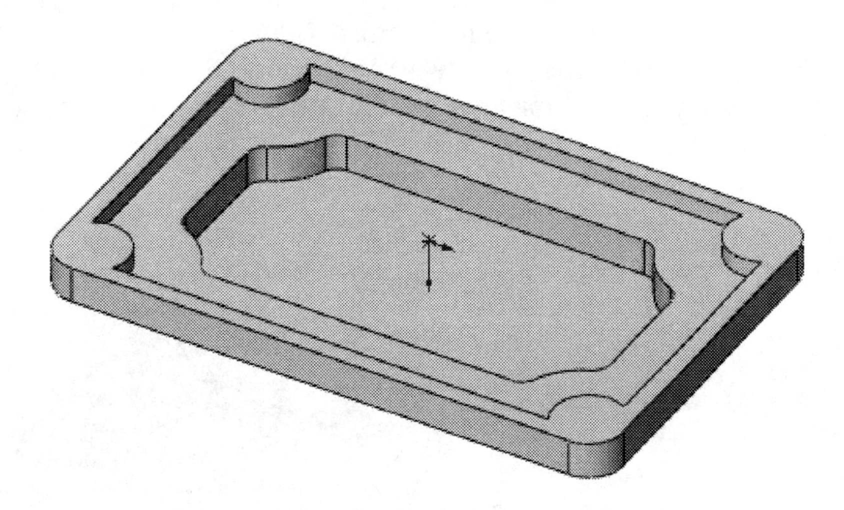

99. - Add a 0.031" Fillet to the three edges indicated; notice the fillet propagates through the tangent edges.

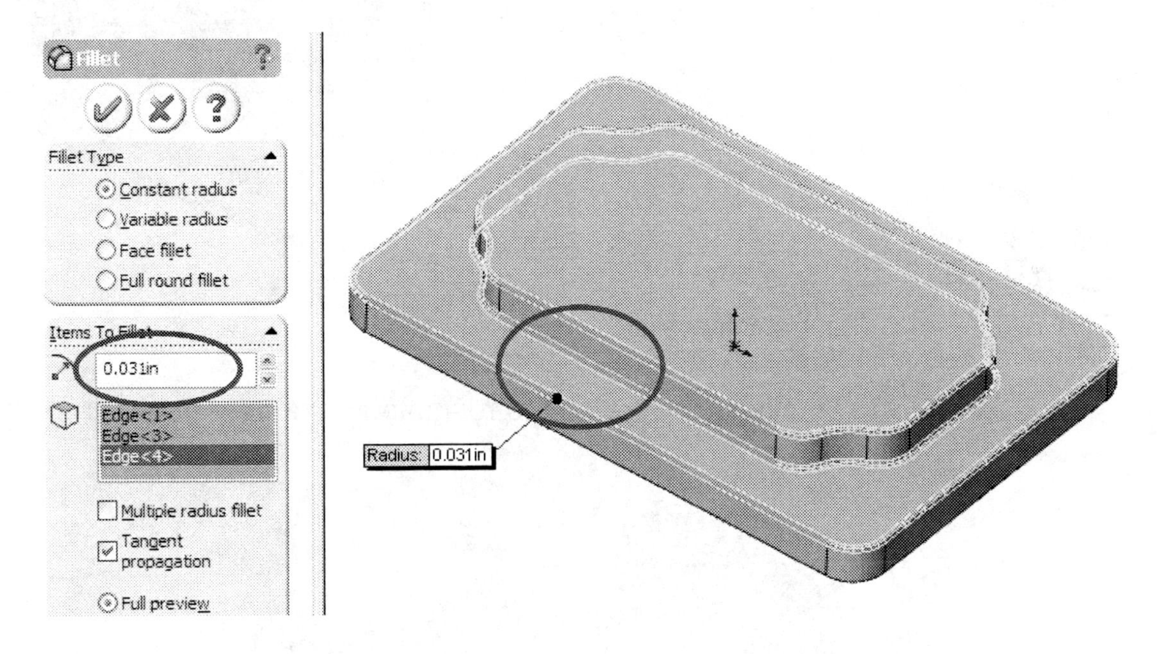

100. - To add fillets to the entire inside edges of the part we will use a different technique. Instead of individually selecting the inside edges, we will select only the "Shell1" feature from the fly-out Feature Manager, with a 0.031" radius. Adding the fillet like this rounds every edge of the "Shell1" feature, making it much faster and convenient, not to mention that it maintains the design intent better.

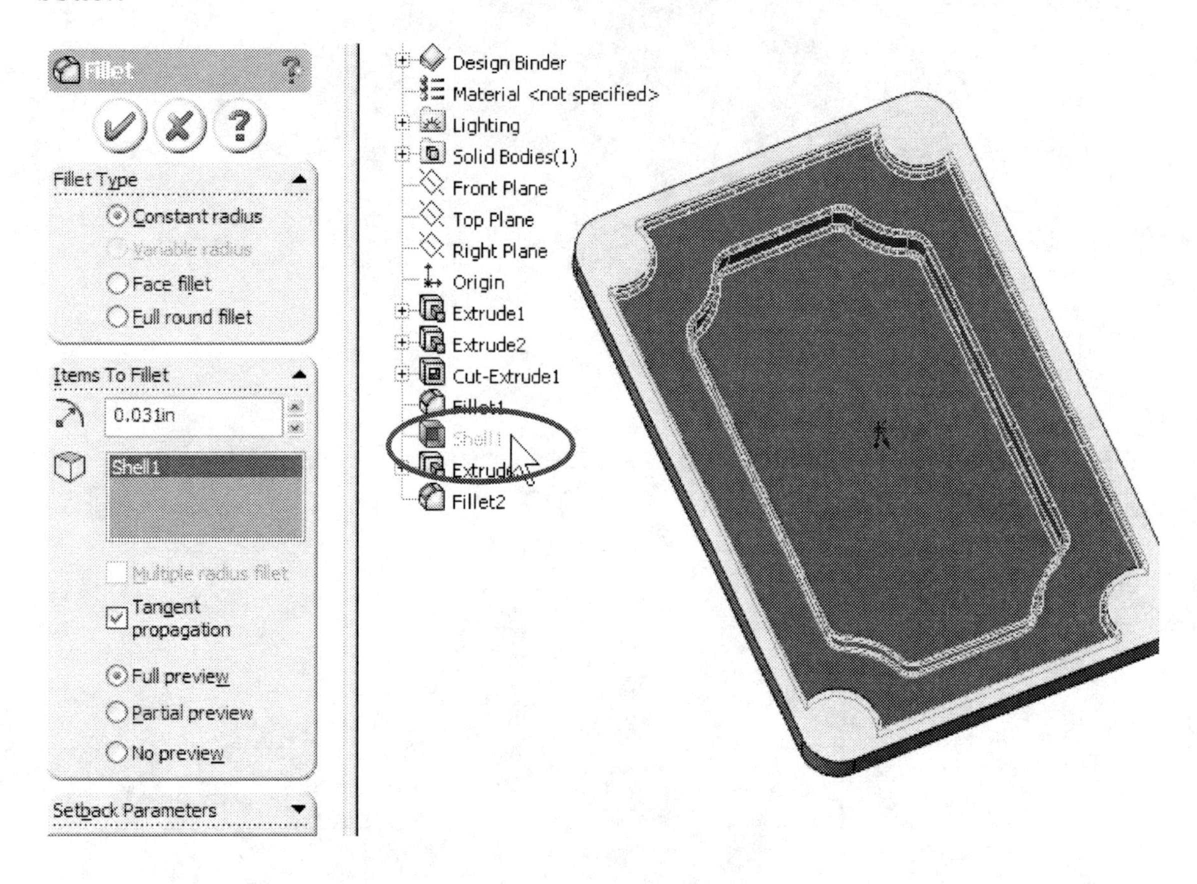

Now every internal edge of the part is rounded in one operation with only one selection.

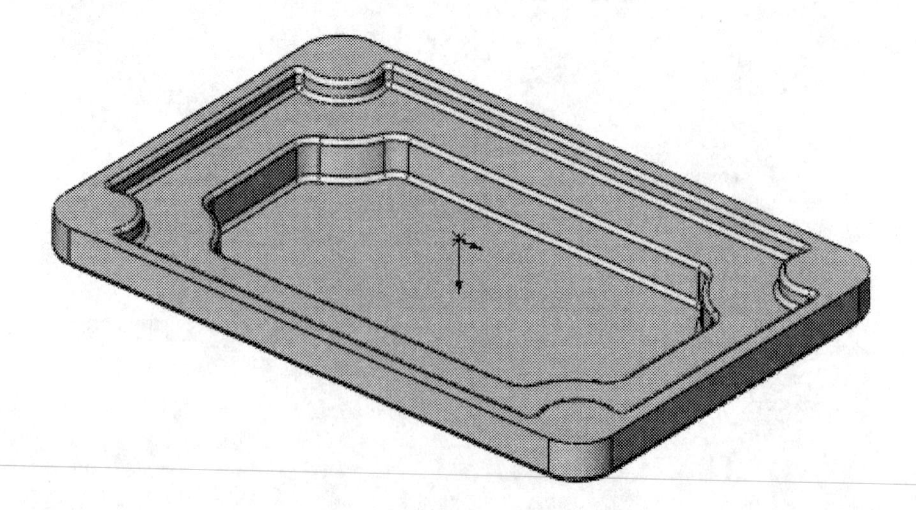

101. - We are now ready to add the four clearance holes for the #6-32 screws on the top cover. We'll use the Hole Wizard for this feature. Switch to a Top view and select the top face. Remember to pre-select the face before initiating the Hole Wizard.

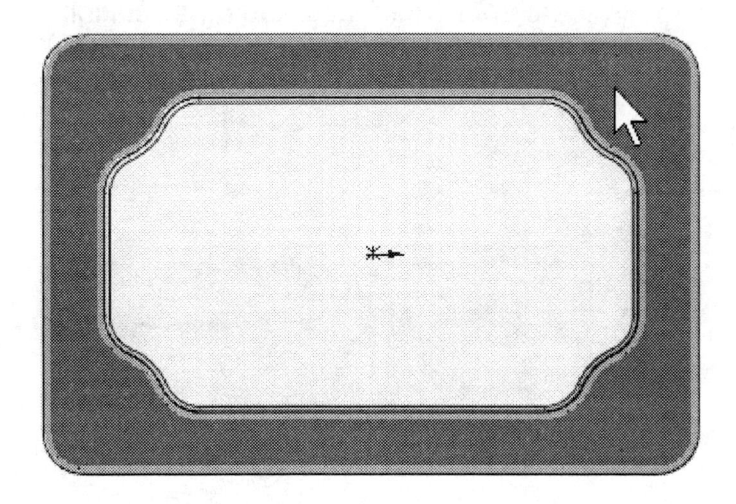

102. - In the first step of the Hole Wizard, select the "Hole" tab at the top. From the "Screw Type" drop down list select "**Screw Clearances**". From the "Size" selection list pick "# 6" and from "Hole Type & Depth" select "Through All". Click "Next" to move to the second step to locate the holes.

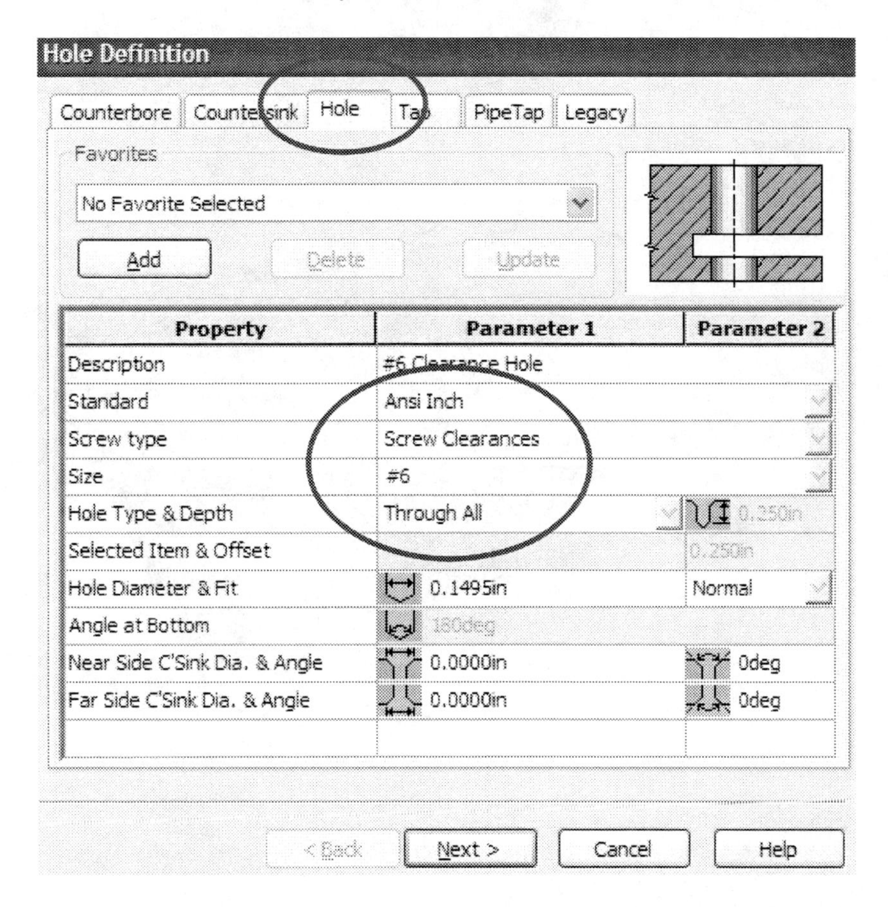

103. - In the second step notice we already have one point in the sketch. Remember that SolidWorks adds one point where we pre-selected the face. We'll now add three more points to create three more holes. Notice that the Point Tool is active, simply click were we want to draw the three extra points, making sure they are concentric to the corner fillets, and adding a relation to the existing point to make it concentric to the last fillet. Click Finish to complete the Hole Wizard.

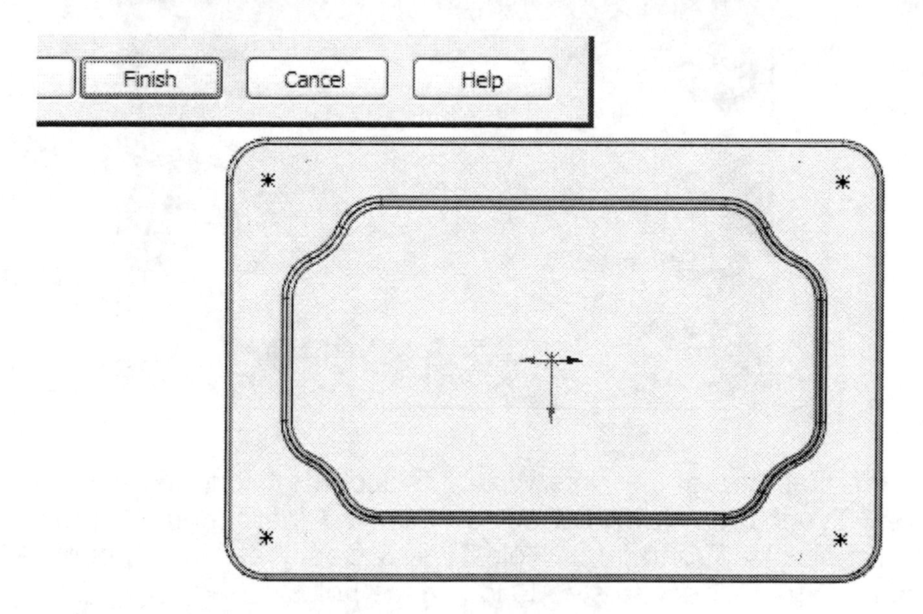

104. - Save the finished part as "Top Cover" and close the file.

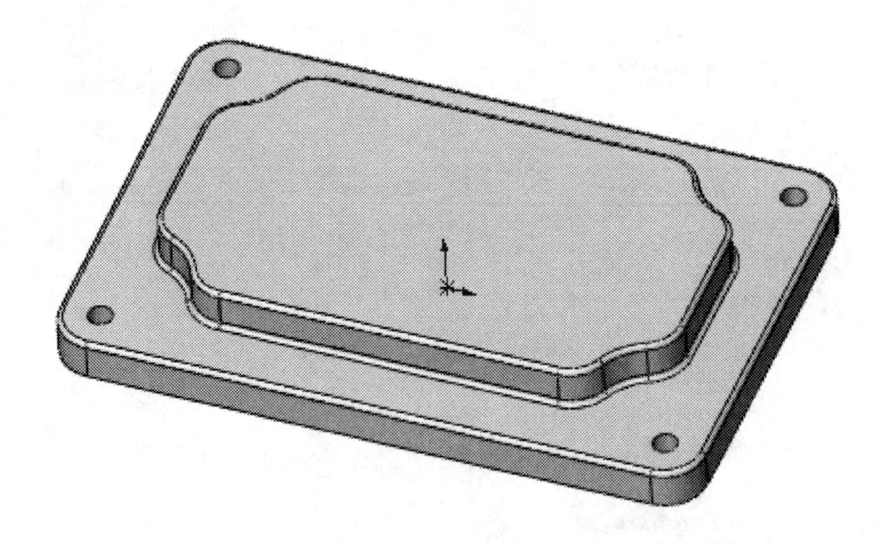

The Offset Shaft

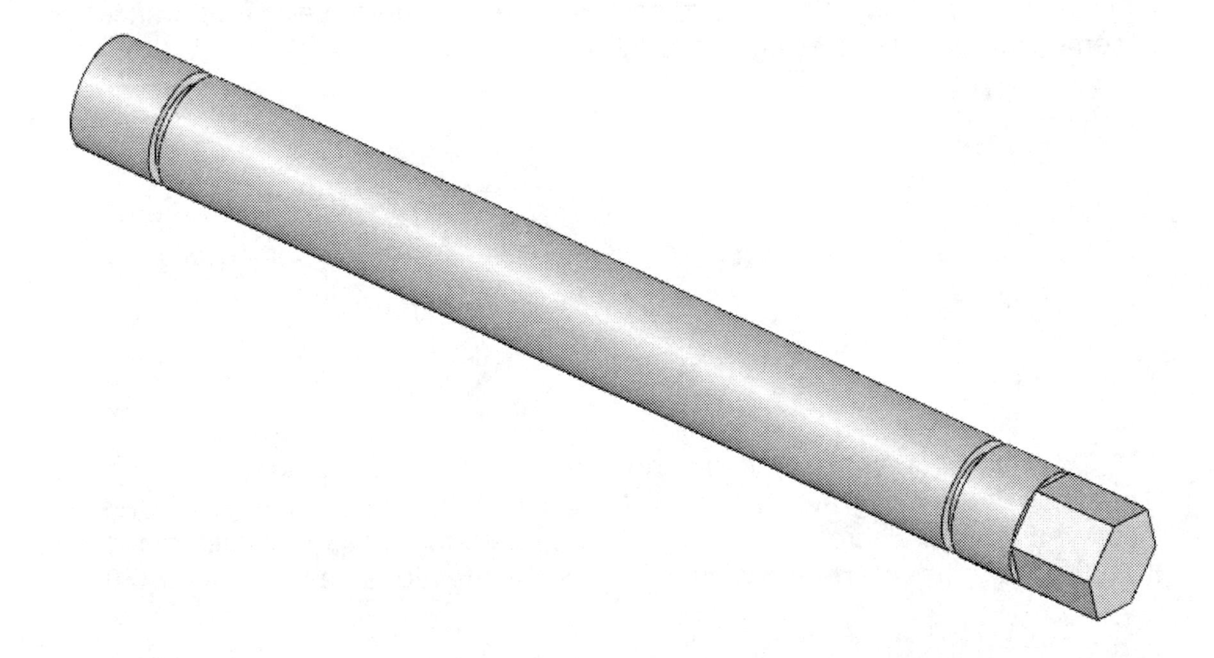

For the Offset Shaft we'll follow the next sequence of operations. In this part we will only need three features and we'll learn how to make polygons in the sketch, a new option for the Cut Extrude feature and a Revolved cut.

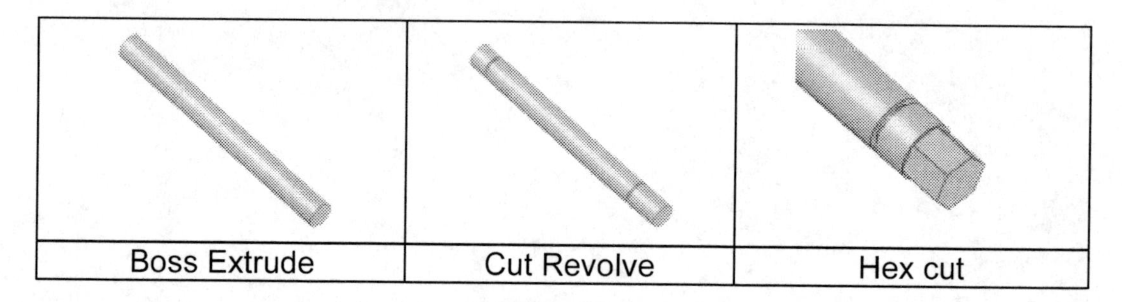

105. - For the first feature in this part we'll make a sketch in the Right plane and dimension it as shown. Since this shaft will need to meet certain tolerances for assembly, we will give a tolerance to the diameter.

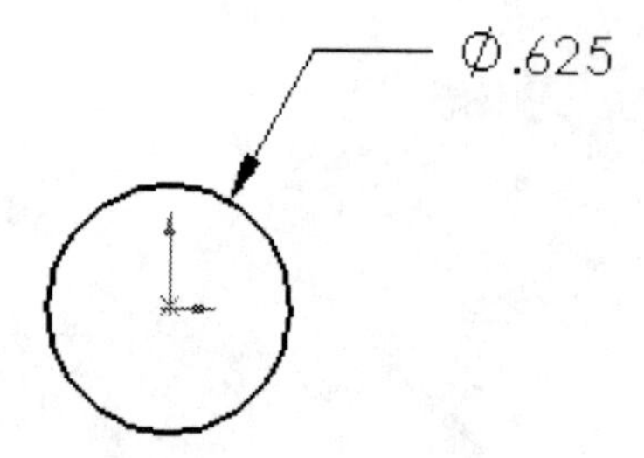

106. - To add (or change) the tolerance of the shaft's diameter, select the dimension from the graphics area. Notice that the dimension properties are displayed. This is where we can change the tolerance type. For this shaft select from the "**Tolerance/Precision**" options box "Bi-directional" and +0.000"/-0.005"

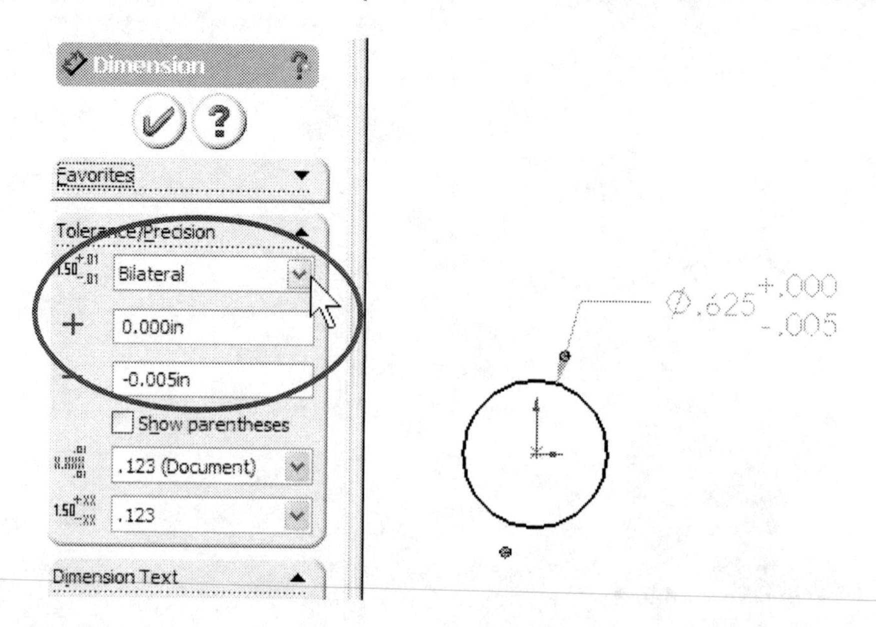

107. - Now extrude the shaft 6.5" as indicated.

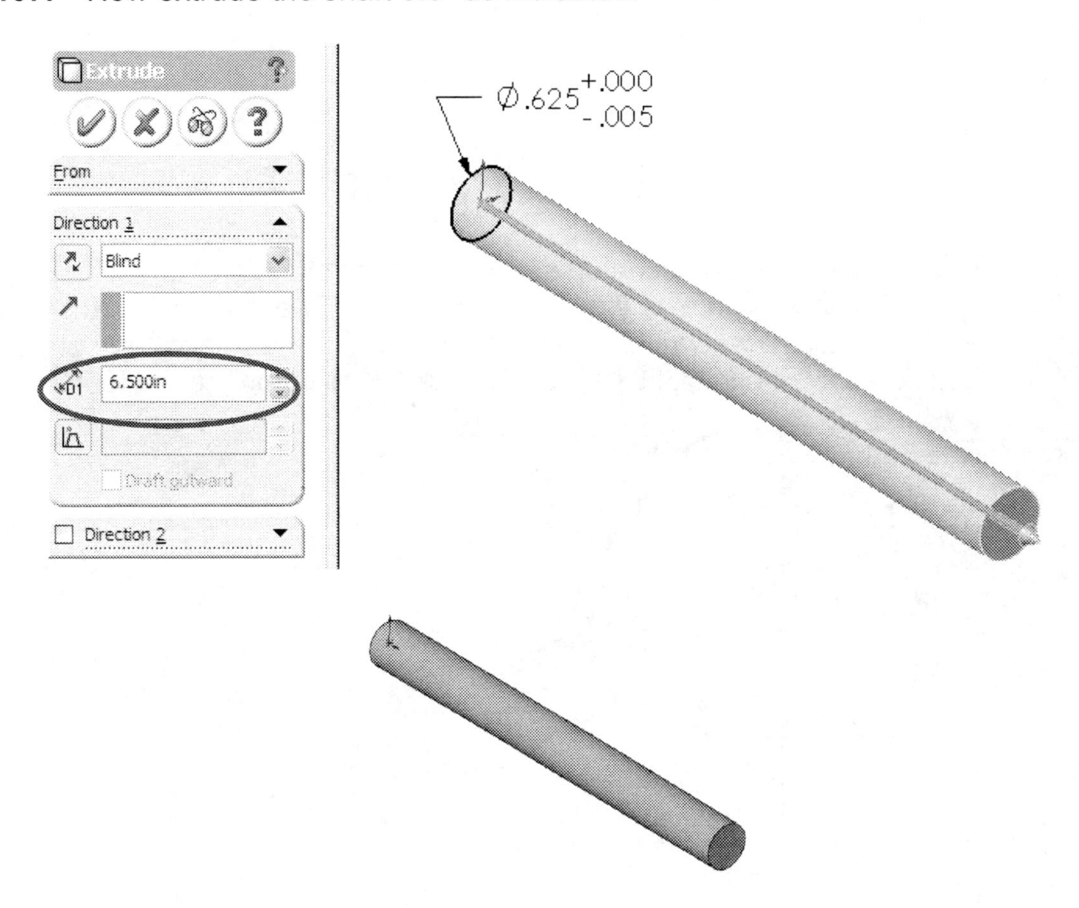

108. - For the second feature we'll make a revolved cut. As its name implies, we'll remove material from the part similar to a turning operation. Switch to a Front View and select the Front Plane from the Feature Manager, *then* click in the Sketch icon. Draw the following sketch and be sure to add the centerline; this will be the profile that will be used as a "cutting tool". It's good practice to add a centerline to the sketch to use it as a center of rotation for the cut. Notice that we can make a Revolved Cut with multiple profiles.

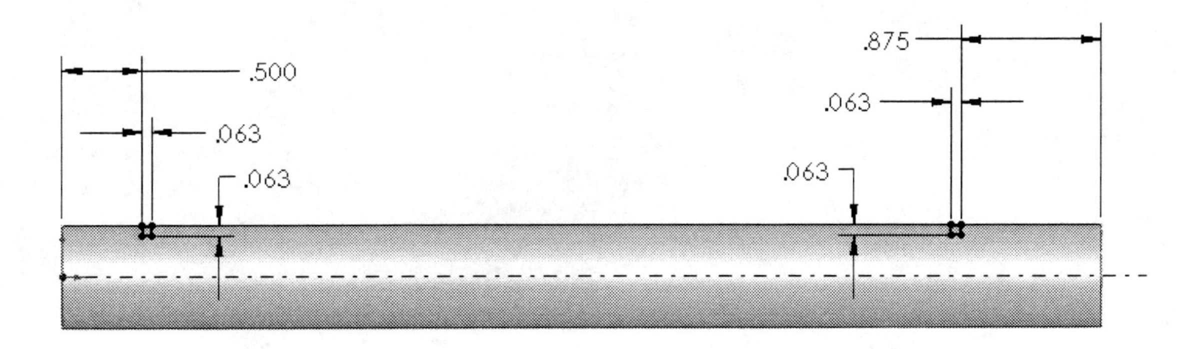

109. - Now that we are done with the sketch, select the "**Revolved Cut**" icon from the Features Toolbar.

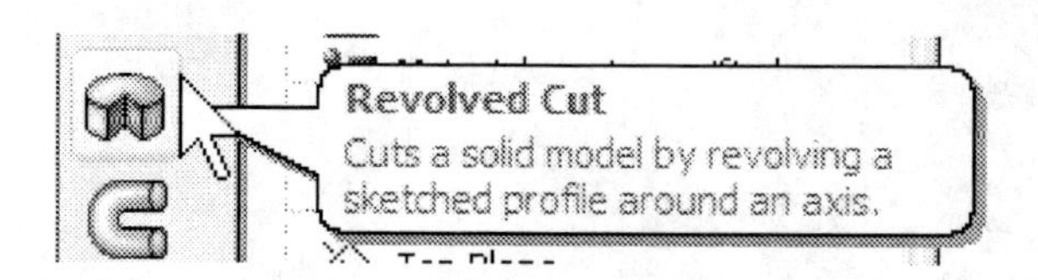

110. - Since we only have one centerline in the sketch, SolidWorks automatically selects it as the axis to make the Revolved Cut; if we had more than one centerline in the sketch, we would be asked to select one to make the cut about it. The default revolved cut is 360 degrees. Click OK to complete the feature.

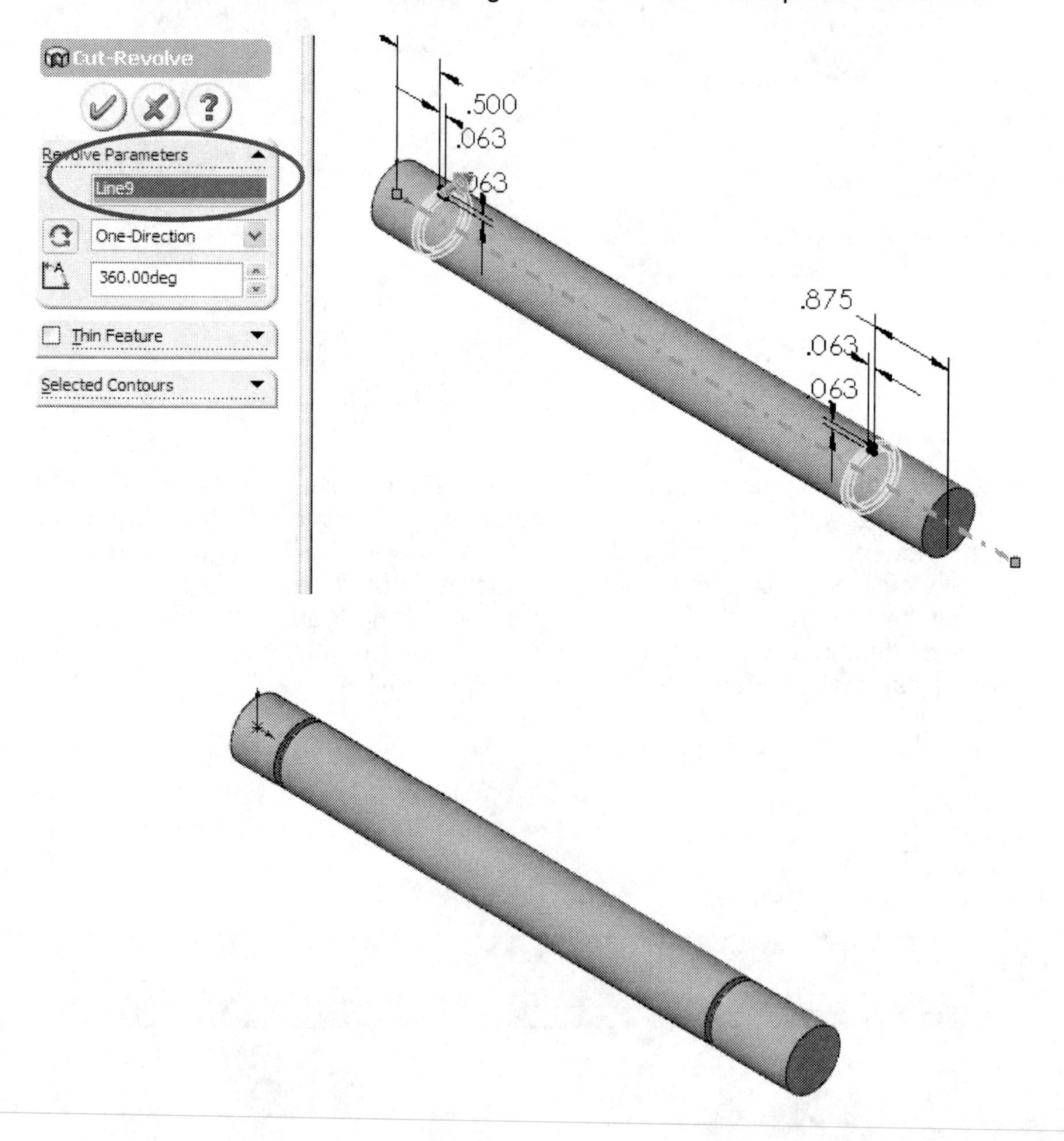

111. - For the last feature, we'll add a hexagonal cut at the right end of the shaft. Switch to a Right view, and insert a sketch in the rightmost face of the shaft.

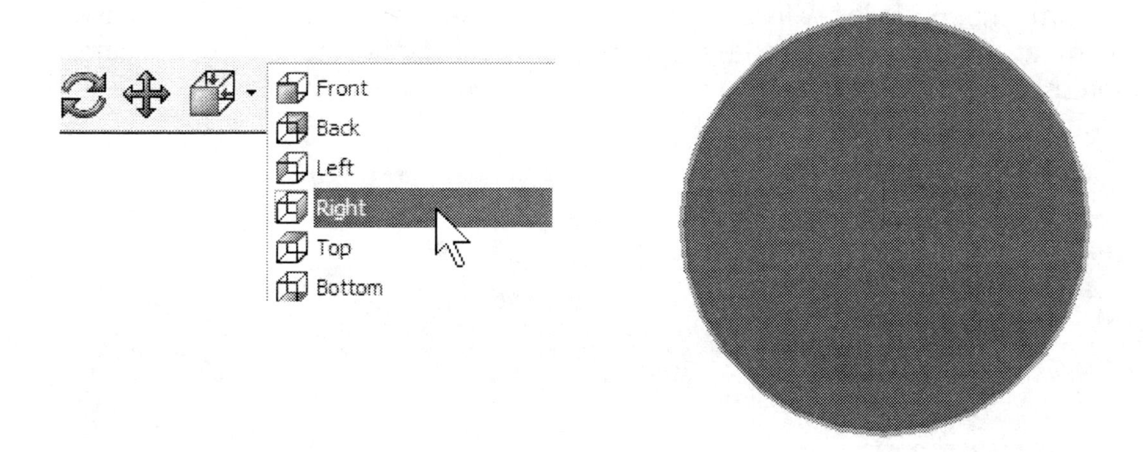

112. - We can make the hexagon utilizing dimensions and geometric relations, but we really want to do it easily, so we'll use the "**Polygon**" tool in SolidWorks. Go to the "Tools" menu, "Sketch Entities, Polygon"

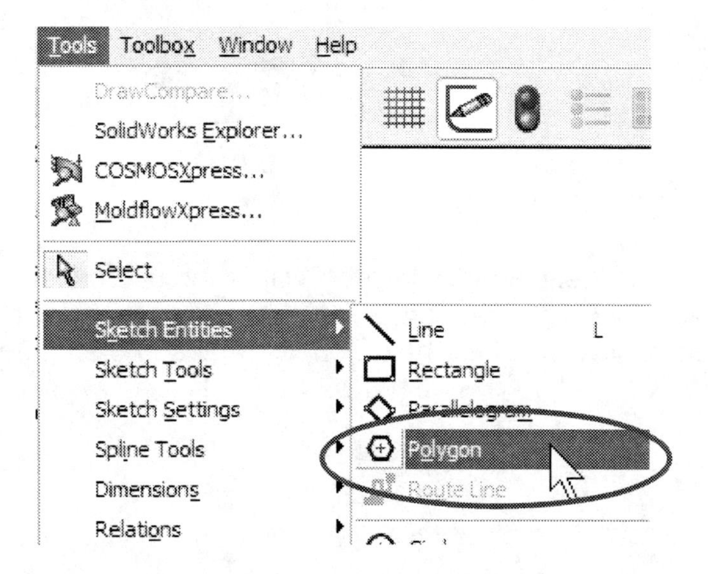

...or select the Polygon icon from the Sketch Toolbar, if available. (Remember that SolidWorks Toolbars can be customized to add or remove icons as needed.)

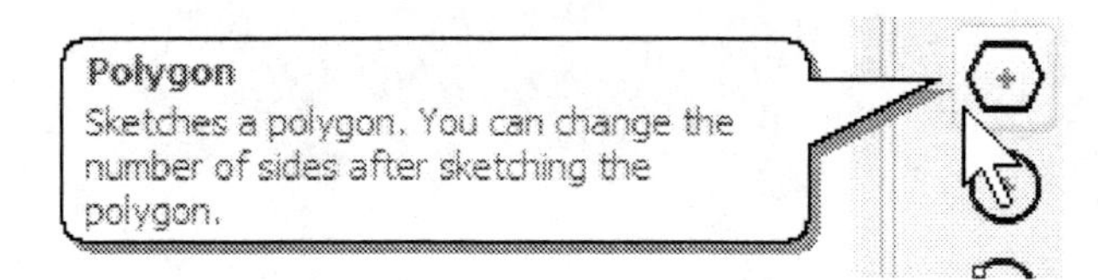

113. - After we select the "Polygon" tool, we are presented with the options in the Property Manager. This tool helps us to create polygons by making it either inscribed or circumscribed to a circle. For this exercise we'll select the "Circumscribed circle" with 6 sides from the "Parameters" options. Don't worry too much about the rest of the options, as we'll define the hexagon using two more geometric relations.

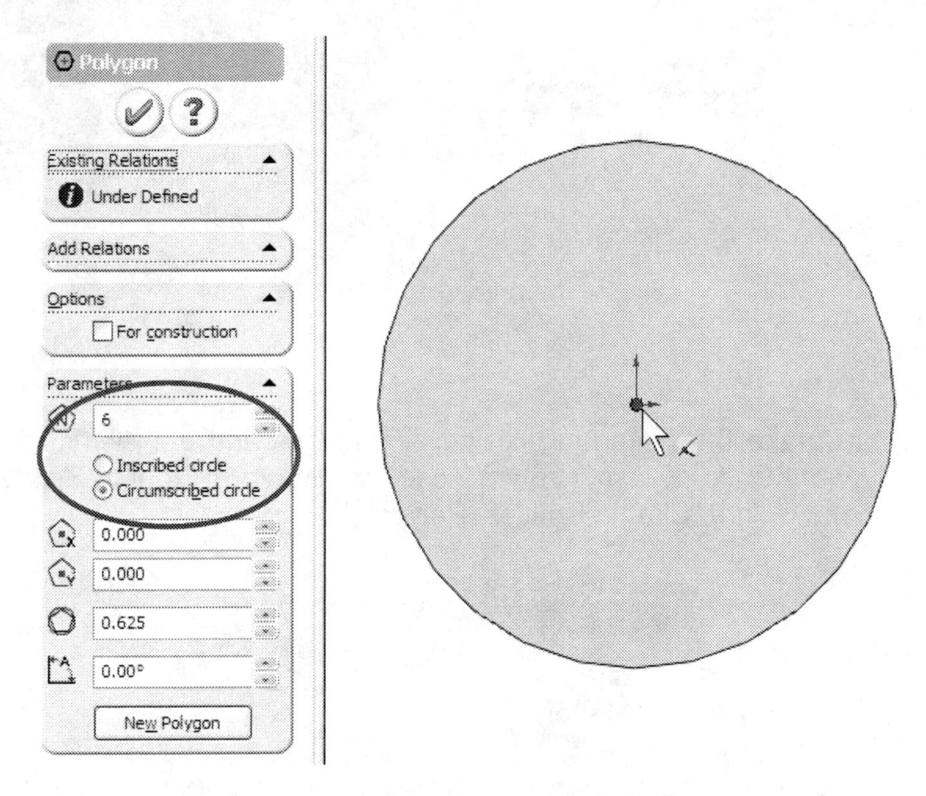

114. - Since we are defining the polygon with a circumscribed circle, we have to draw a circle. Start at the center of the shaft as shown; notice that we immediately get a preview of the hexagon, its radius and angle of rotation.

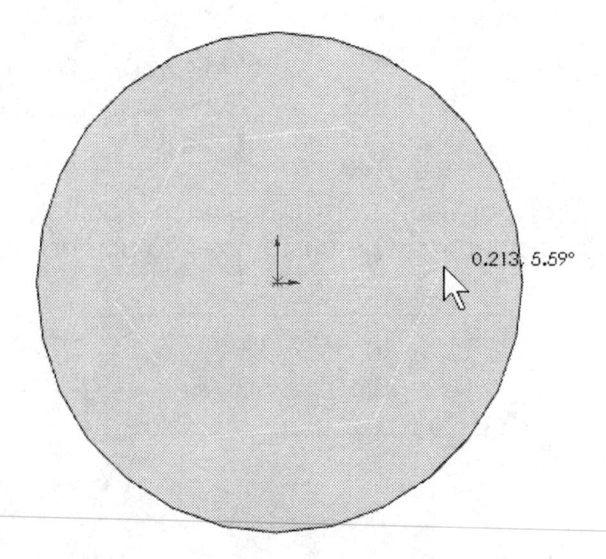

115. - Draw the circle a little smaller or larger than the shaft, the idea is to make the circumscribed construction circle the same size as the shaft using a geometric relation. Hit "ESC" or turn off the polygon tool when done.

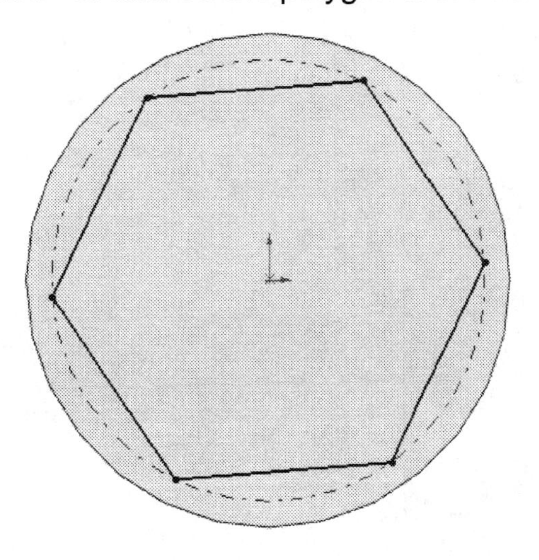

116. - Now select the "Add Relation" tool and select the polygon's construction circle and the edge of the shaft; add an "Equal" relation to make them the same size.

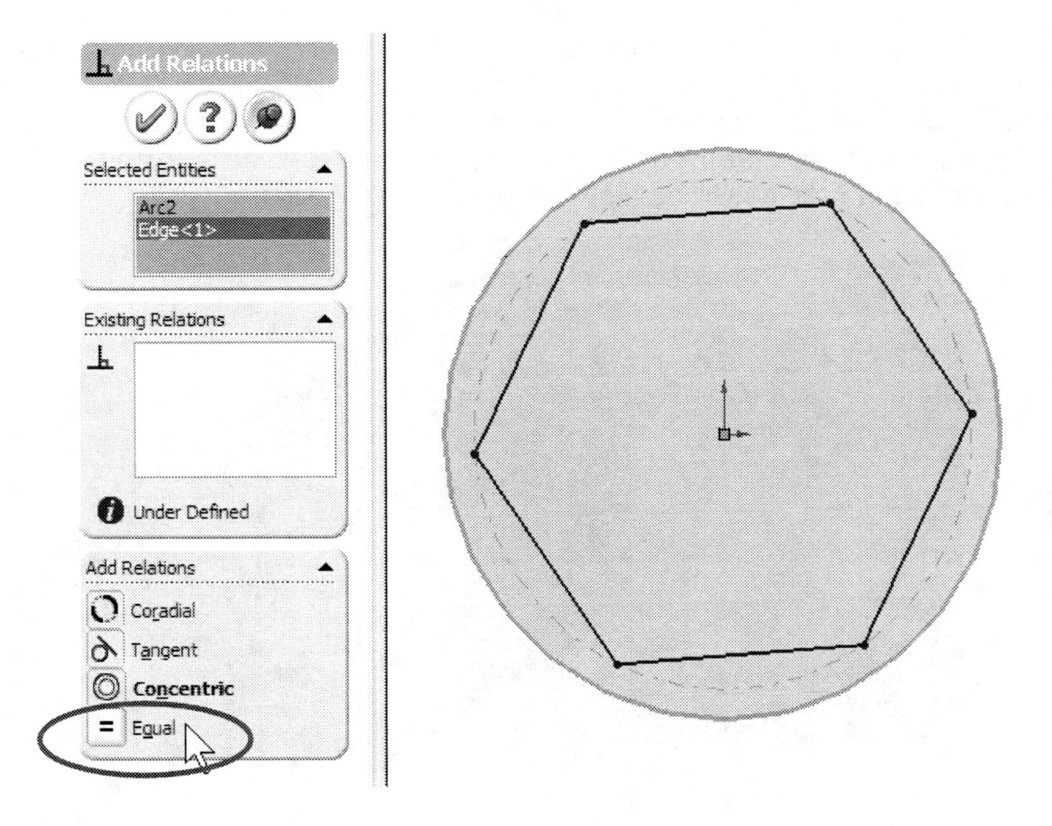

117. - Now select the center of the shaft (the origin) and one endpoint in the hexagon. Add a "Horizontal" relation to fully define the sketch.

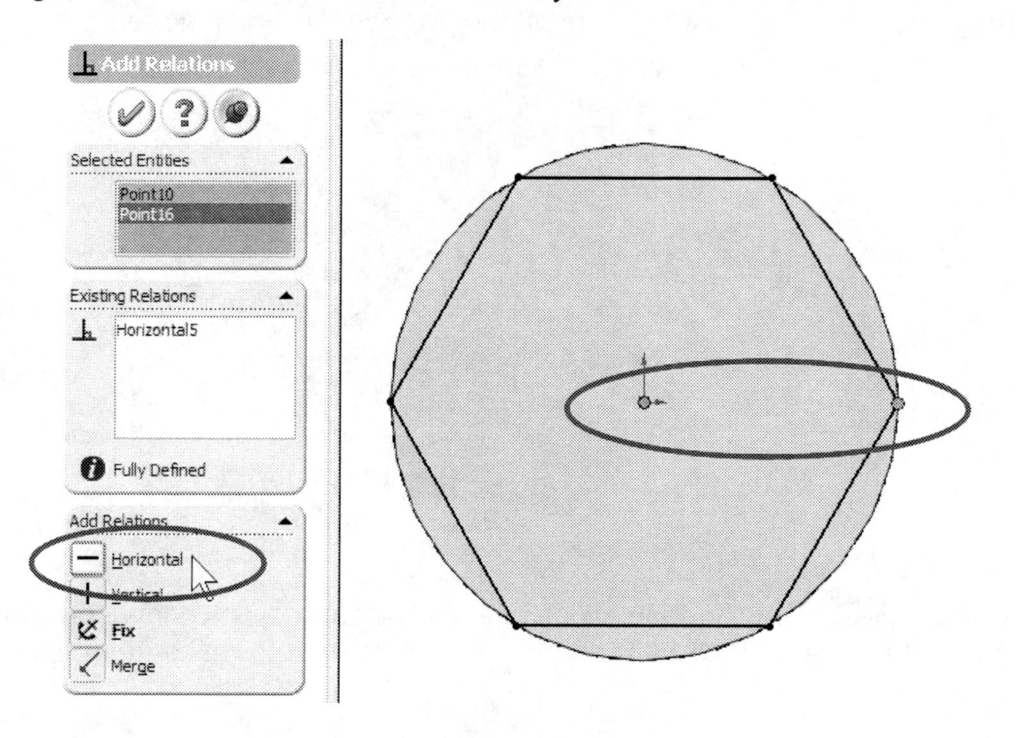

118. - Now we are ready to make the cut. For this operation we'll use a little used but very powerful option in SolidWorks. Select the "Extrude Cut" icon as we have done before, but now activate the checkbox "**Flip side to cut**". This option will make the cut *outside* of the sketch, not inside. Notice the arrow indicating which side of the sketch will be cut. Make the cut 0.5".

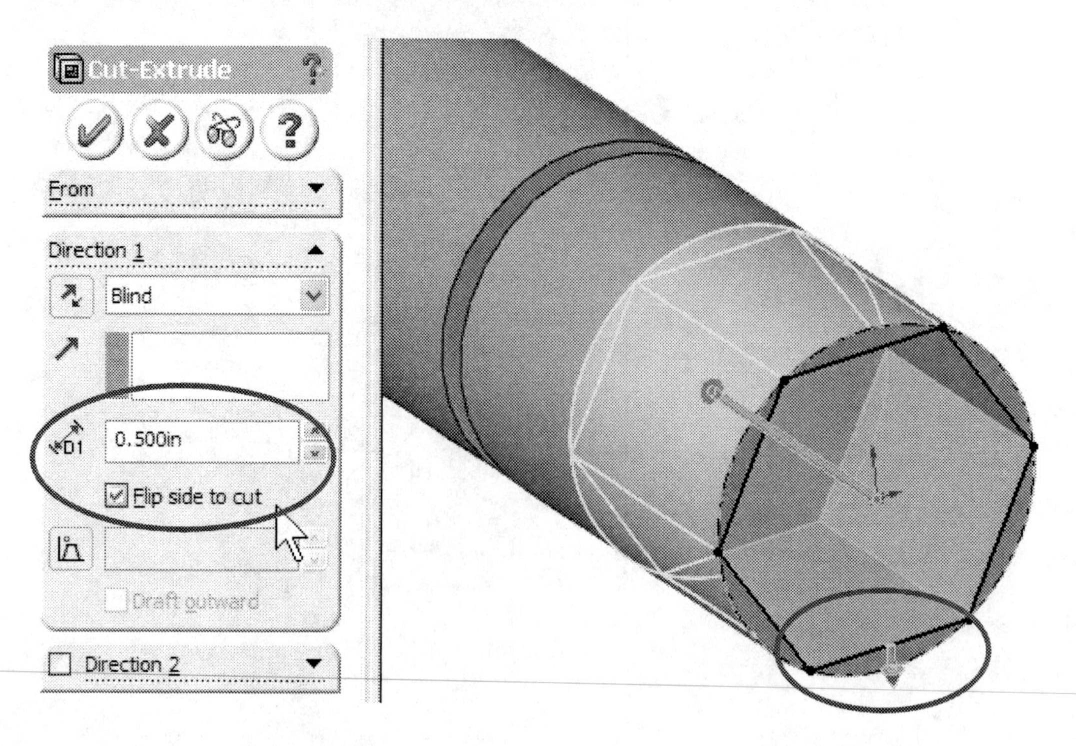

119. - Save the part as "Offset Shaft" and close the file.

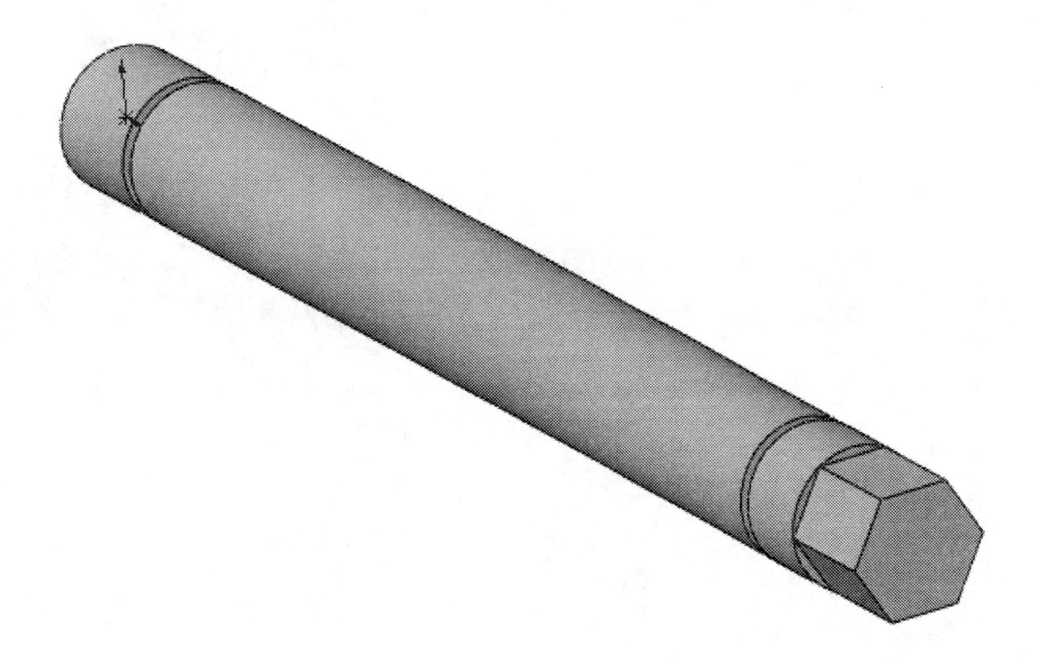

Notes:

The Worm Gear

For the Worm Gear we will make a simplified version of the gear. The intent of this book is not to go into gear design, but rather to help the user understand and learn how to use basic SolidWorks functionality. With this part we'll learn a new extrusion (or cut) end condition called Mid Plane, how to chamfer the edges of a model, add diameter dimensions to a sketch used for a revolved feature, and dimension to a quadrant of a circle or circular edge. We'll also practice previously learned features. For the Worm Gear we will follow the next sequence of features.

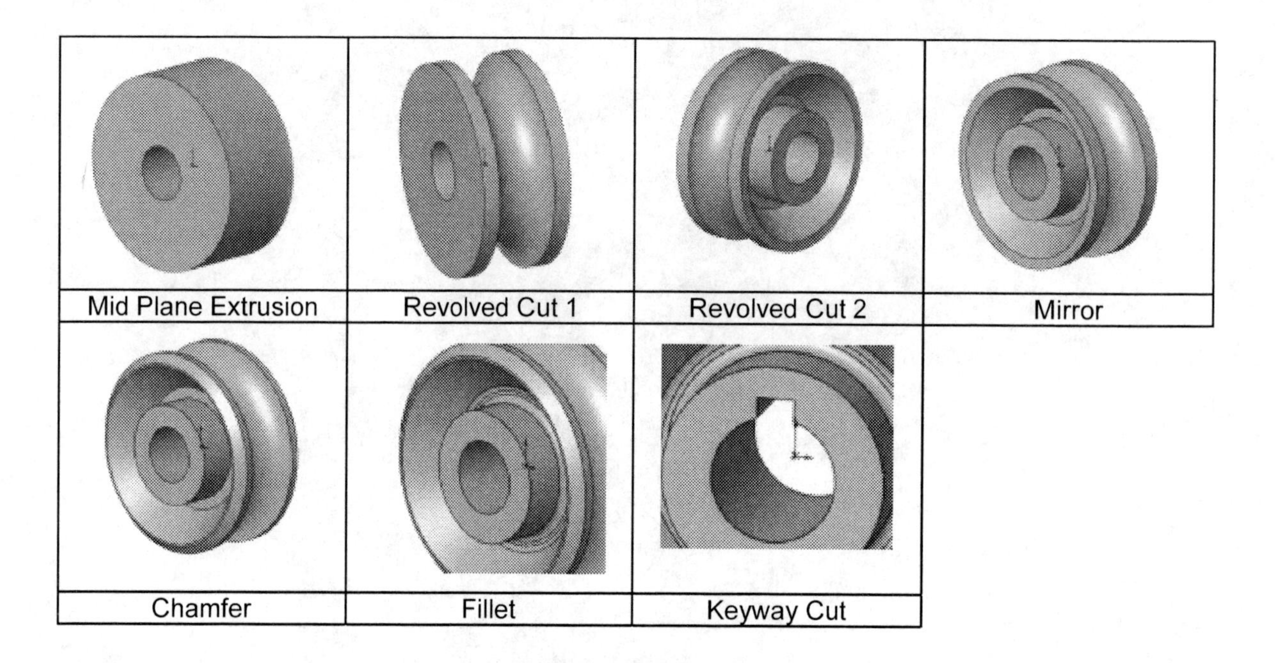

120. - For the "Worm Gear" start by making a new part and draw the following sketch in the Front plane.

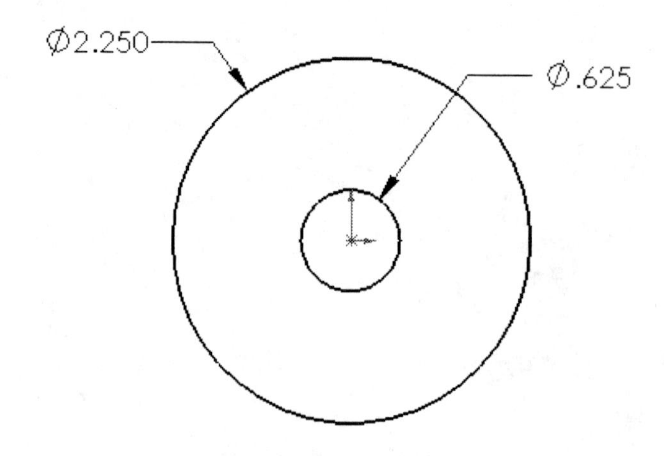

121. - In this case we want the part to be symmetrical about the Front Plane. To achieve this, we'll extrude it with the "**Mid Plane**" end condition; this condition extrudes half of the distance in one direction, and half in the second direction. Change the end condition to "Mid Plane" and extrude it 1".

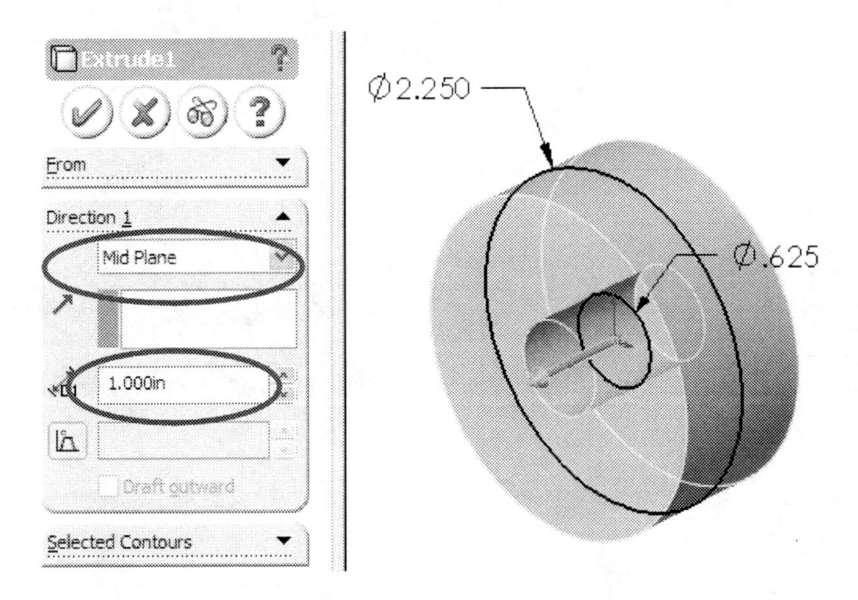

122. - For the second feature we will use a Revolved Cut to make the rounded slot around the outside perimeter. Switch to a Right View and create a Sketch in the Right Plane as shown (don't forget the centerline). Make sure to add a Midpoint relation to the top edge of the cylinder projection.

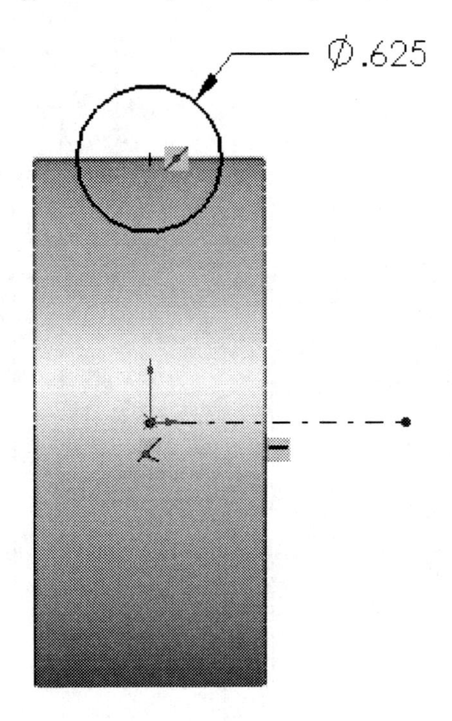

123. - Make a Revolved Cut to complete the feature.

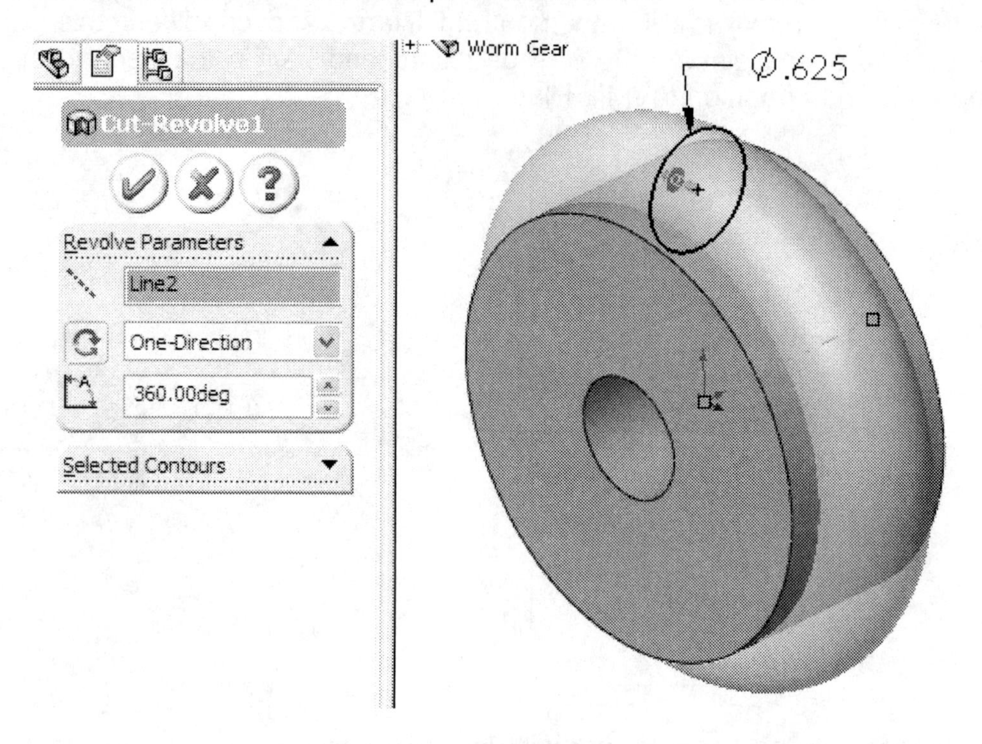

124. - We will now make a second revolved cut to remove material from one side. Switch to a Right view, and create a new Sketch in the Right Plane as shown. Change to "Hidden Lines Visible" or "Wireframe" view mode for clarity.

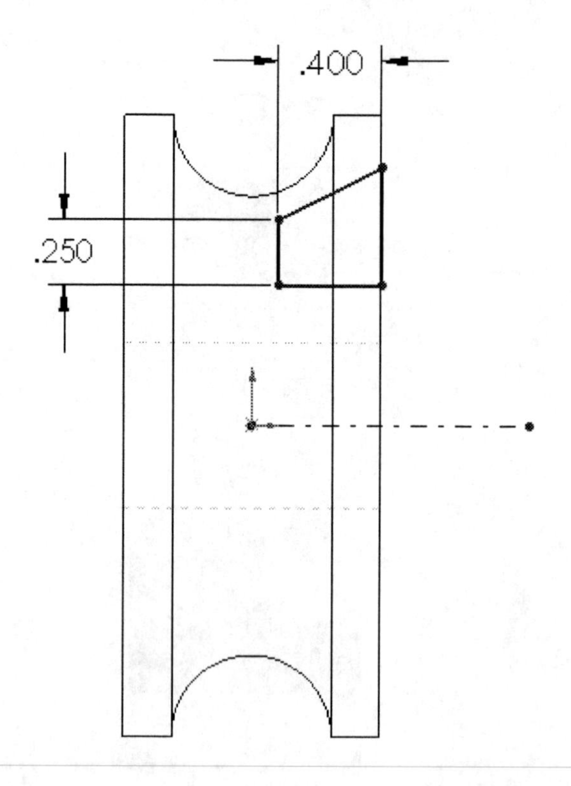

125. - We'll now use a new dimensioning technique to add diameter dimensions for Revolved Features. First select the dimension tool, and add a dimension <u>from the Centerline</u> (NOT an endpoint) to the top endpoint of the sketch. Before locating the dimension, cross the centerline and notice how the dimension value doubles. Now locate the dimension, and change it to 2". Repeat and add a 1" diameter dimension to the indicated line.

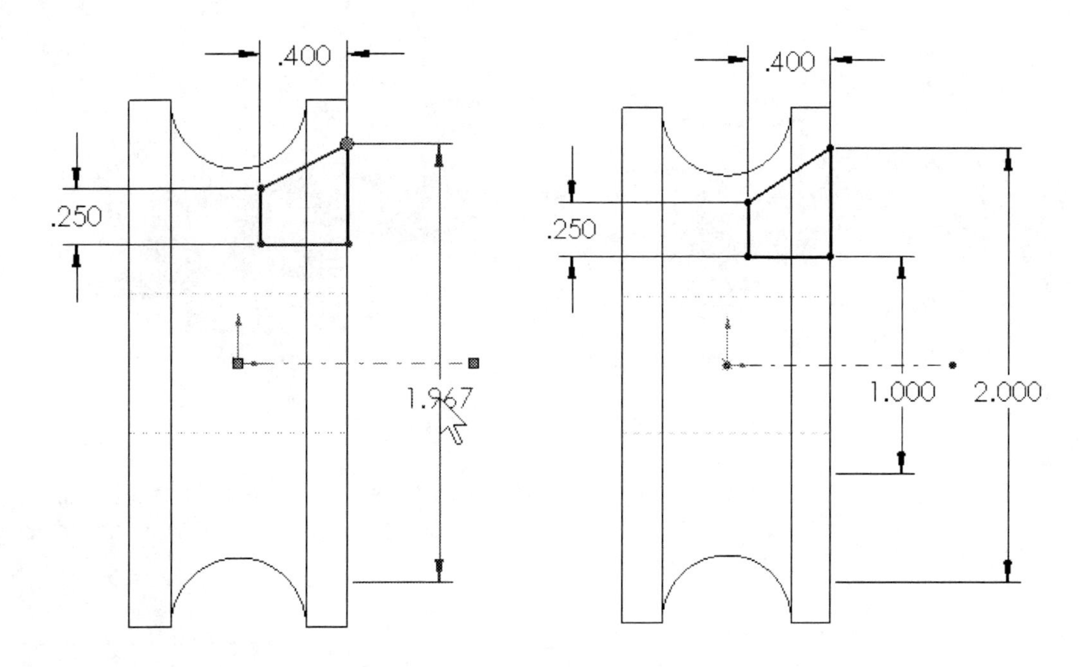

126. - Make a Revolved Cut about the centerline to complete the feature.

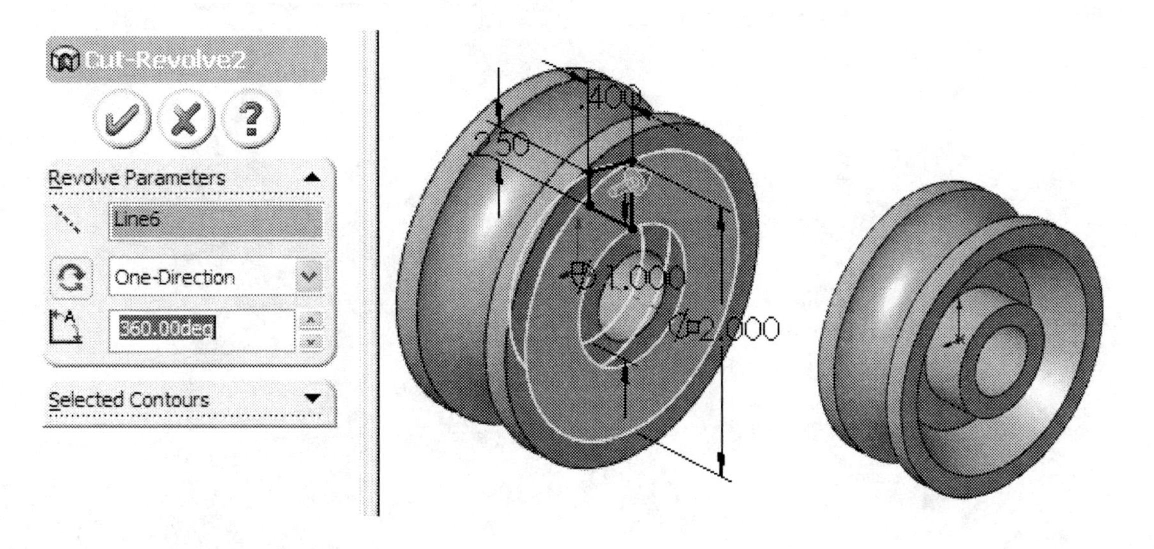

127. - The cut has to be on both sides of the part; to add the second cut, we'll Mirror the previous Cut Revolve about the Front Plane as we've done before.

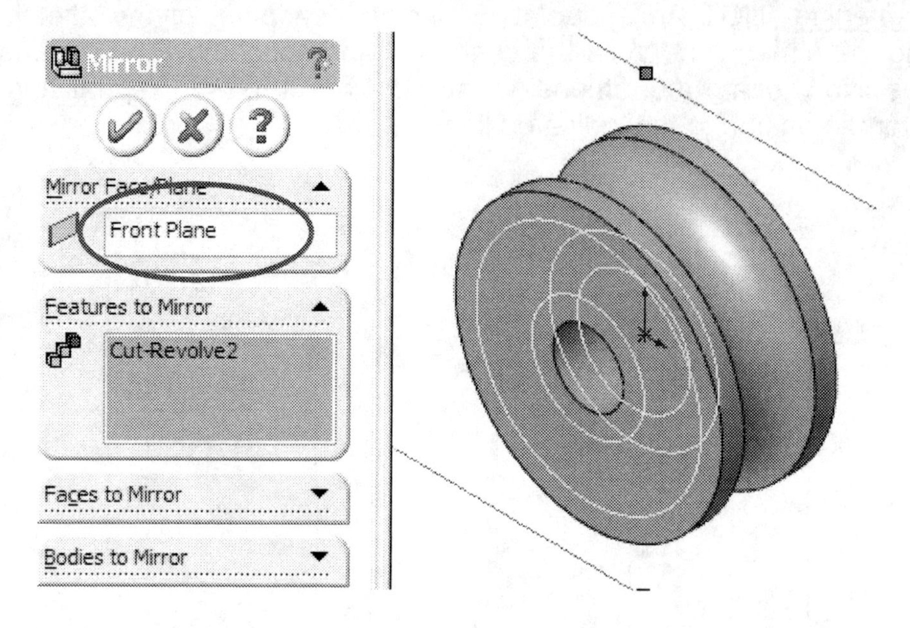

128. - To eliminate the sharp edges on the outside perimeter we'll add a 0.1" x 45° **Chamfer**. The chamfer is applied similar to the fillet, but instead of rounding the edges, it ads a bevel to it.

Select the "Chamfer" icon from the Features Toolbar, set the options and values and select the two edges indicated. Click OK when finished.

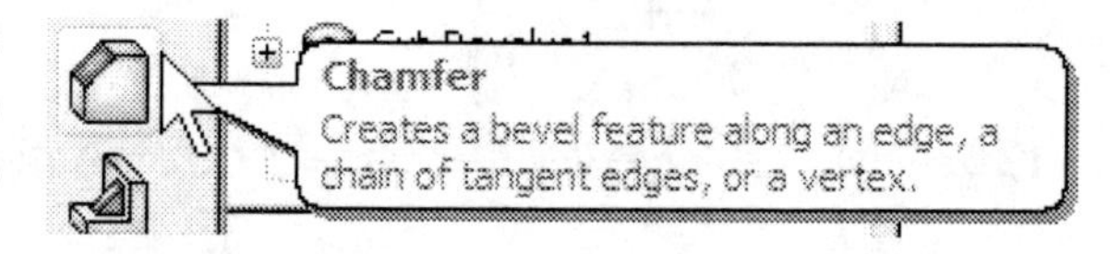

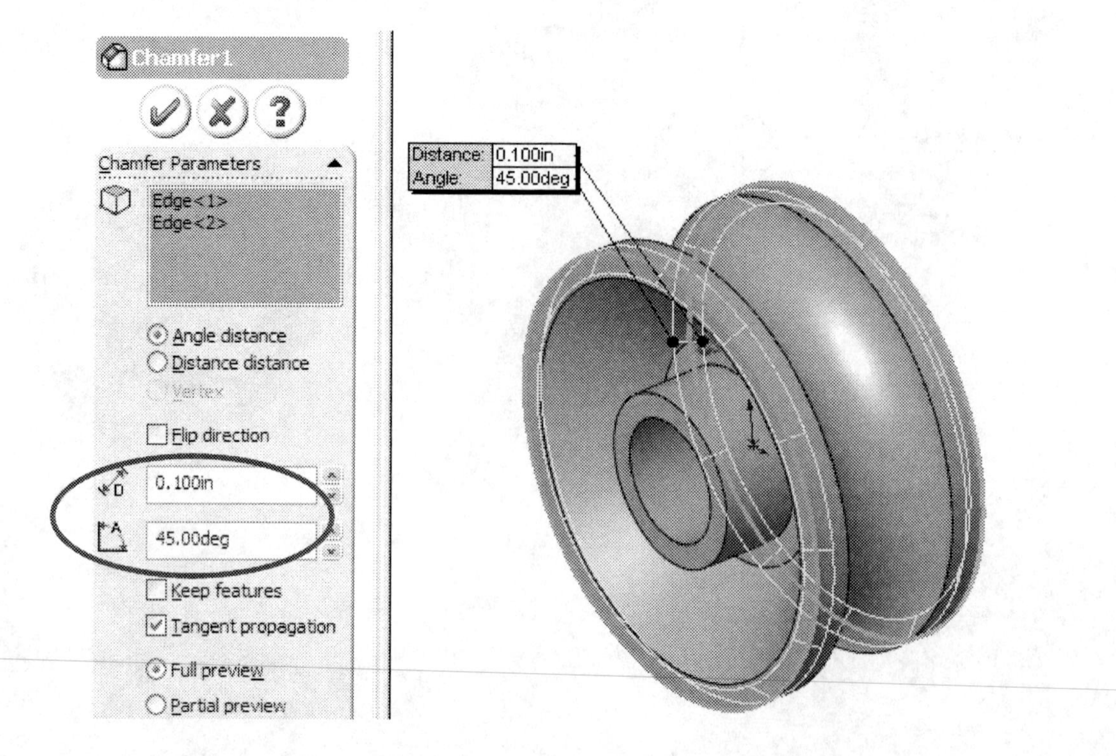

129. - Add a 0.0625" radius fillet to the four inside edges indicated.

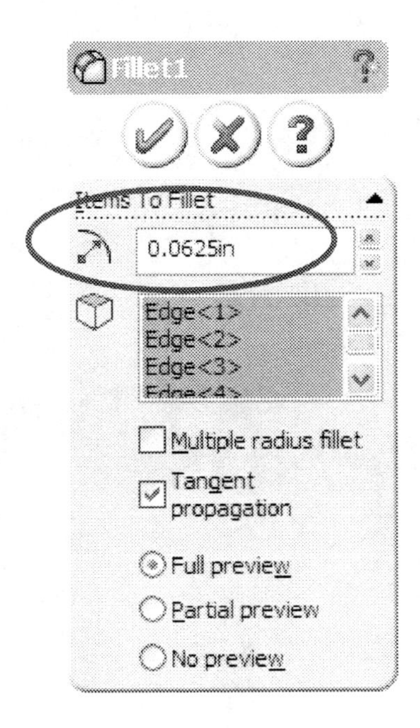

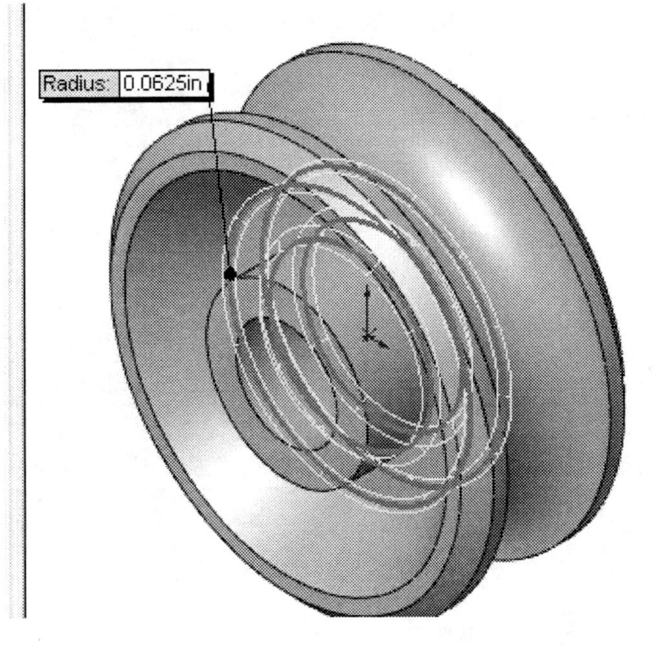

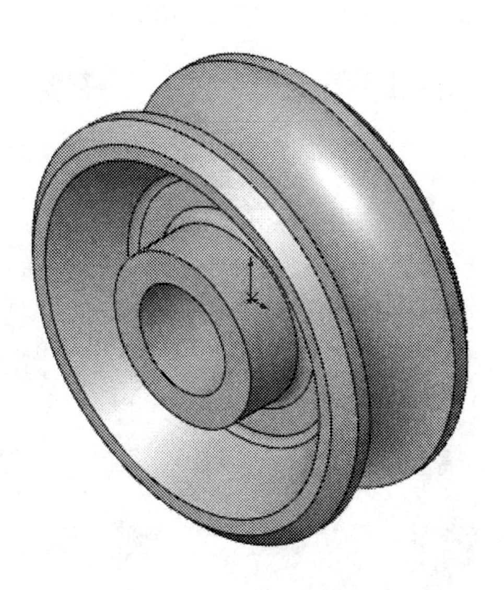

130. - For the last step, we'll make the keyway. Switch to a Front View and make a sketch in the frontmost face. Draw a rectangle and add a Midpoint relation between the bottom line and the origin.

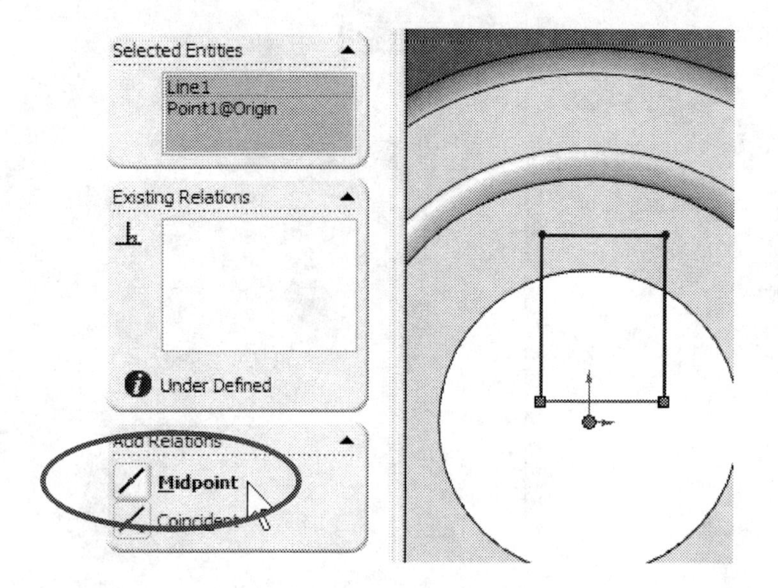

131. - By adding the Midpoint relation, the rectangle will be centered about the origin. Add a 0.188" width dimension as indicated.

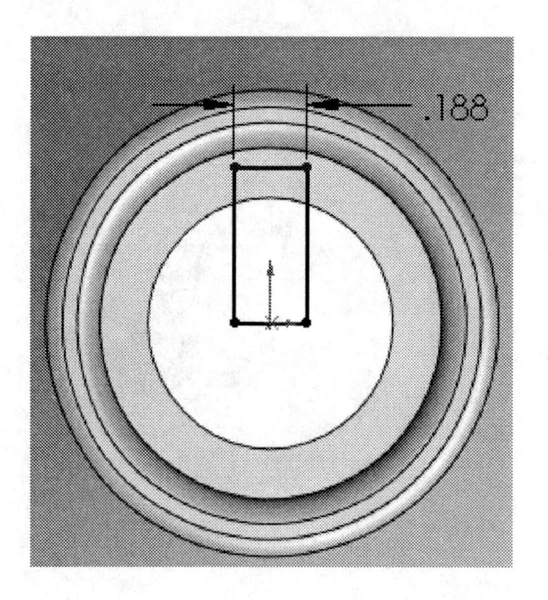

132. - Now add a dimension from the top of the circular edge to the horizontal line. By selecting the topmost part of the circular edge, we'll add a dimension to the top quadrant. Locate the dimension and change its value to 0.94".

NOTE: With SolidWorks 2005 SP2, it will be required to hold the "Shift" key to activate the quadrant dimension; otherwise you will add a dimension to the center of the circle.

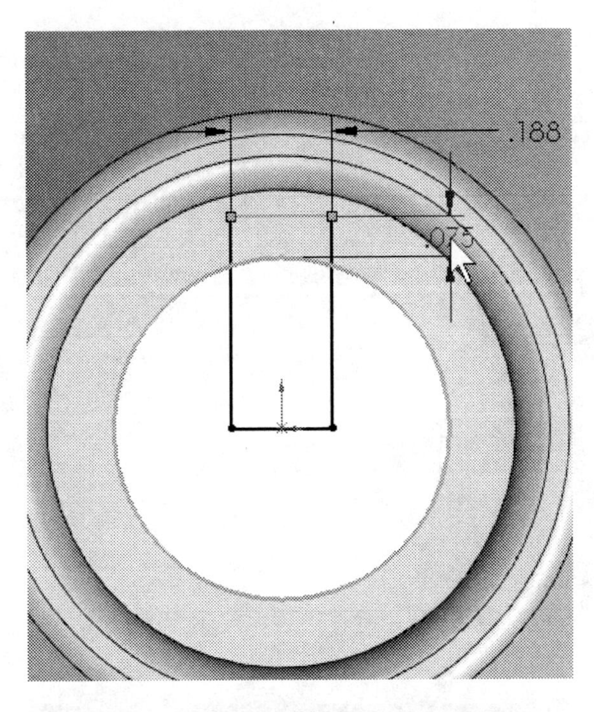

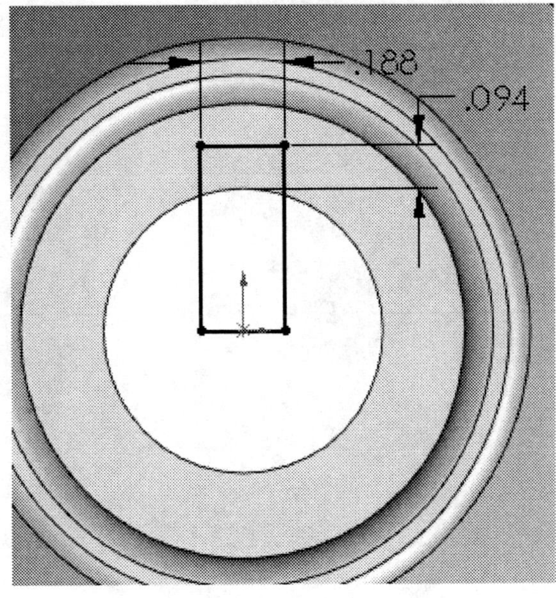

133. - Select the Cut Extrude icon and select the "Through All" option to finish the part.

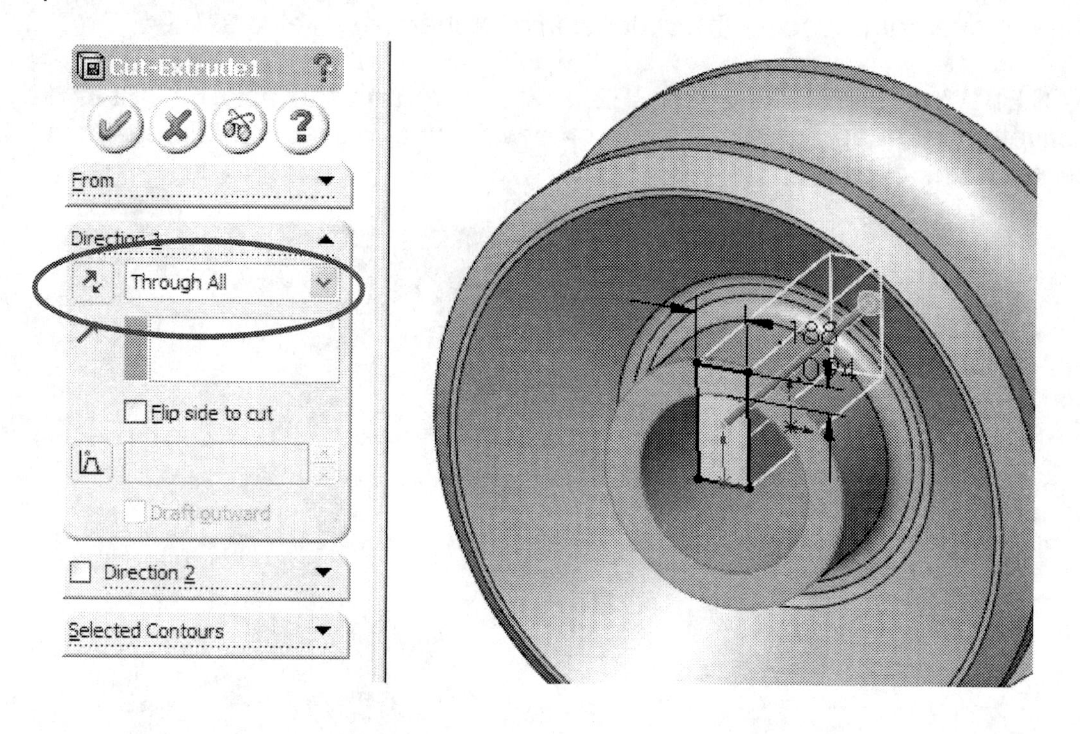

134. - Save the part as "Worm Gear" and close the file.

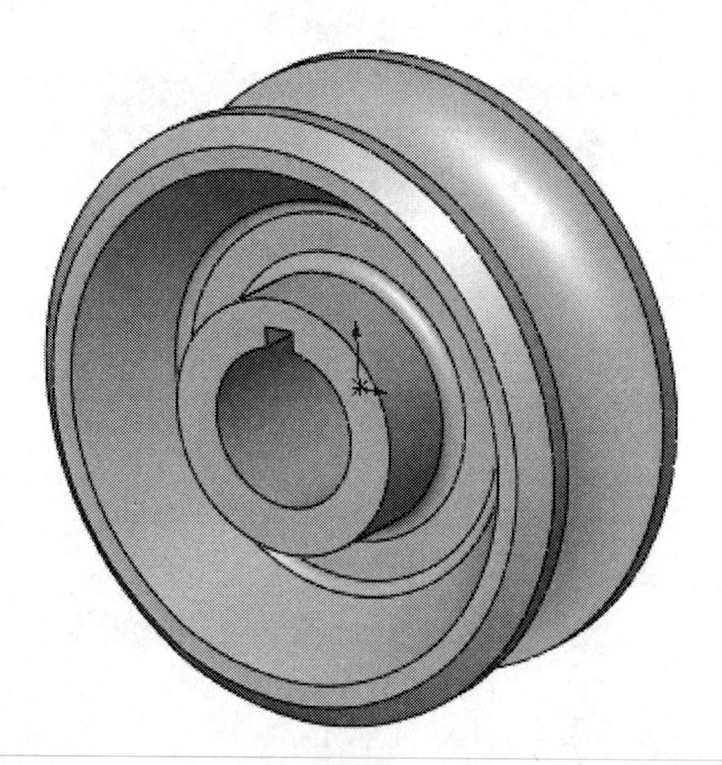

The Worm Gear Shaft

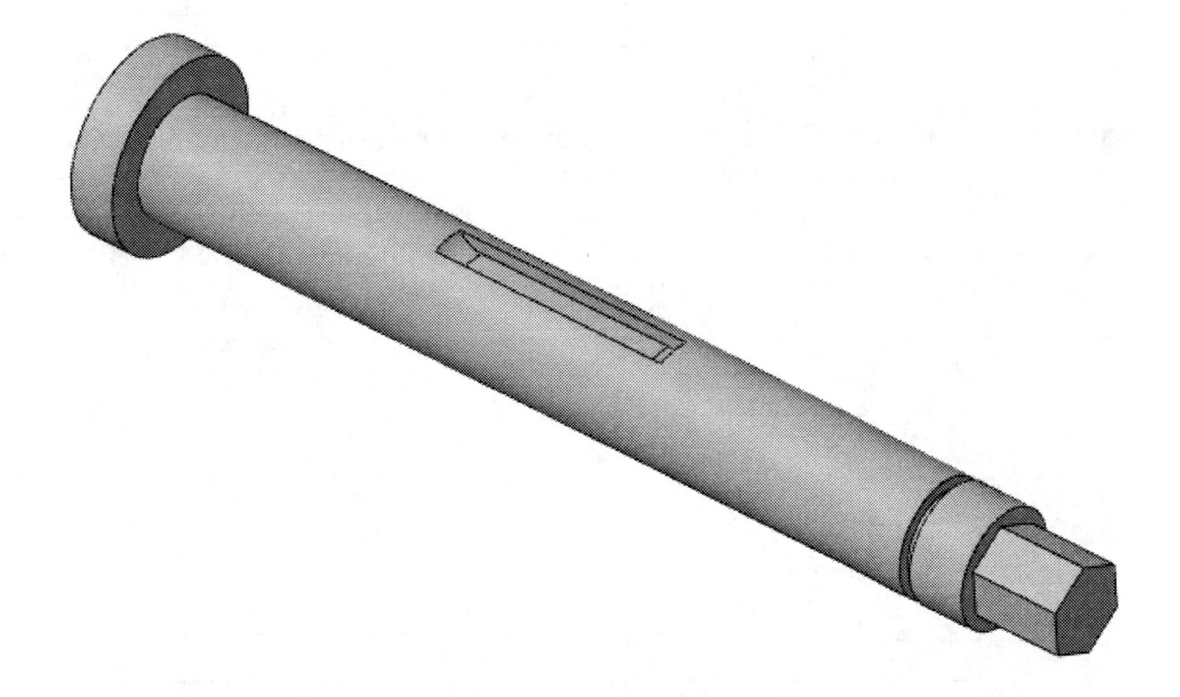

For the Worm Gear Shaft we will also make a simplified version of it. This part will review the Revolved Feature previously learned, the sketch Polygon tool and a Mid Plane cut.

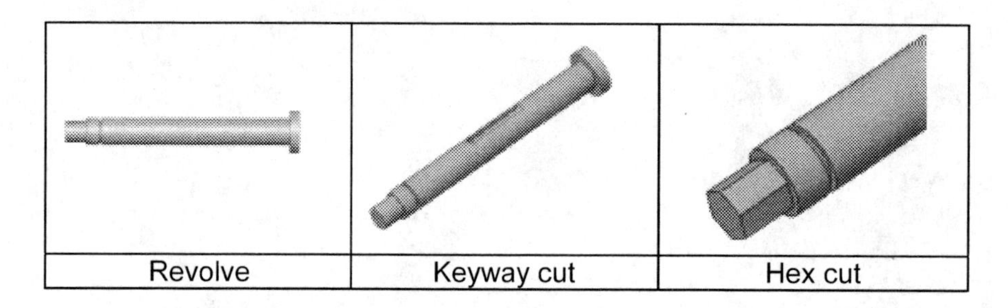

Revolve	Keyway cut	Hex cut

135. - In this part we only need to make three features. The first feature will be a revolved extrusion. Make a new part file and insert a sketch in the Front plane as shown next. It's very important to add the centerline, as we'll need it for the diameter dimensions as we did in the previous example. Select the "Revolved Boss" icon and complete the first feature.

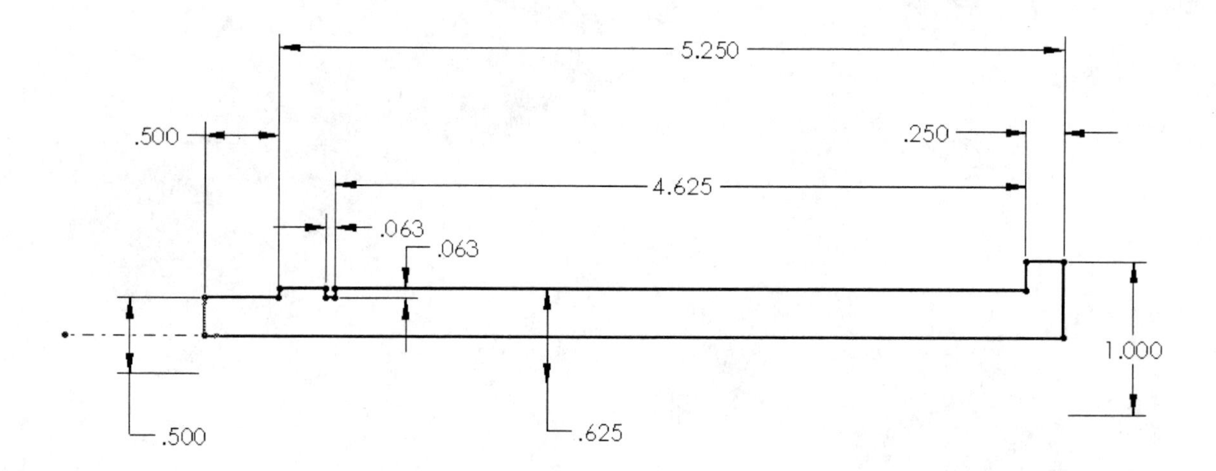

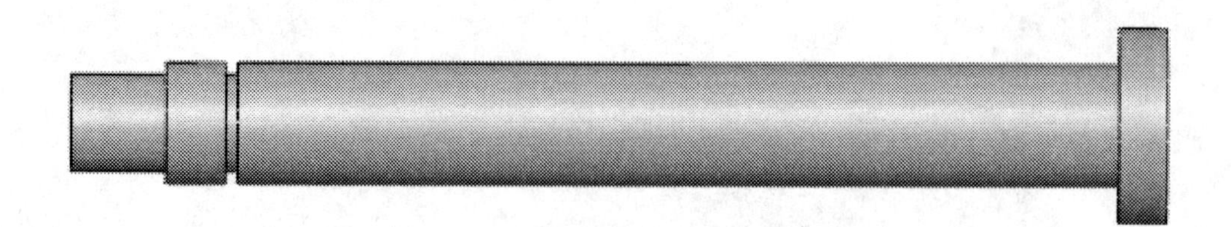

136. - The second feature is the keyway. Select the Front Plane from the feature manager and make a sketch as shown in the next image. It is OK to leave the top line under defined. What we'll do is a cut using the "Mid Plane" end condition, and the way the sketch is cutting the part, it will "cut air". This is a common practice and is fine as long as we make sure the sketch is big enough to accommodate possible future changes.

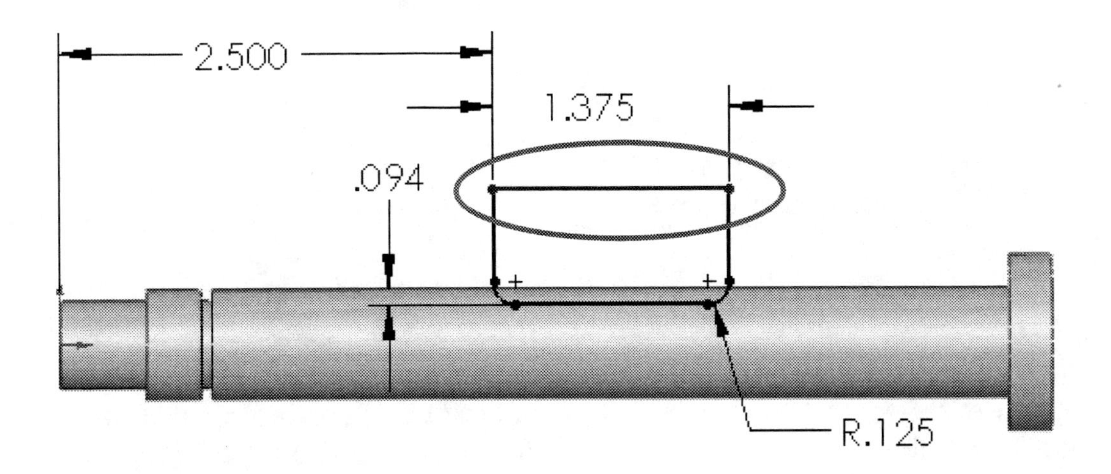

137. - Select the "Cut Extrude" icon and use the "**Mid Plane**" end condition with a dimension of 0.1875".

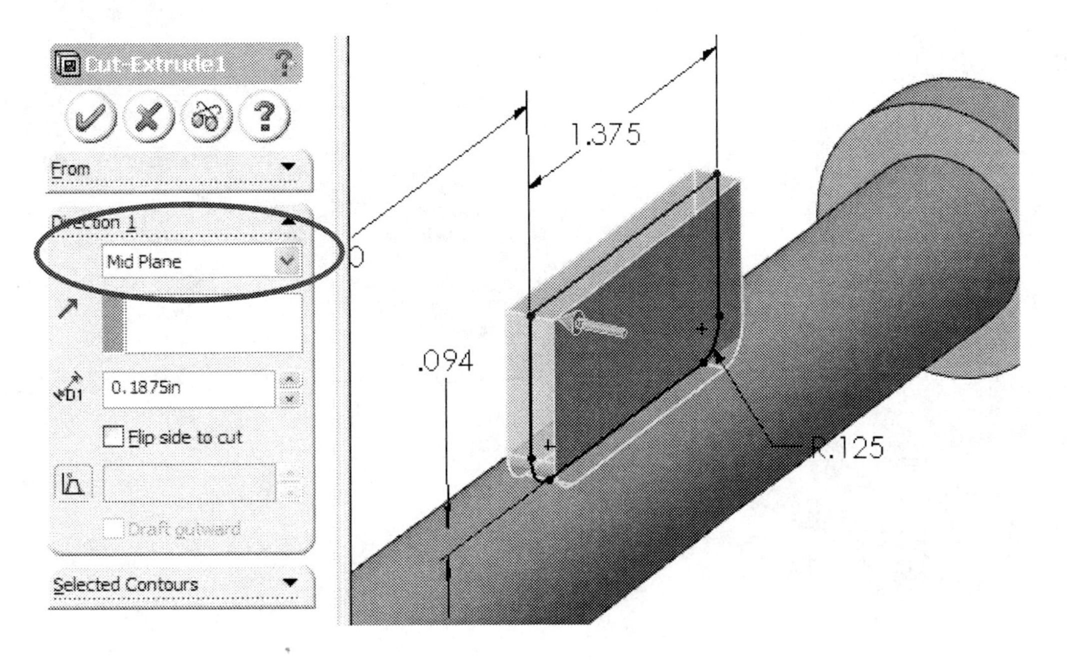

138. - Finished keyway cut.

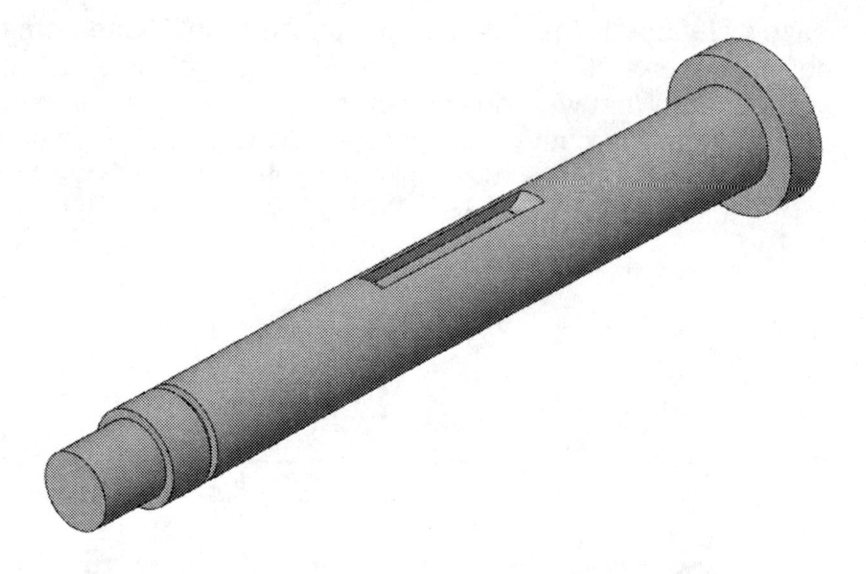

139. - For the last feature we'll do the exact same steps we did with the "Offset Shaft", with the exception that we are making the hexagonal cut in the left most face of the part. Repeat steps 111 to 118 to complete the component. Save the part as "Worm Gear Shaft" and close the file.

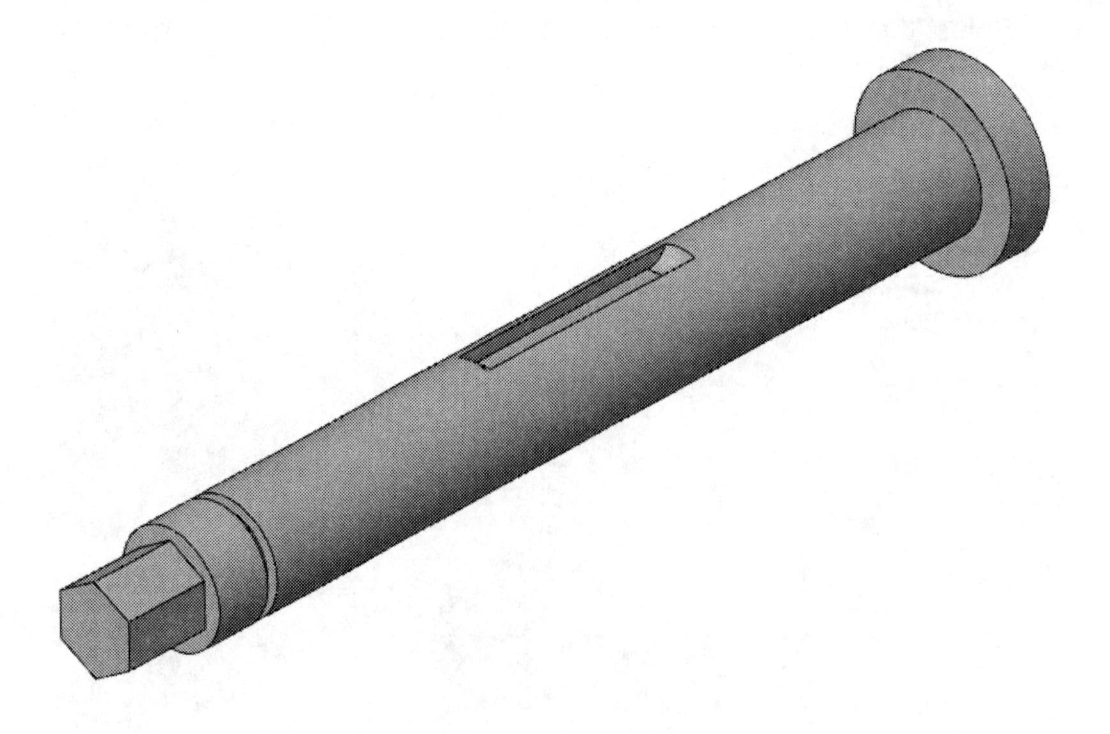

Detail Drawing

Now that we have completed modeling the components, it's time to make the 2D detail drawings for manufacturing. In SolidWorks, first we have to make the 3D models and from them make the 2D drawings. By deriving the drawing from the solid model, the 2D drawing is linked and associative to the part. This means that if the part is changed, the drawing will be updated and vice versa. Drawing files in SolidWorks have the extension *.slddrw and each drawing can contain multiple sheets, each corresponding to a different printed page.

SolidWorks offers a very simple to use environment where we can easily create 2D drawings of parts and assemblies. In this section we'll cover Part drawings only. Assembly drawings will be covered after the assembly section. We will add different views, annotations, dimensions, sections and details necessary for manufacturing to fabricate the component without missing any detail.

SolidWorks allows us to make the 2D drawings using any of the many dimensioning standards available. In this book we will use the ANSI standard. Once in a drawing, the dimensioning standard can be easily changed to a different one by going to the "Tools, Options" menu and selecting the "Document Properties" tab. In the Detailing section we can simply select the desired standard. It is important to note that after changing a standard, SolidWorks will change the dimension styles, arrow head type, etc. accordingly. For more information, look at the Appendix.

The detailing environment of SolidWorks is a true What-You-See-Is-What-You-Get interface. When we make a new drawing, we are asked what size of drawing sheet we want to use (unless we are using a template with a predefined sheet size). This size corresponds to the printed sheet size. Do not be too concerned about selecting the right sheet size, as we can easily change to a larger or smaller sheet if we determine that our drawing will not fit the current sheet.

Notes:

Drawing One: The Housing

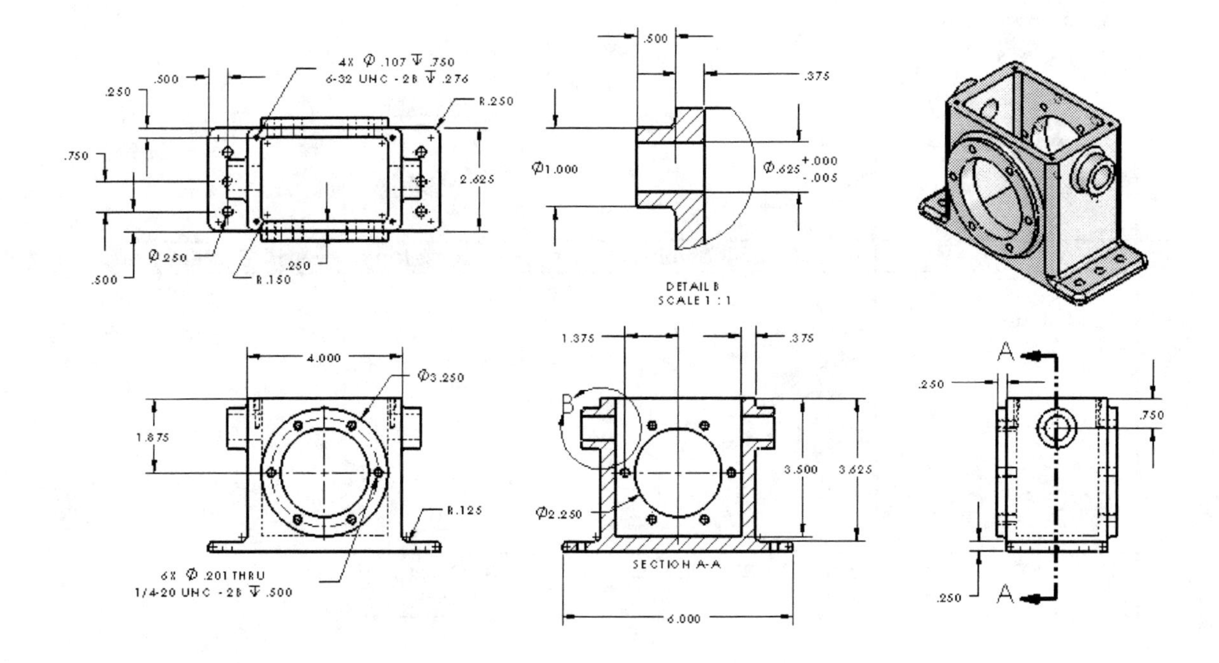

In this lesson we'll learn how to make a drawing from an existing part, add different views including an isometric, change the view display style, move the views, make section and detail views, import model dimensions and manipulate them. The detail drawing of the Housing part will follow the next sequence.

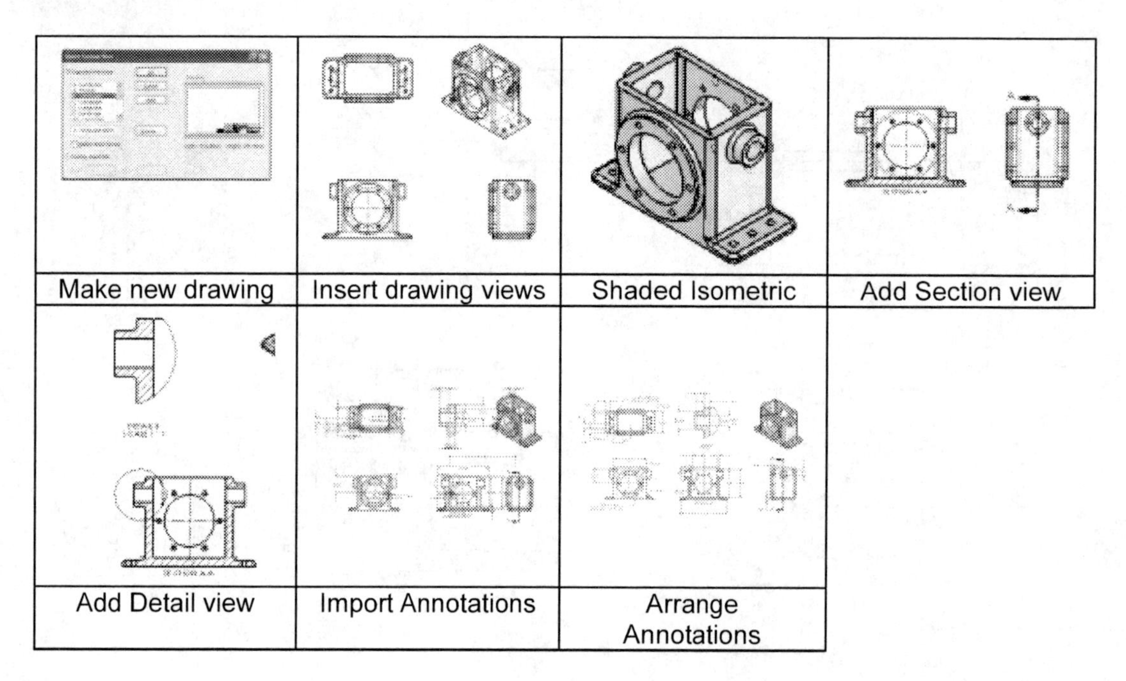

140. - In order to make the detail drawing of the Housing part, open the Housing part in SolidWorks, and then select the "**Make Drawing from Part/Assembly**" icon from the main toolbar.

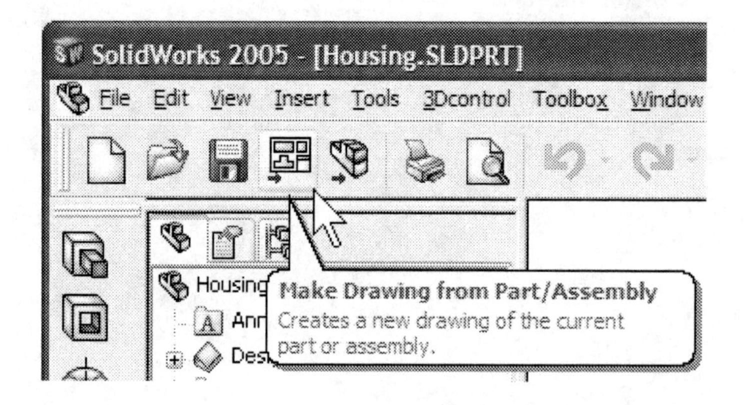

141. - Right away we are presented with the New Document dialog, but this time it is filtered to show only Drawing templates. Select the "Drawing" Template and click OK to continue.

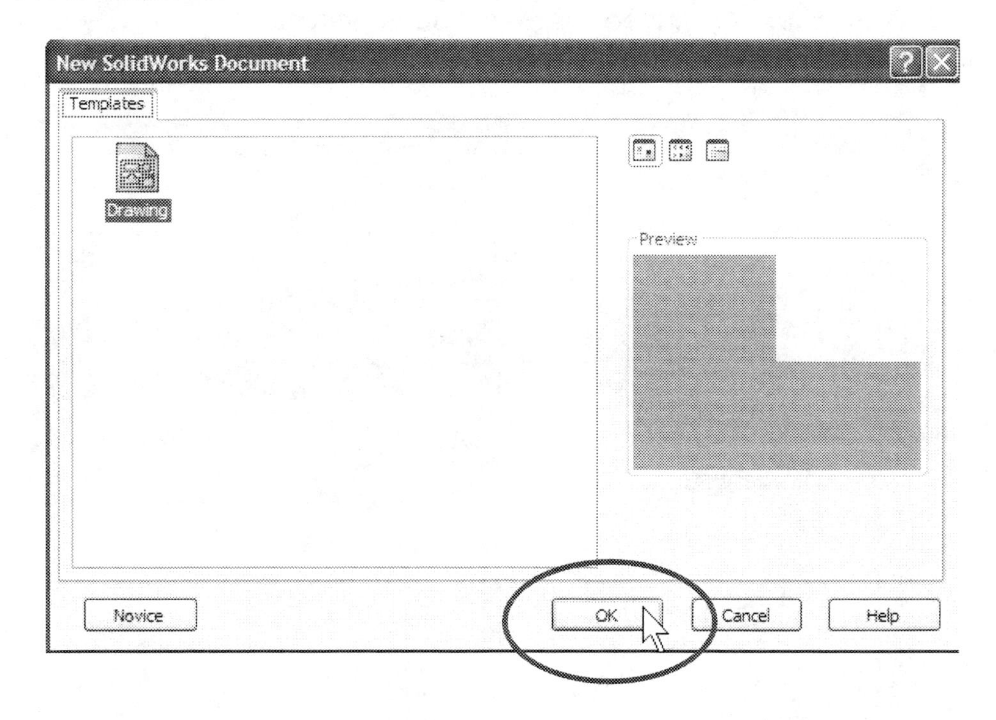

142. - Now select the standard sheet size that we want to use for this drawing. Select Paper Size "B-Landscape", and turn off the "**Display Sheet Format**" option; we will talk about title blocks in the Appendix.

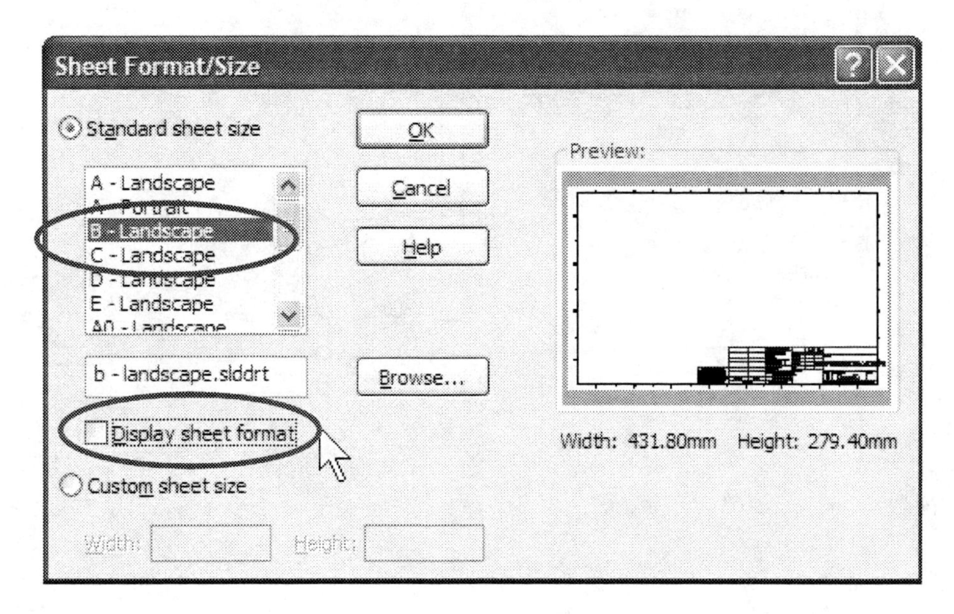

143. - After selecting the paper size, a new drawing is opened and SolidWorks is ready for us to choose the options for the first view. Select "Front" from the "Orientation" box, and "Hidden Lines Visible" from the "Display Style" box. Make sure the "**Auto-Start projected view**" option is activated; this way we'll be able to quickly add all the views to the drawing.

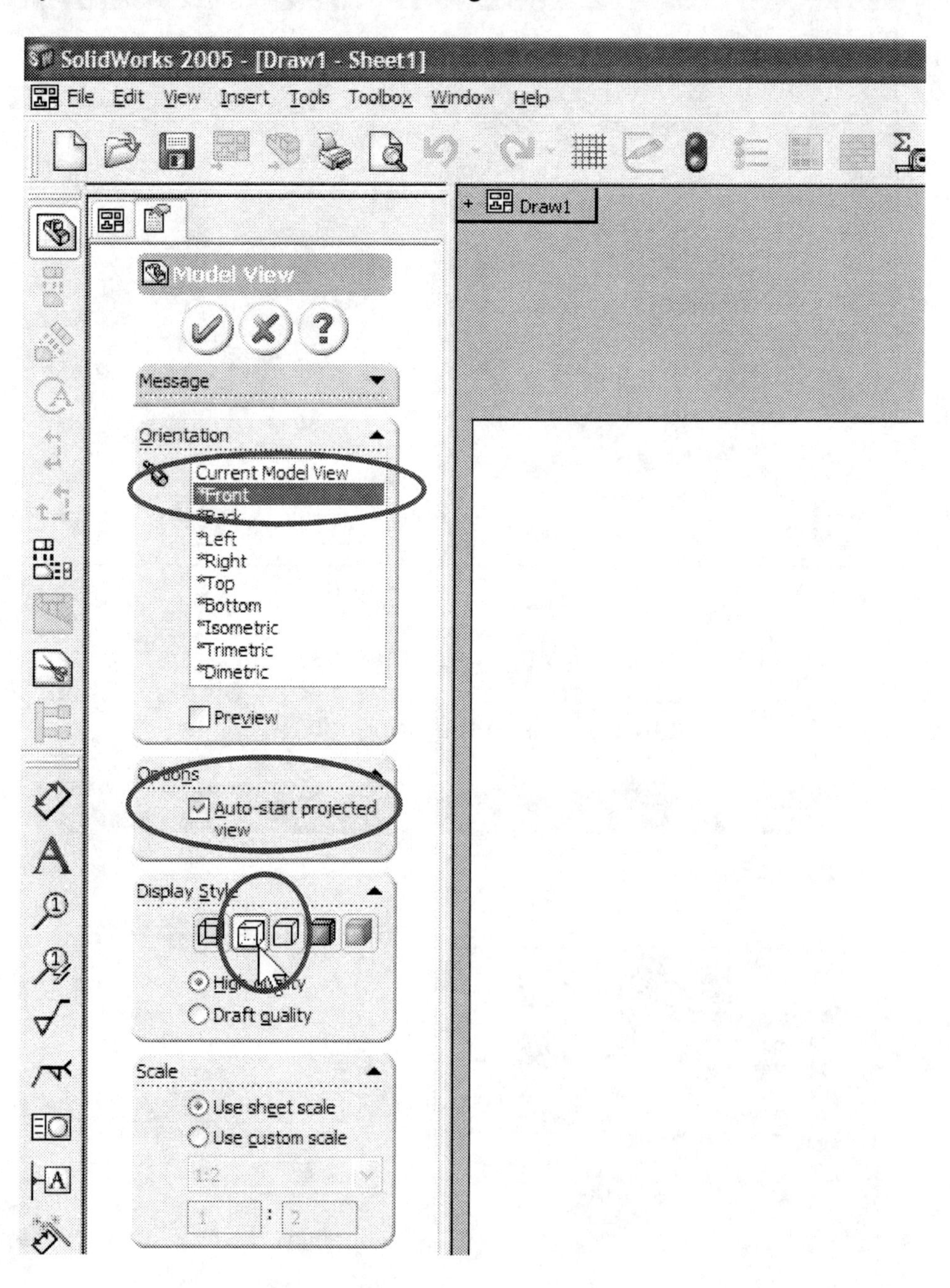

144. - Move the cursor to the graphics area where we'll see the outline of the Front view; simply click on the page to locate the view.

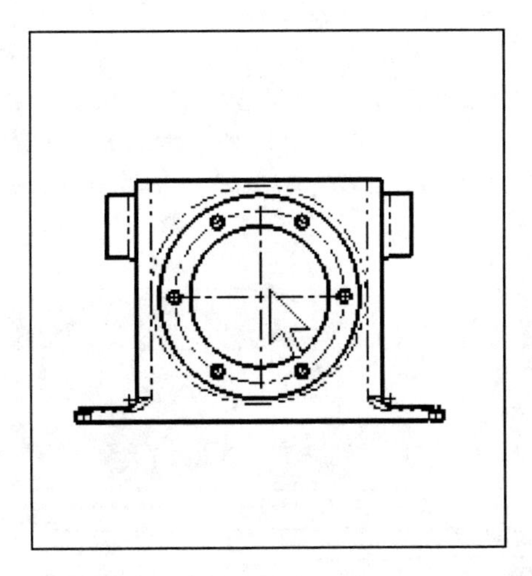

145. - After locating the Front view, SolidWorks automatically starts the **Projected View** command (remember we set this option before locating the Front view); simply move the mouse in the direction of the projected view required, and click to locate it. Add Top, Right and Isometric views as shown. When we're done adding the views, click "OK" to finish. Your drawing should now look like this:

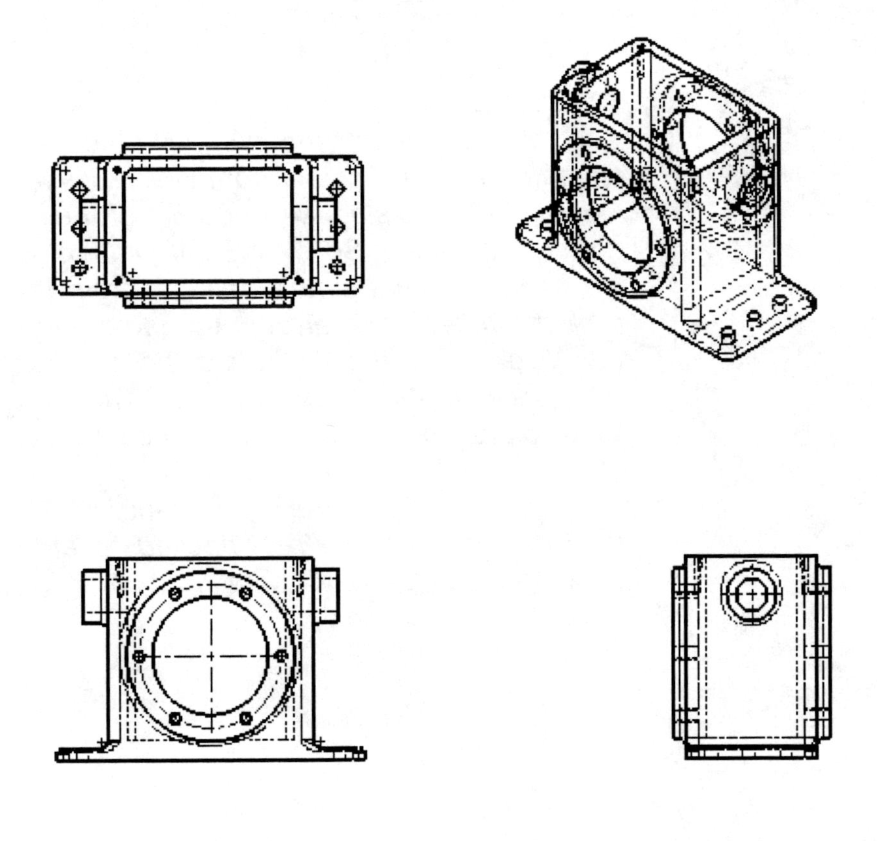

NOTE: Adding standard and projected drawing views

*When we make a new drawing from a part using the "Make Drawing from Part/Assembly" icon, SolidWorks automatically starts the "**Model View**" command to start adding views to the drawing sheet. However, if we make a new drawing using the "New Document" icon, we don't get this behavior and we have to add the views ourselves. To add model views, select the "Model View" icon from the Drawing Toolbar.*

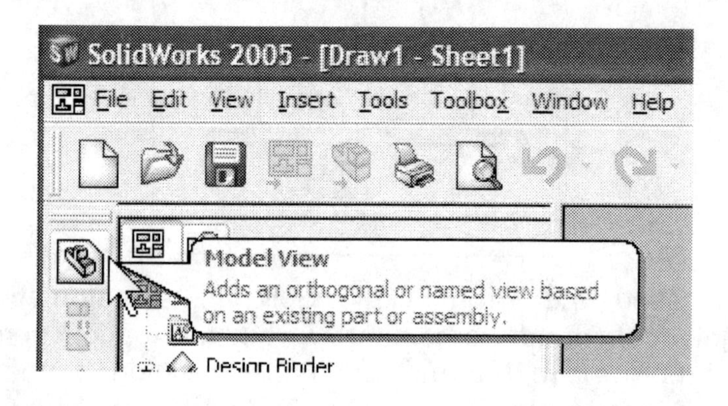

We are then presented with the dialog where we can select the model from which we want to add views. In this dialog we get a list of the open parts and assemblies in SolidWorks. If the component that we need is listed, simply select it from the list and click the blue "Next" arrow. If it is not open, click in the "Browse" button to locate the component, and then click in the "Next" arrow. Then the "Model View" dialog is presented as before.

The option "Start command when creating new drawing" has to be checked to automatically show the "Model View" command when we make a new drawing.

*To add a projected view, select an existing drawing view and click in the "**Projected View**" icon. After that, it's simply a matter of moving the mouse to get the projected view that we need, just like we did before. The behavior is the same.*

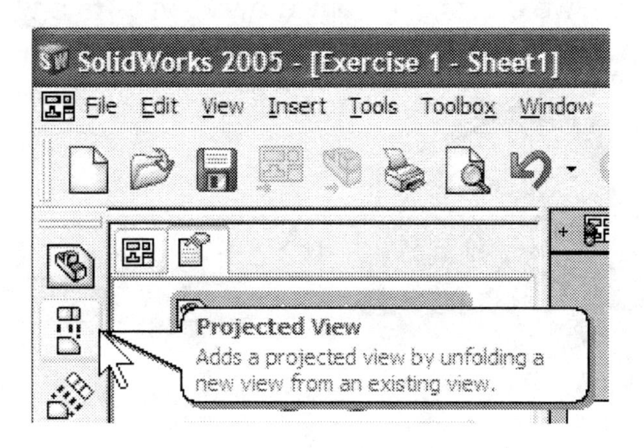

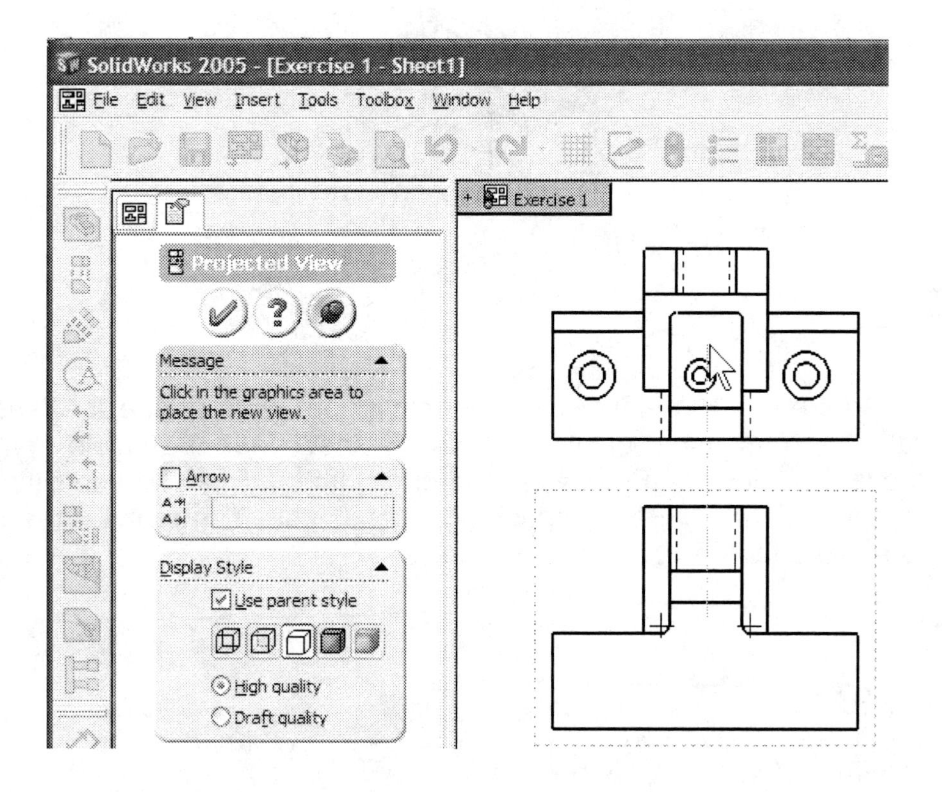

146. - Now that we have the views in place, we want to show the Isometric view in "Shaded With Edges" mode. Select the Isometric view on the screen, and click the Shaded With Edges icon in the view toolbar. Notice the green dotted line around the view; this is indicating to us that the view is selected. Using the same procedure, we can change any drawing view to any display mode.

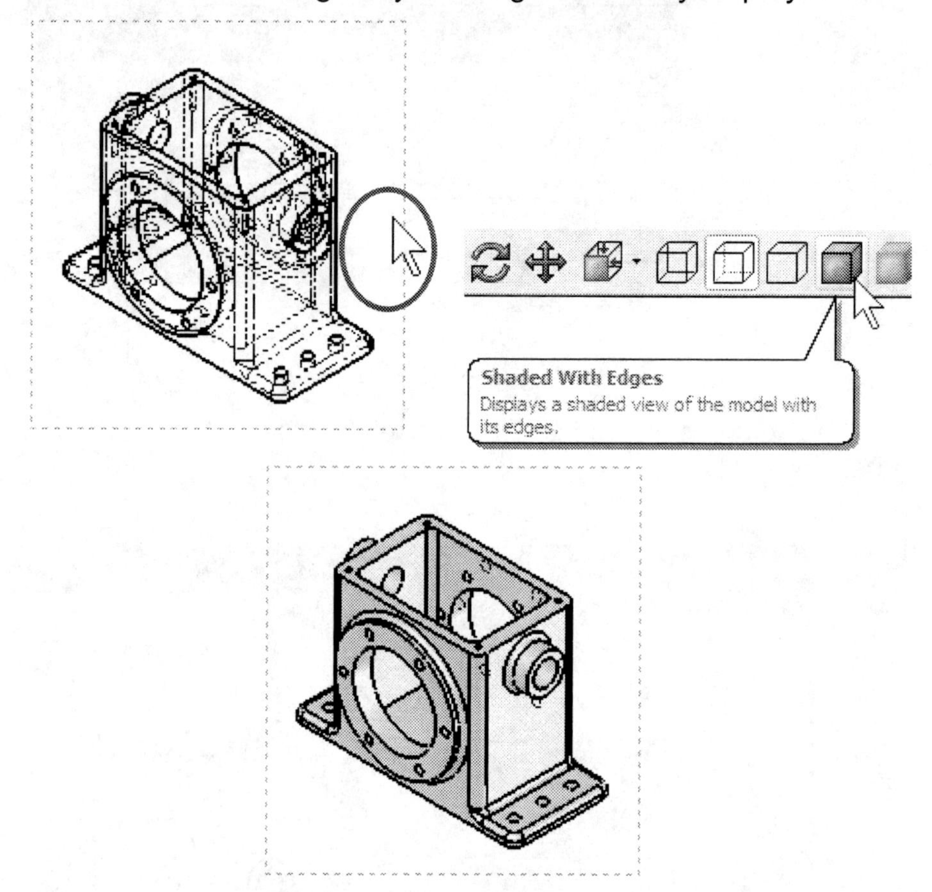

147. - In the drawing views we may or may not want to see the tangent edges. SolidWorks has three different ways to show them: Visible, With Font, or Removed. Select the Front view and click with the Right Mouse Button; from the Drop down menu select the option "Tangent Edge, **Tangent Edges Removed"**. Do the same for the Top and Right views.

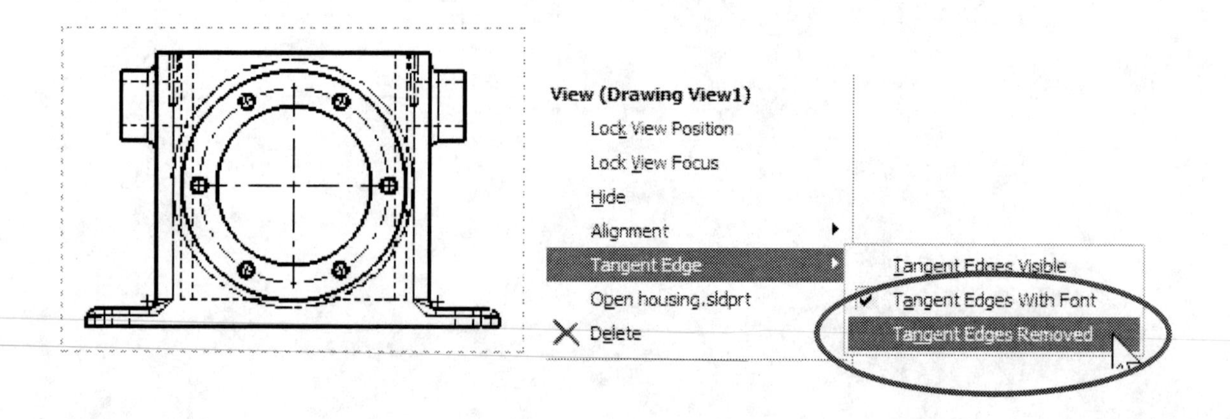

148. - The differences between the three types of edge display are shown in the following table. "Tangent Edges Visible" shows all the model edges with a solid line, "Tangent Edges with Font" shows the tangent edges with a dashed line, and "Tangent Edges Removed" eliminates the tangent edges from the view.

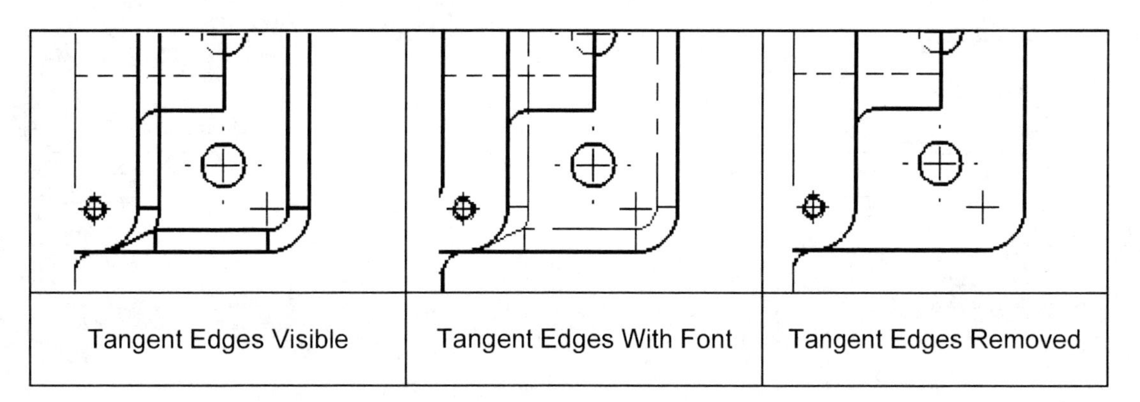

149. - The next thing we want to do in the drawing is to move the views to arrange them on the sheet. To move a view, click and drag the view either from the View border or any model edge.

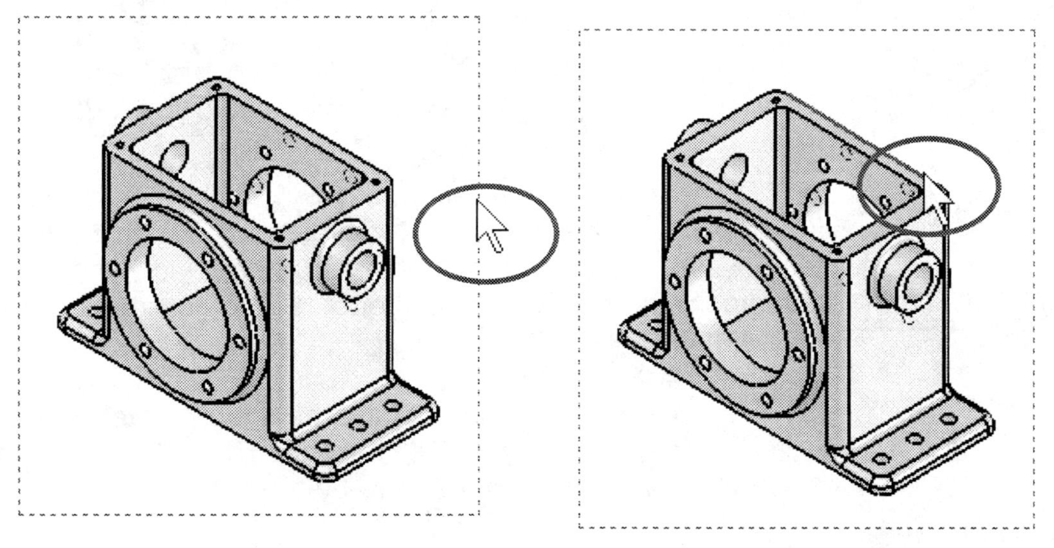

150. - Click and drag the drawing views in the drawing sheet and arrange them as shown, with "Tangent Edges Removed" for all views, and the Isometric in Shaded with Edges.

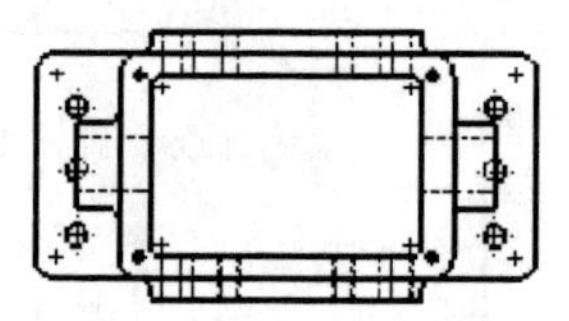

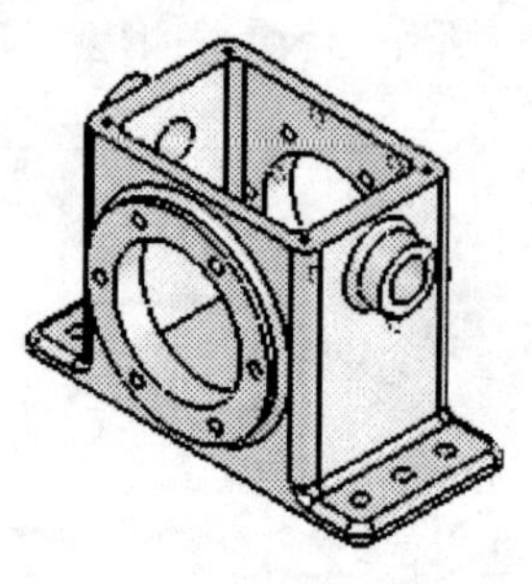

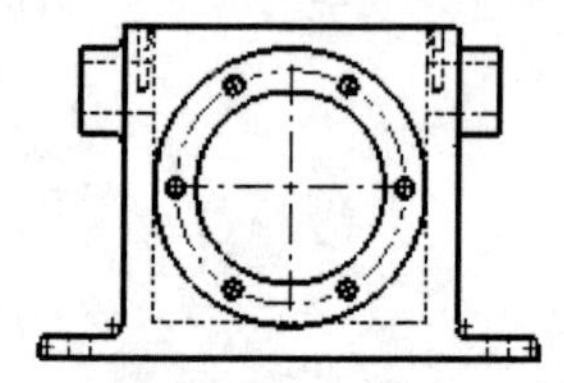

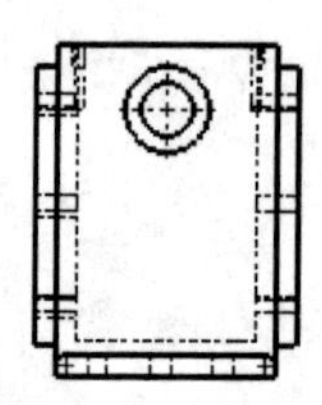

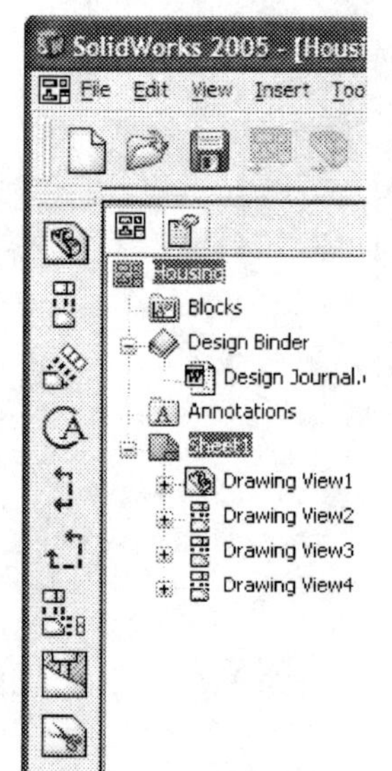

NOTE: The toolbars were automatically changed to match the drawing environment in which we are now. The Features Toolbar was replaced by the Drawing and Annotation toolbars, with detailing tools better suited for the drawing environment.

151. - To help us better understand the drawing, we'll make a section of the Right view. To make the section, select the "**Section View**" icon from the drawing toolbar.

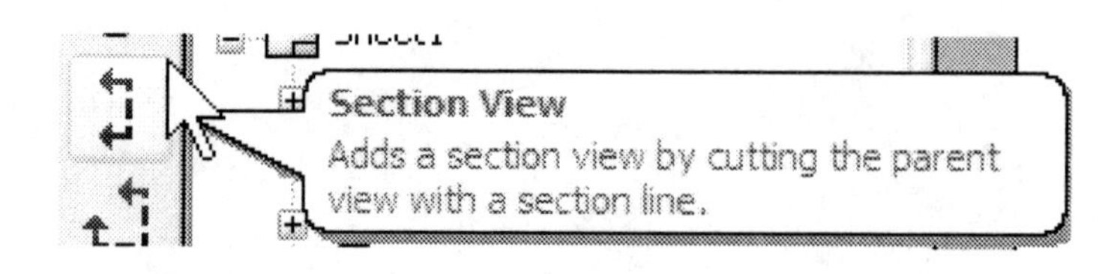

152. - SolidWorks automatically activates the line tool and we are now ready to draw the section line. Since we want a section at the center of the Right View, we'll move the cursor to the edge of the circular boss as shown, but don't click on it! We are only "waking up" the center of the circle to use it as a reference; notice the yellow icons at the quadrants and the center.

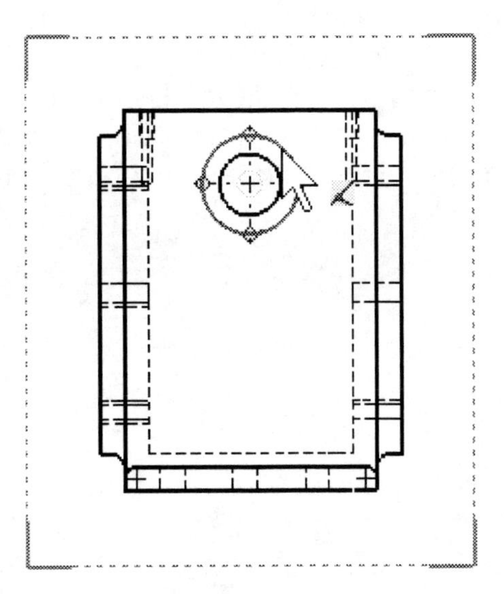

153. - Now, move the pointer to the top of the view, and using the center of the edge as a reference, draw a line starting above the view and all the way across through the view as shown.

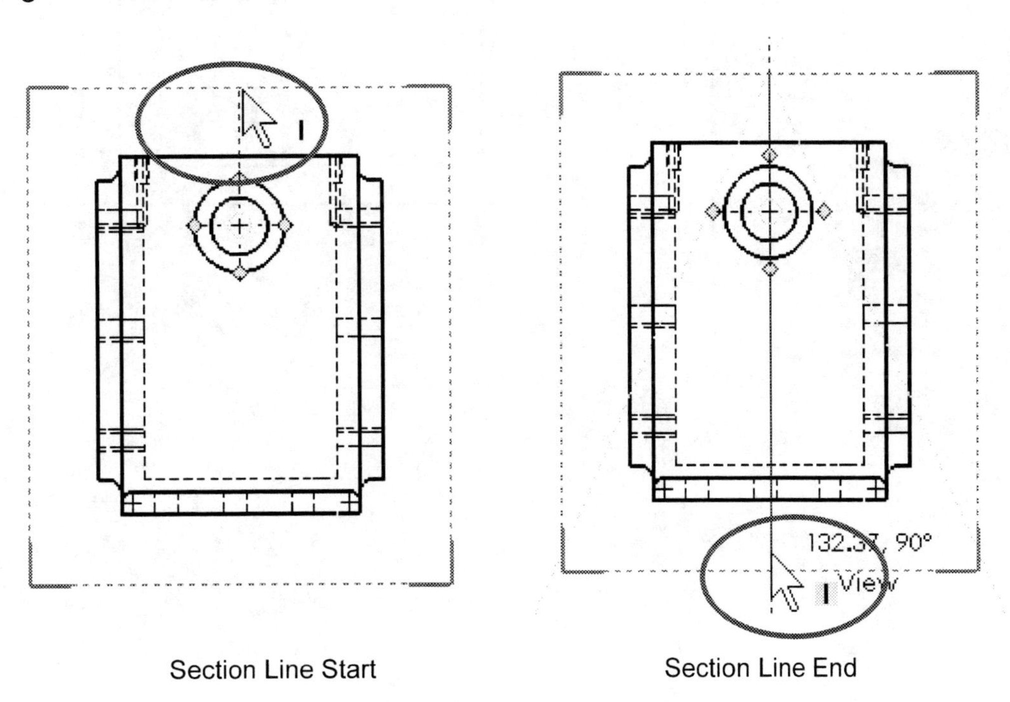

Section Line Start

Section Line End

154. - Immediately after we finish drawing the line, move the mouse and we'll see a dynamic preview showing the section of the Housing; the last thing to do is to locate the section. Move the mouse to the left and click to locate it between the Front view and the Right view.

Now we can see the Section View's Property Manager, where we can change different options such as the Section Label, reverse the section direction ("Flip Direction"), Display Style, Scale, etc. Change the Section View's display style to "Hidden Lines Removed" for clarity. By default, the section will inherit the display style of the view it was made from. To show a view's Property Manger later on, simply select the view, and the Property Manager will be displayed automatically.

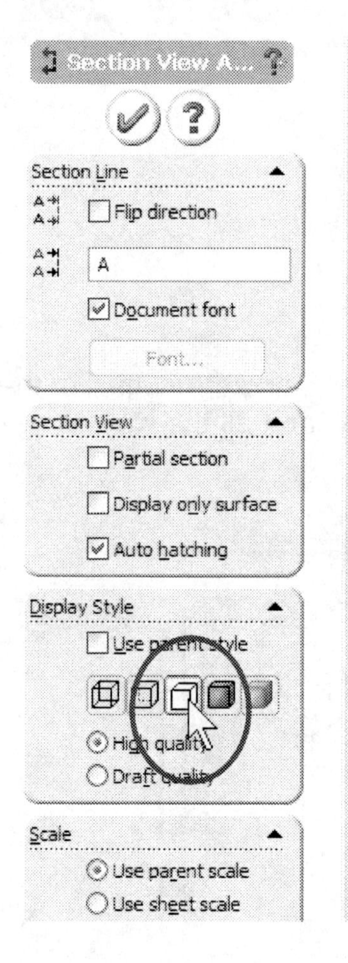

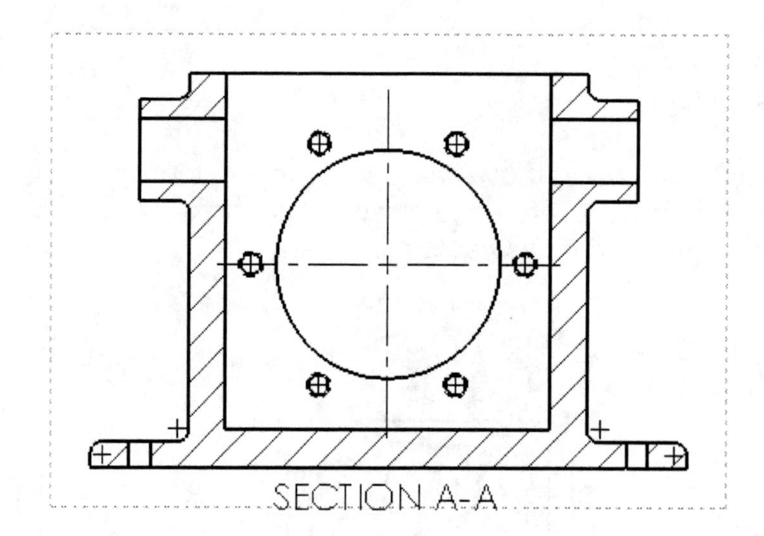

155. - The next step is to add a **Detail View**. From the Drawing toolbar, select the Detail View icon. Similarly to the Section View command, the Detail View activates the Circle tool for us to draw the detail circle.

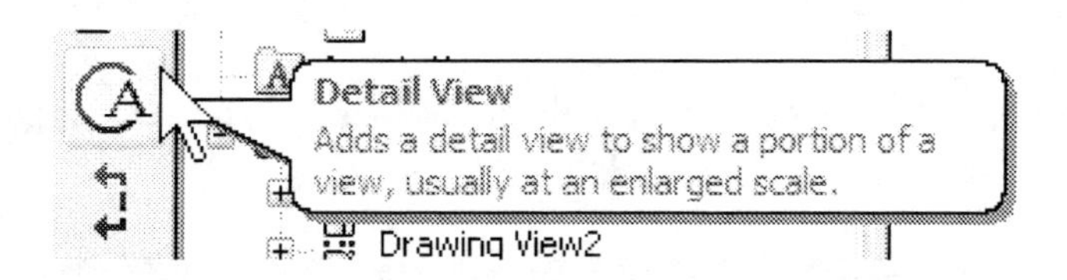

Draw the detail circle at the upper left area of the Section View…

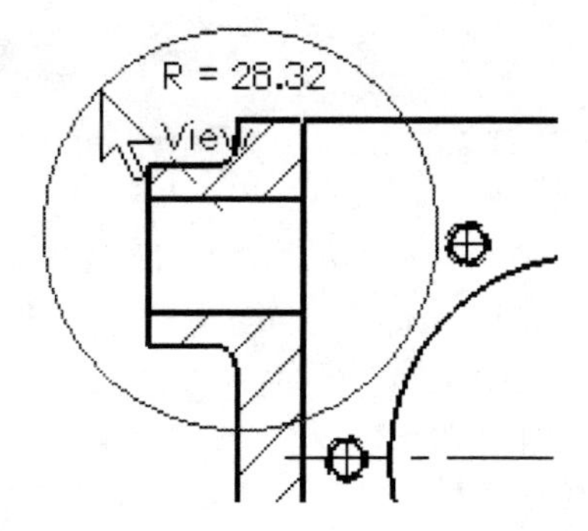

156. - …and just like the section view, move the mouse to locate the Detail above the section view using the dynamic preview. By default, detail views are two times bigger than the view they came from. This option can be changed in the "Tools" menu, "Options"; under the "System Options" tab select the "Drawing" section and change the "**Detail view scaling**" factor here to be "X" number of times bigger than the view it came from.

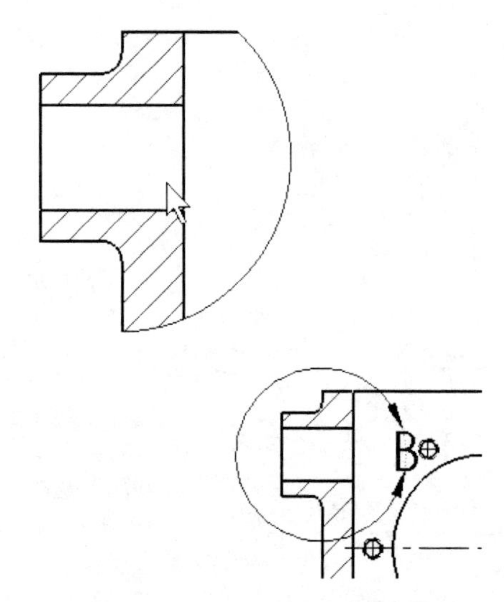

NOTE: If the detail area is not as big as needed, click and drag the detail's circle and/or its center as needed to resize and move the detail area; the Detail View will update dynamically.

157. - Now that we have all the views that we need in the drawing, the next step is to import the Housing's dimensions from the part into the drawing. If you remember, we added all the necessary dimensions to the part when it was modeled, and now we can bring those dimensions to the detail drawing.

Go to the "Insert" menu and select "**Model Items**" to display the dialog to import dimensions and annotations from the model.

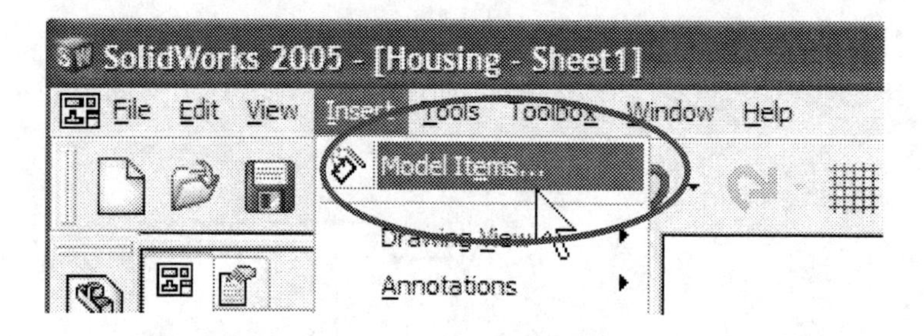

158. - In the Model Items Property Manager, select the type of dimensions and annotations we want to import to the drawing. For this exercise select the options listed below. Remember to select "Import From: Entire Model" to import the dimensions of the entire Housing, and "Import items into all views" to add the dimensions to all the views. "Dimensions Marked for Drawing" is selected by default, turn on the "Hole Wizard Locations" to import the location of holes made with the Hole Wizard, and "Hole Callout" to add the annotations for the machine shop to make the holes.

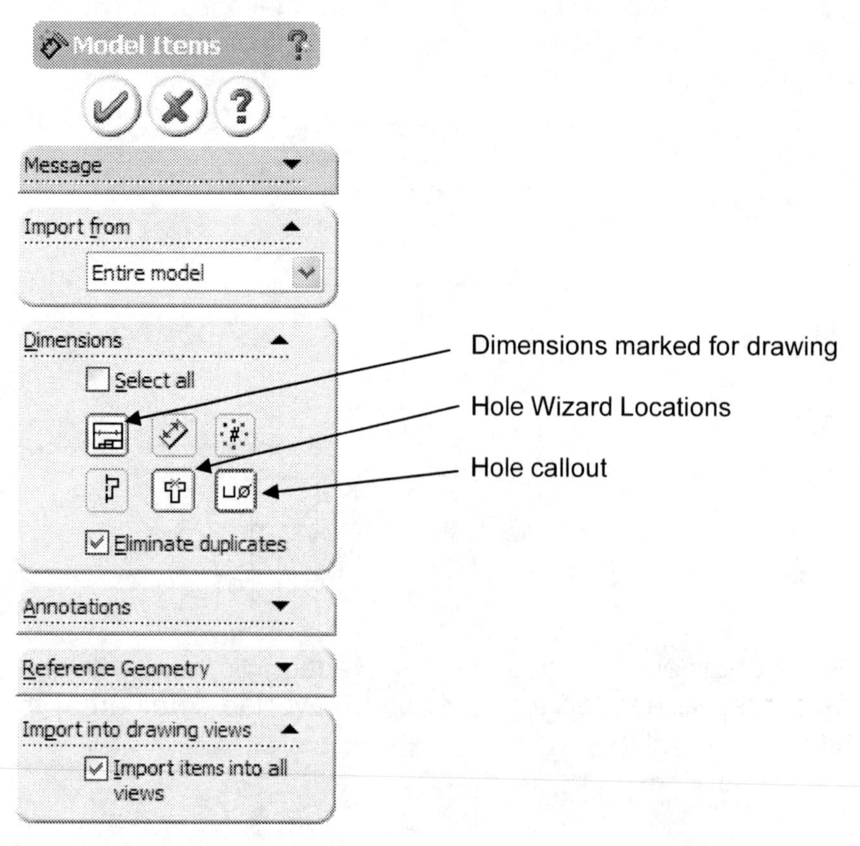

159. - After selecting OK, SolidWorks imports the dimensions and annotations to all views, arranging them automatically. Dimensions are added first to Details, then to Sections, and finally projected views. While SolidWorks makes a good job at adding dimensions to views, they are not always added to the view that best shows it, and here is where we have to do some work.

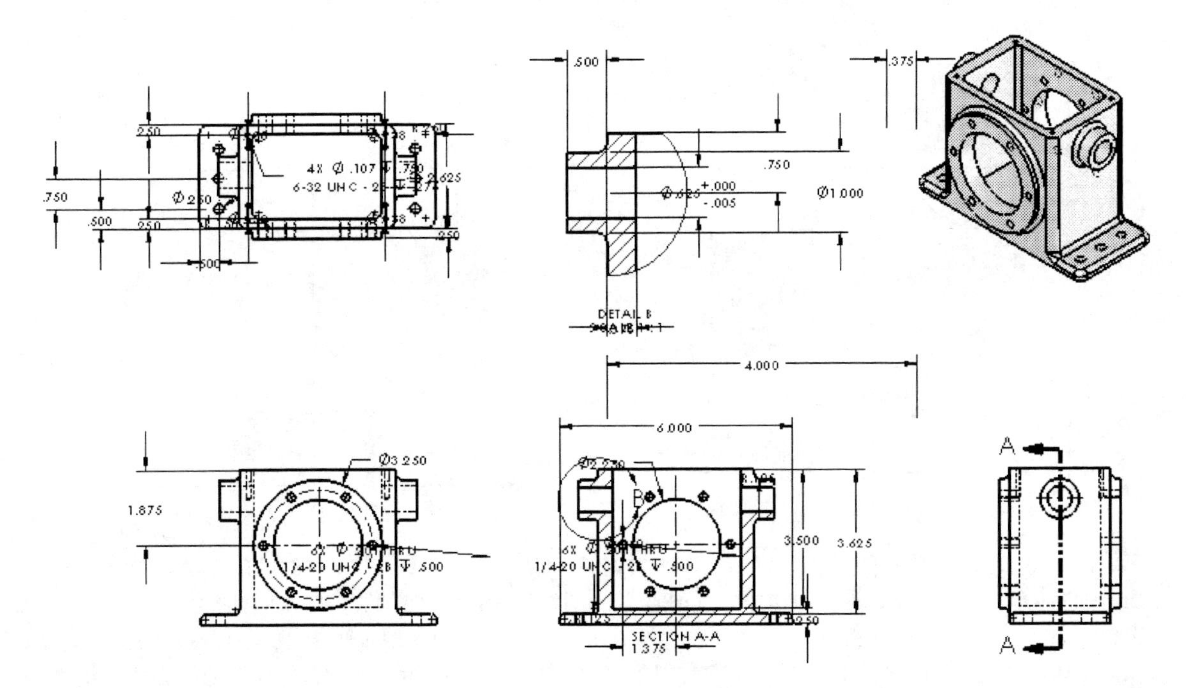

160. - The only thing for us to do now is to arrange the dimensions in order to make them easier to read. To move dimensions and annotations simply Click and Drag them as needed. To move a dimension from one view to another, hold down the "SHIFT" key, <u>and while holding it down</u>, DRAG the dimension to the view where we want it. Notice that the dimension has to be dragged inside the view border to relocate it. Arrange the annotations as needed to make the drawing easy to read.

Some of the dimensions added to the Detail view are incorrectly assigned to it, so we have to move them to a different view. A good way to tell which dimensions are attached to a view is to simply move the view a little. All of the dimensions attached to it will move with it, and you'll be able to put them where you need them.

161. - To complete the drawing, add the center marks using the **"Center Mark"** icon from the Annotations Toolbar. Select the "Center Mark" icon, and click on the circular edges to add a Center Mark as needed. Click OK to finish the command.

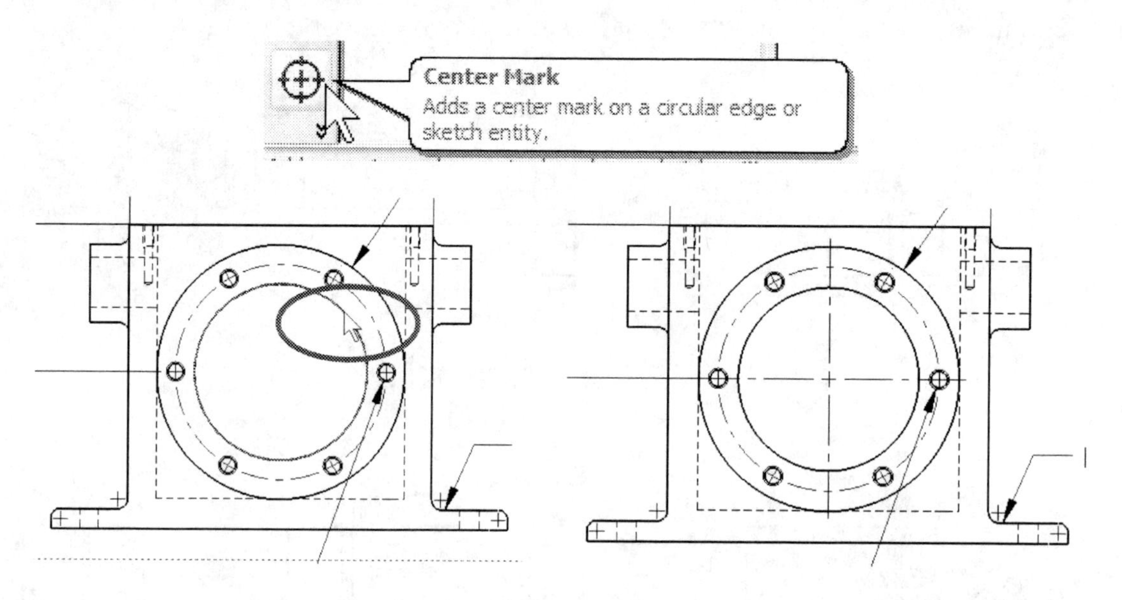

162. - Another thing that you may want to do is to reverse the arrows of a dimension. To do this, simply select the dimension and click in the green dots in the arrow heads to reverse them. To delete a dimension that is duplicated or you don't want, select it, and hit the "Delete" button in your keyboard. You only delete it from the drawing, but not from the part. If you re-import the dimensions to the drawing, SolidWorks will only bring back the ones that are missing. We'll cover more detailing and annotations in the following drawings; there is a lot more than you think to discover in SolidWorks.

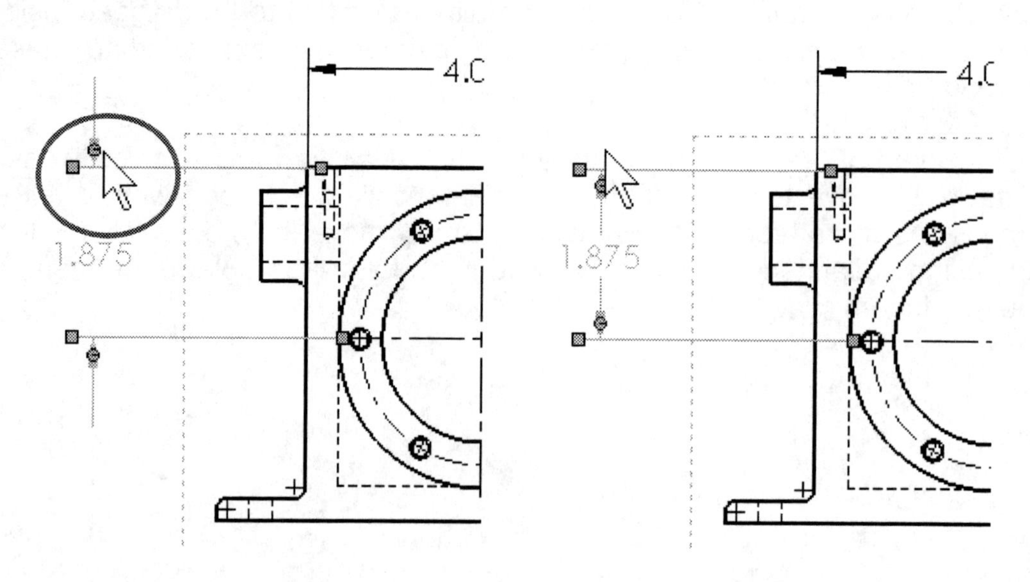

163. - Notice the Hole Callouts added have the drill sizes needed to make the holes specified, as well as the number of holes of each type. Your finished drawing should now look something like this (we broke it in 2 parts for clarity):

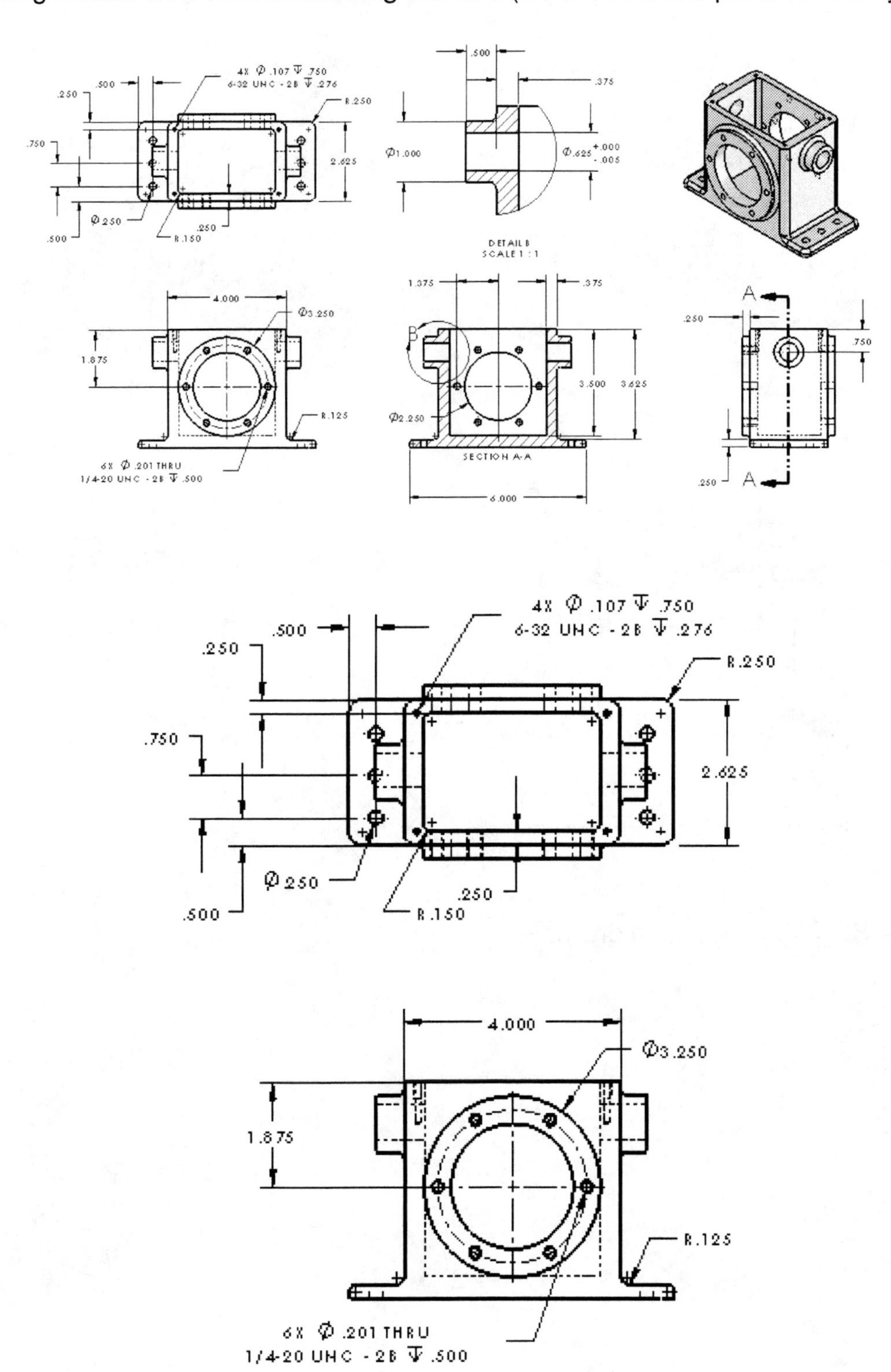

164. - Notice the tolerance was carried over to the drawing from the part model.

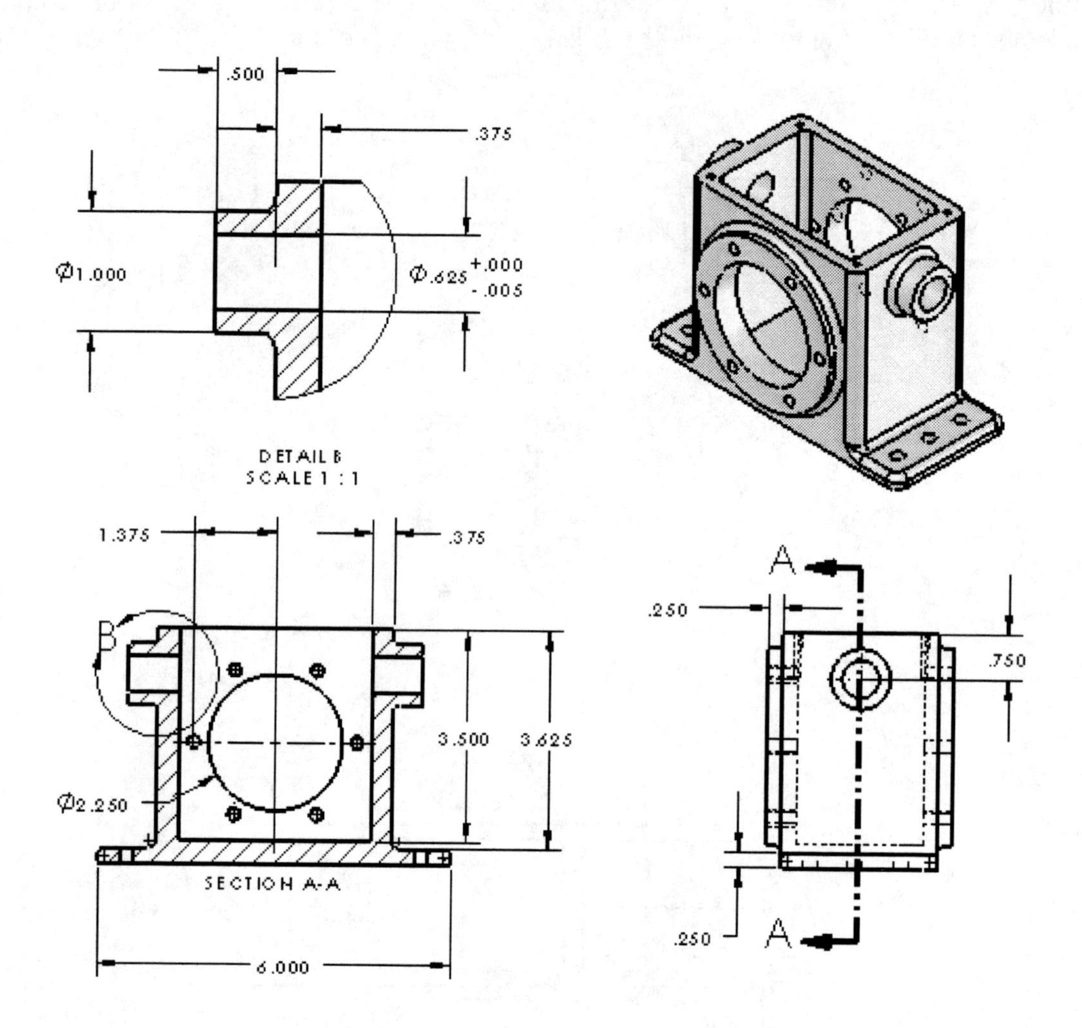

165. - Save and close the drawing file.

Drawing Two: The Side Cover

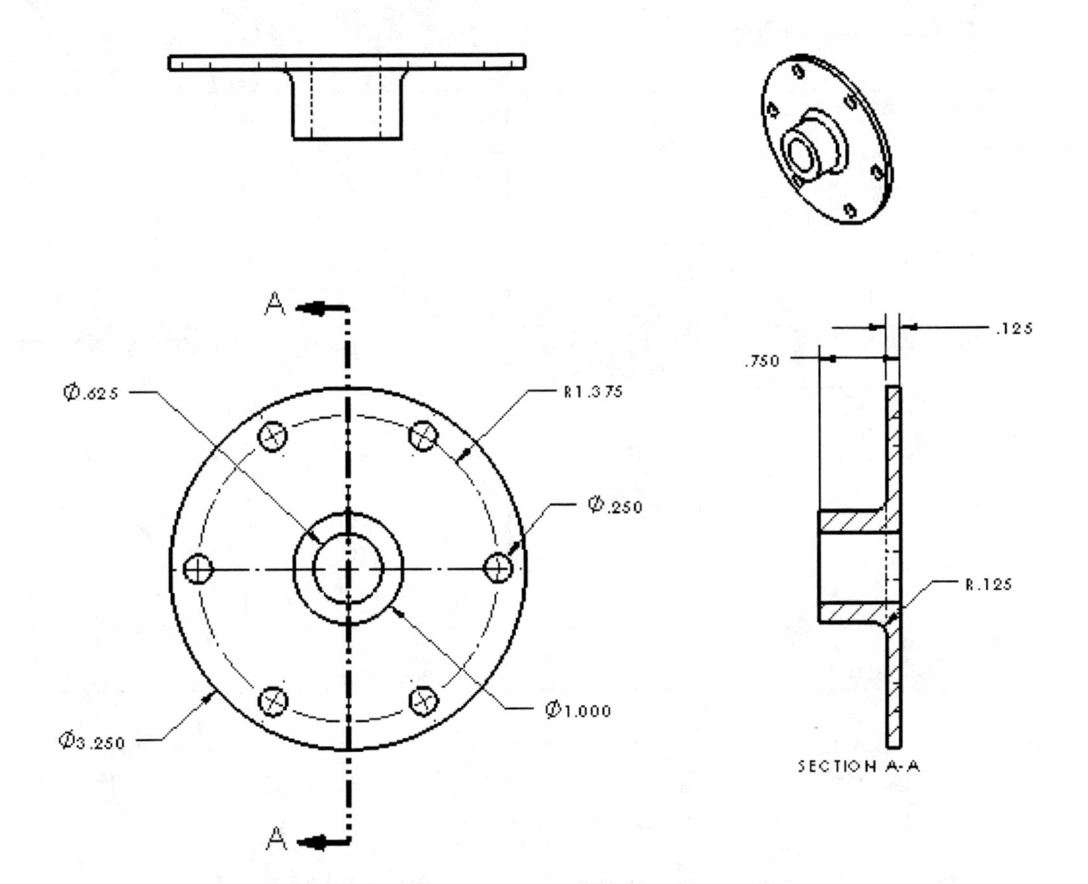

In this lesson we'll review material previously covered, like how to add standard views, sections and importing annotations from the model to the drawing. We'll learn how to add dimensions that are not present or in the desired format, change the sheet scale and a view's scale. The detail drawing of the Side Cover part will follow the next sequence.

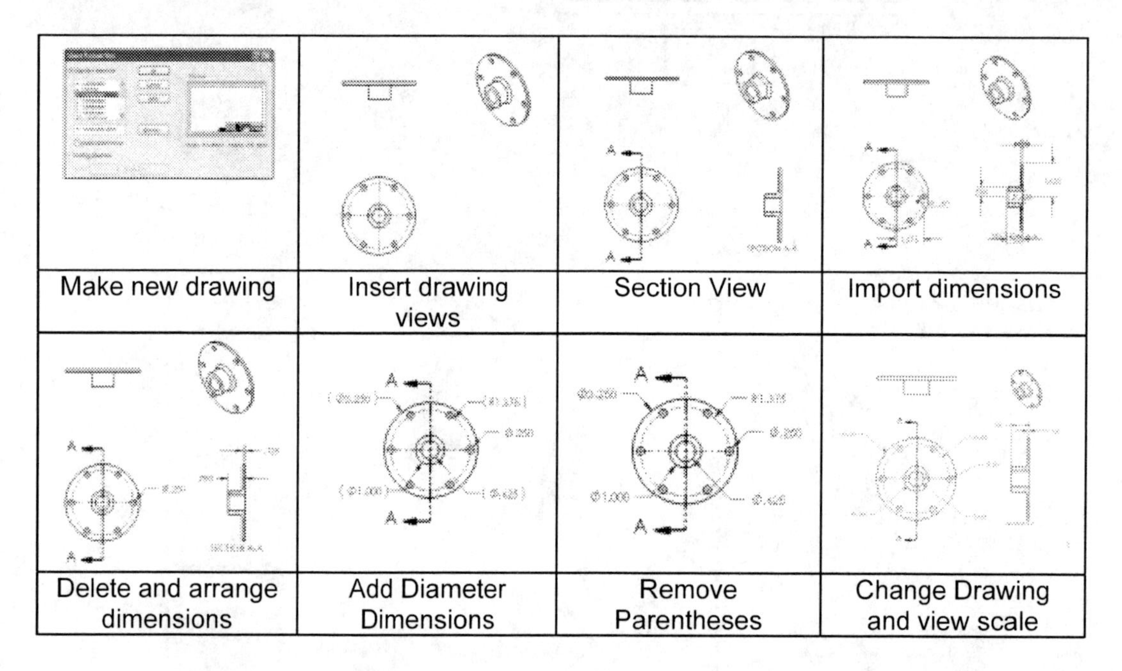

Make new drawing	Insert drawing views	Section View	Import dimensions
Delete and arrange dimensions	Add Diameter Dimensions	Remove Parentheses	Change Drawing and view scale

166. - We will repeat the process as with the Housing part. Open the Side Cover part and select the "Make Drawing from Part/Assembly" icon to start the drawing. Select the "Drawing" template and select OK.

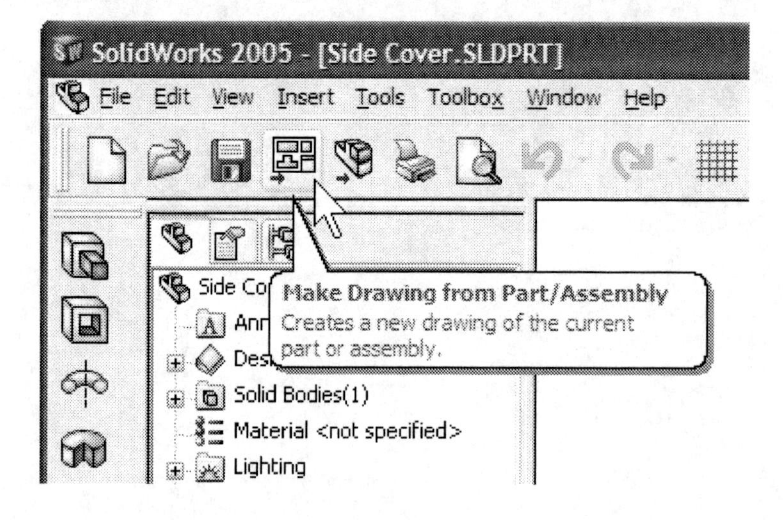

167. - For the Side Cover part we'll use the "A-Landscape" drawing template without sheet format.

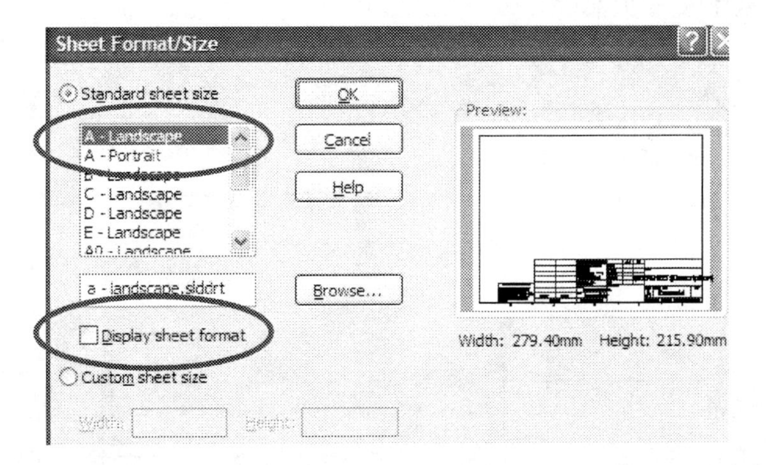

168. - From the Model View command, select "Front" from the "Orientation" box, "Hidden Lines Visible" from the "Display Style" and "Auto-Start Projected View" to add two more views of the Side Cover as we did in the previous exercise.

Click on the drawing to locate the Front view, move the pointer up to add the Top view and up to the right for the Isometric. Change the view properties to "Tangent Edges Removed" for the Front and Top view, and "Tangent Edges Visible" for the Isometric (Step 147).

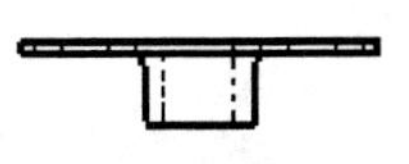

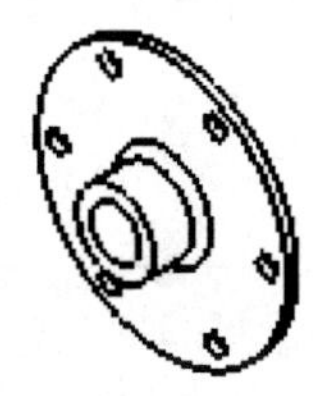

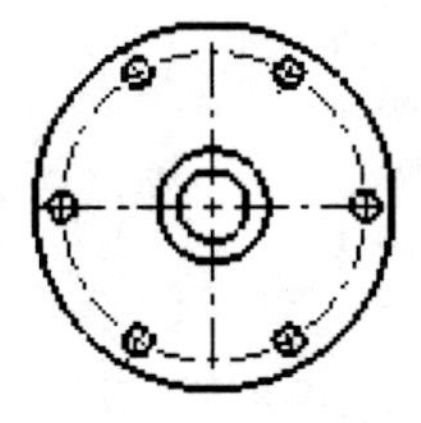

169. - Now make a section of the Front view. Select the "Section View" icon; remember to touch the circular edge to show the center of the part; draw a line through the middle and locate the Section View to the right.

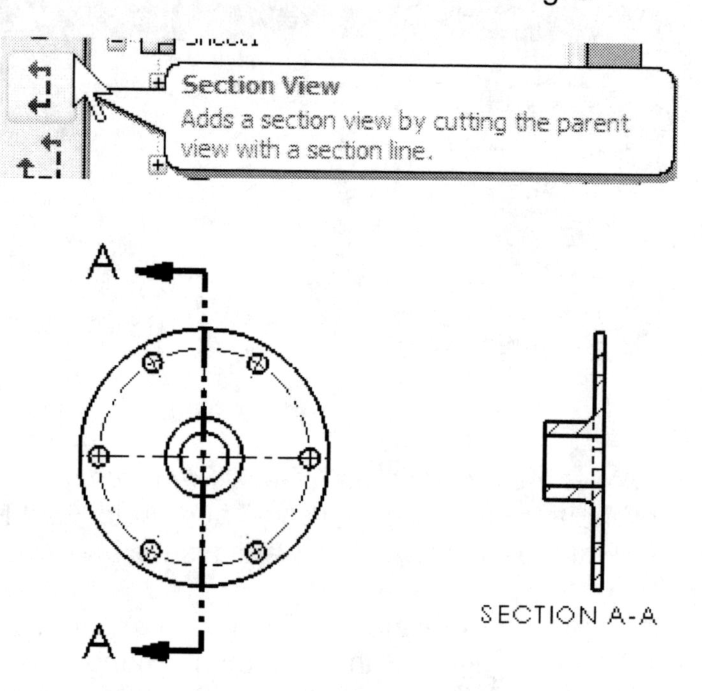

170. - Now we are ready to import the dimensions from the part into the drawing. Go to the "Insert" menu, "Model Items" and select the same options for import as with the Housing in step 157.

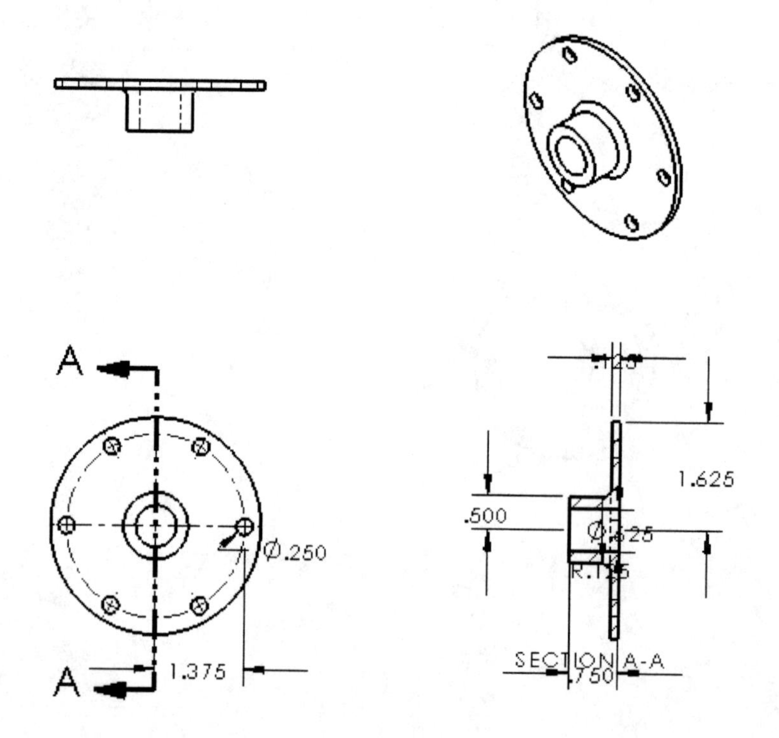

171. - When we modeled the Side Cover, we added radial dimensions, but for the detail drawing we want to show Diameter dimensions instead. We will add those dimensions manually. First delete the dimensions needed to match the image on the bottom and arrange the rest as needed.

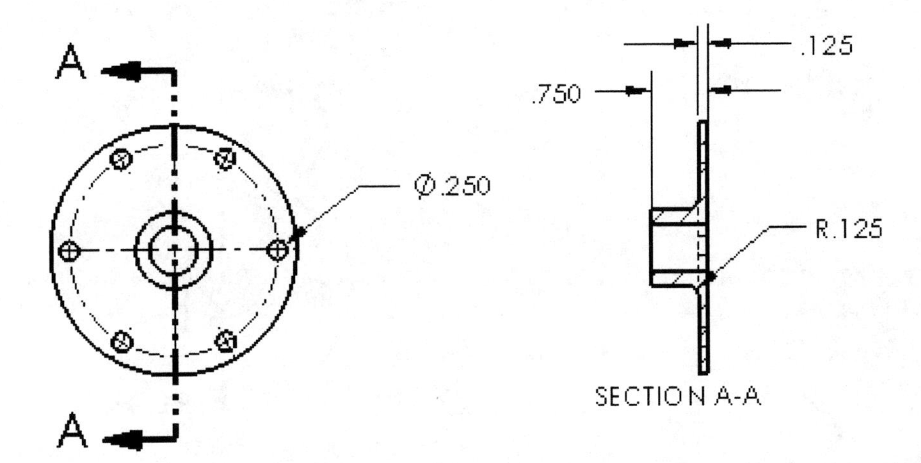

172. - Now select the "Smart Dimension" tool, and manually add the diameter dimensions indicated. Notice the parentheses; this is a way to indicate they are **Reference Dimensions** that were manually added and not imported from the model. The parenthesis is an option set in the document properties. See the Appendix for more details.

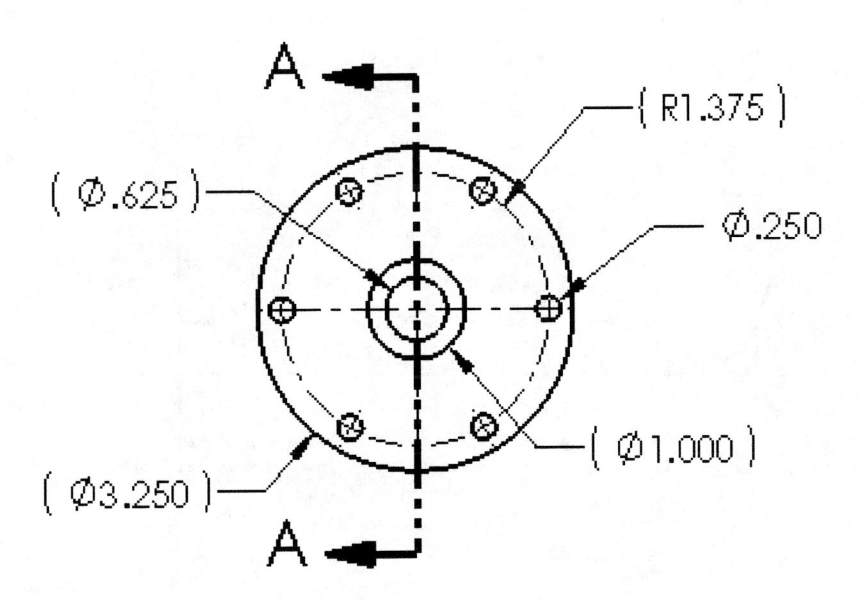

173. - To remove the **Dimension Parentheses**, Ctrl-Select all the dimensions, and from the Dimension's Property Manager, turn off the "Parentheses" option from the "Display Options" box.

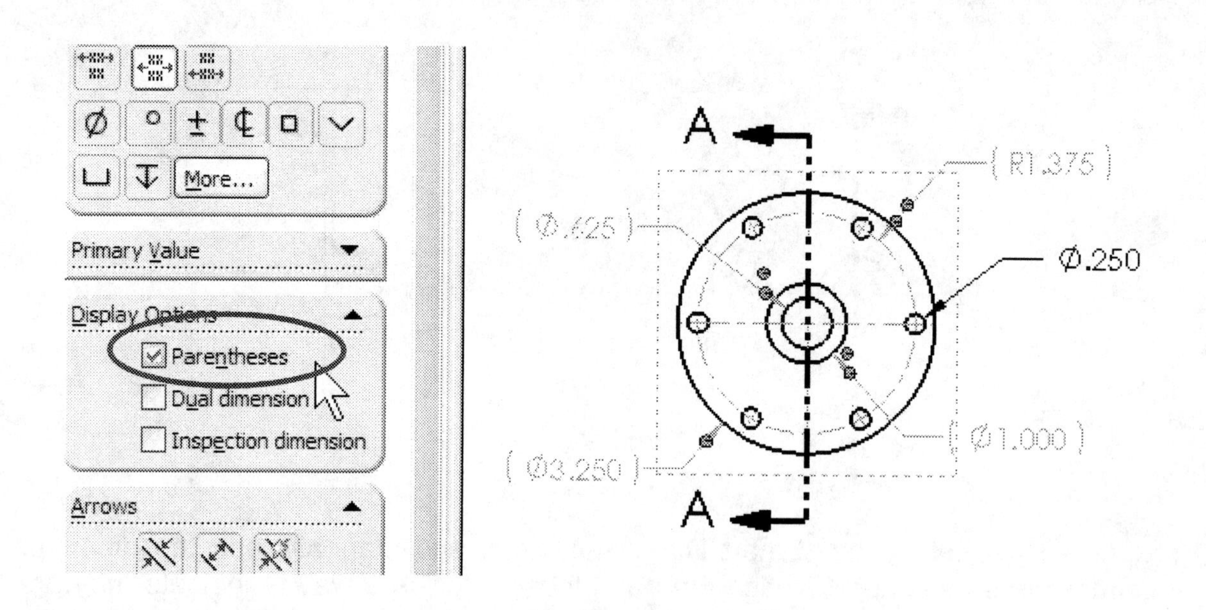

174. - Since the sheet seems a bit too big for the drawing, we'll change the sheet scale. Click with the Right Mouse Button *in the sheet, not a drawing view*, and from the pop-up menu select "Properties".

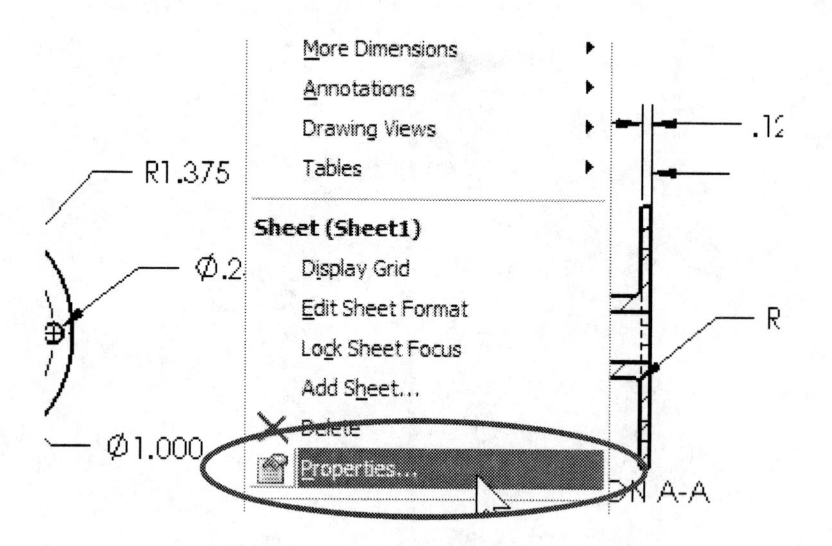

175. - From the options box, select a scale of 1:1 and click OK. After arranging the views and dimensions, we notice that the Isometric is a bit too big to fit nicely in our sheet, so we'll proceed to change the scale of the Isometric view only.

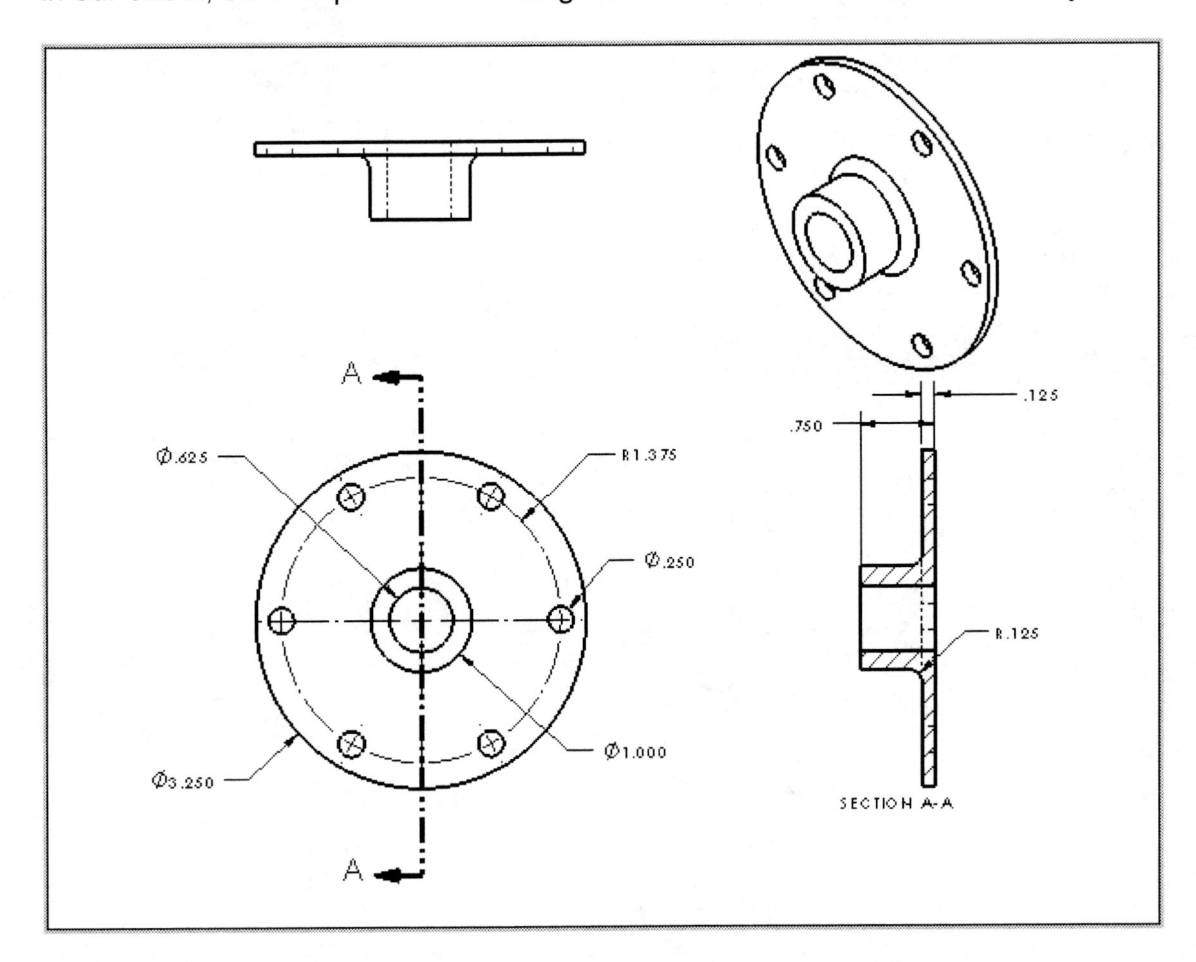

176. - Select the Isometric view, and from the Property Manager, change the view scale to 1:2 from the "**Use custom scale**" in the "Scale" options box.

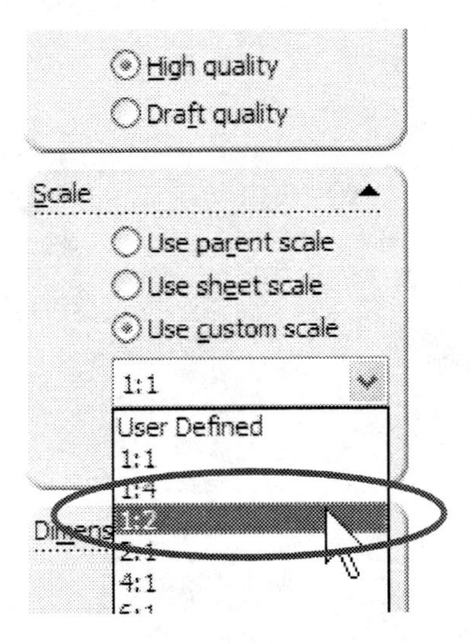

177. - Our drawing is finished, save and close the file.

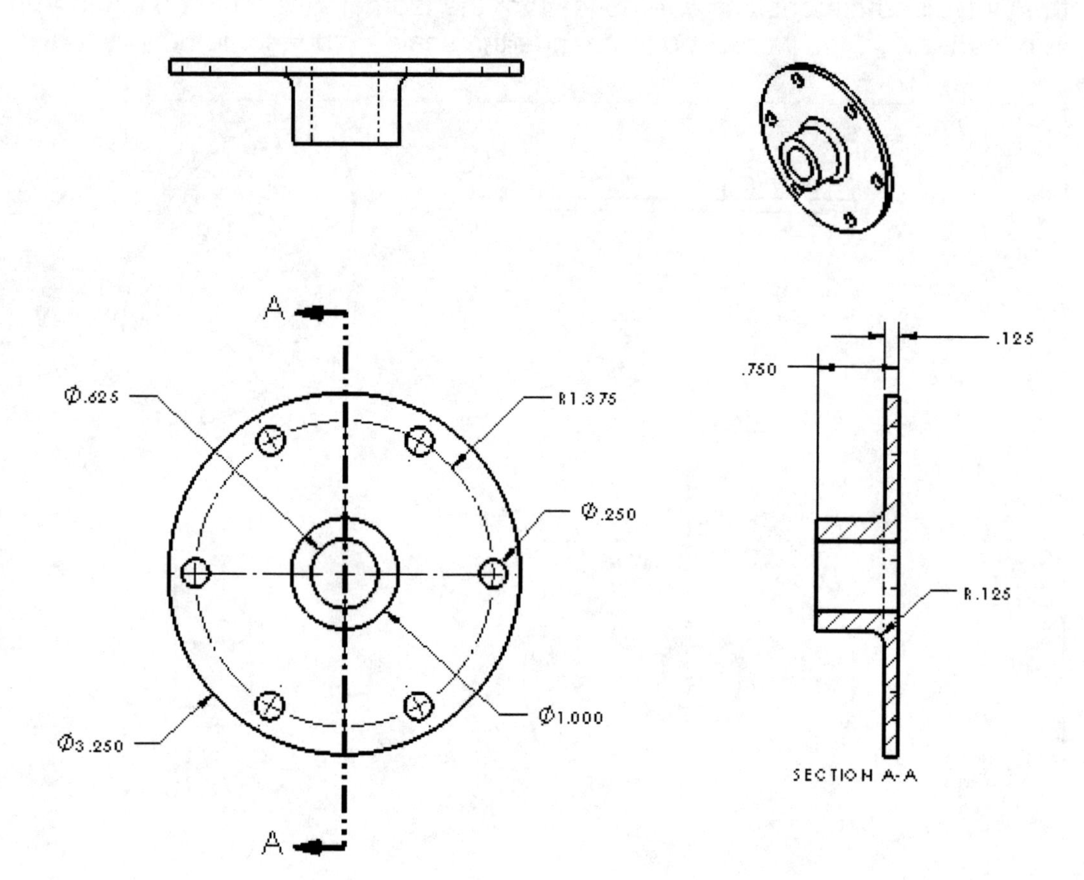

Drawing Three: The Top Cover

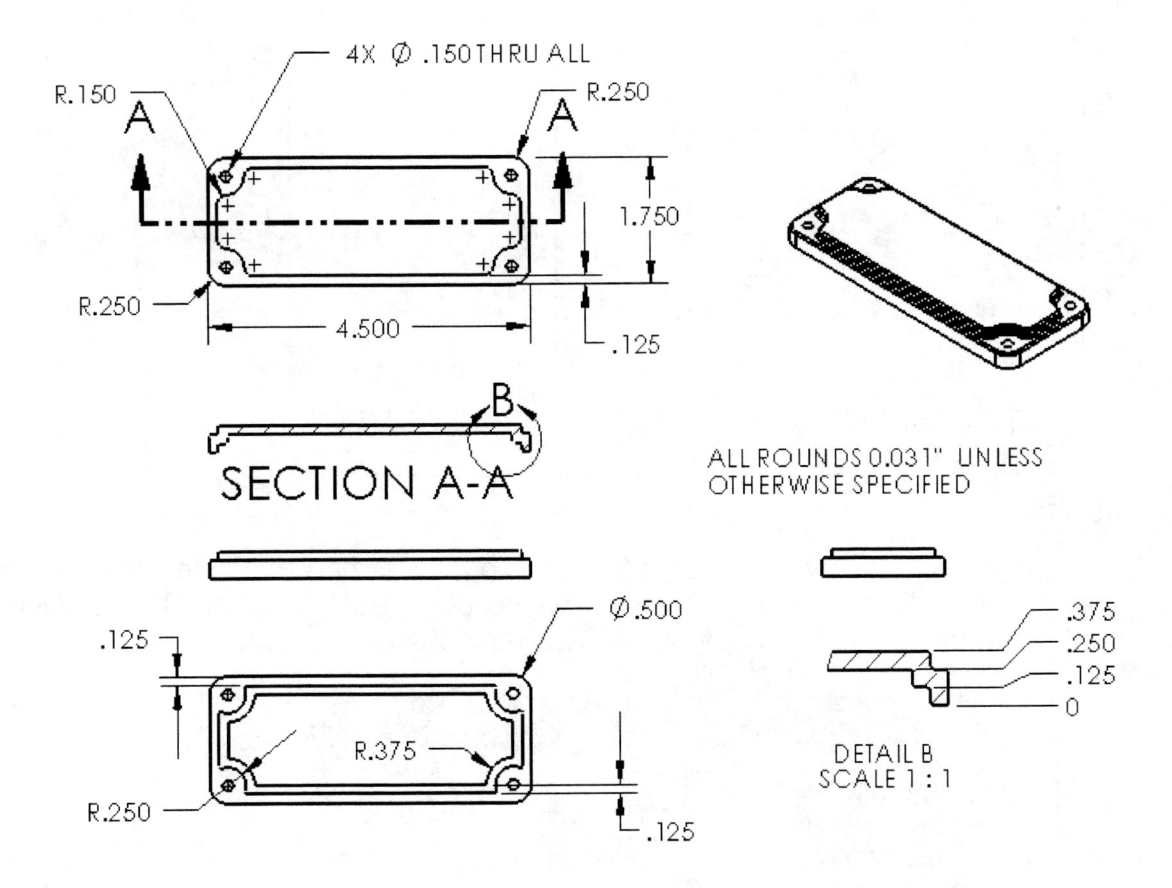

178. - In this lesson we'll review previously covered material as well as new options like "Display only surface" in a section, adding notes and Ordinate dimensions. The drawing of the Top Cover will follow the next sequence.

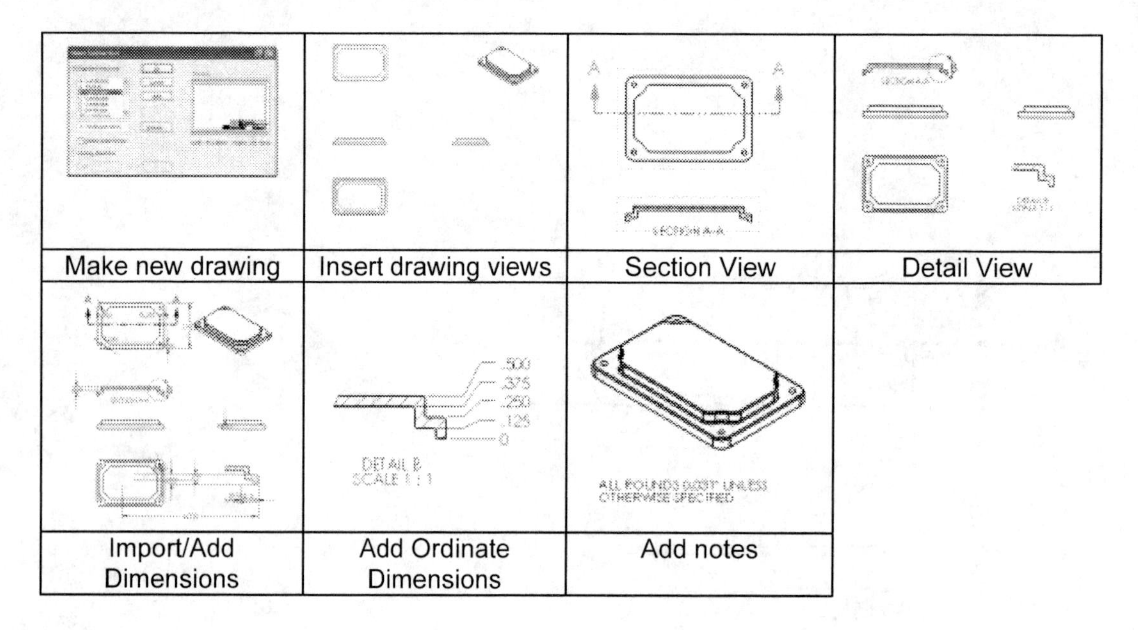

179. - For the Top Cover drawing open the Top Cover part and select the "Make Drawing from Part/Assembly" icon as we've done before. Select the drawing template with an "A-Landscape" sheet size without sheet format.

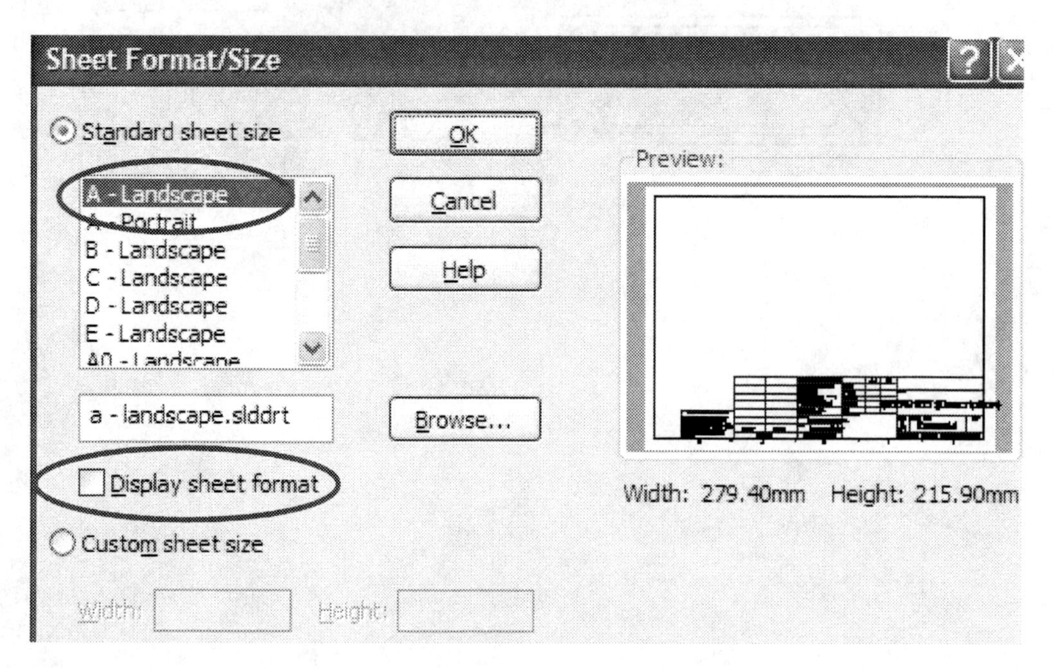

180. - Insert Front, Top, Bottom, Right and Isometric views with "Hidden Lines" display mode and "Tangent Edges Removed". Change the Isometric view display mode to "Tangent Edges Visible" to match the next image.

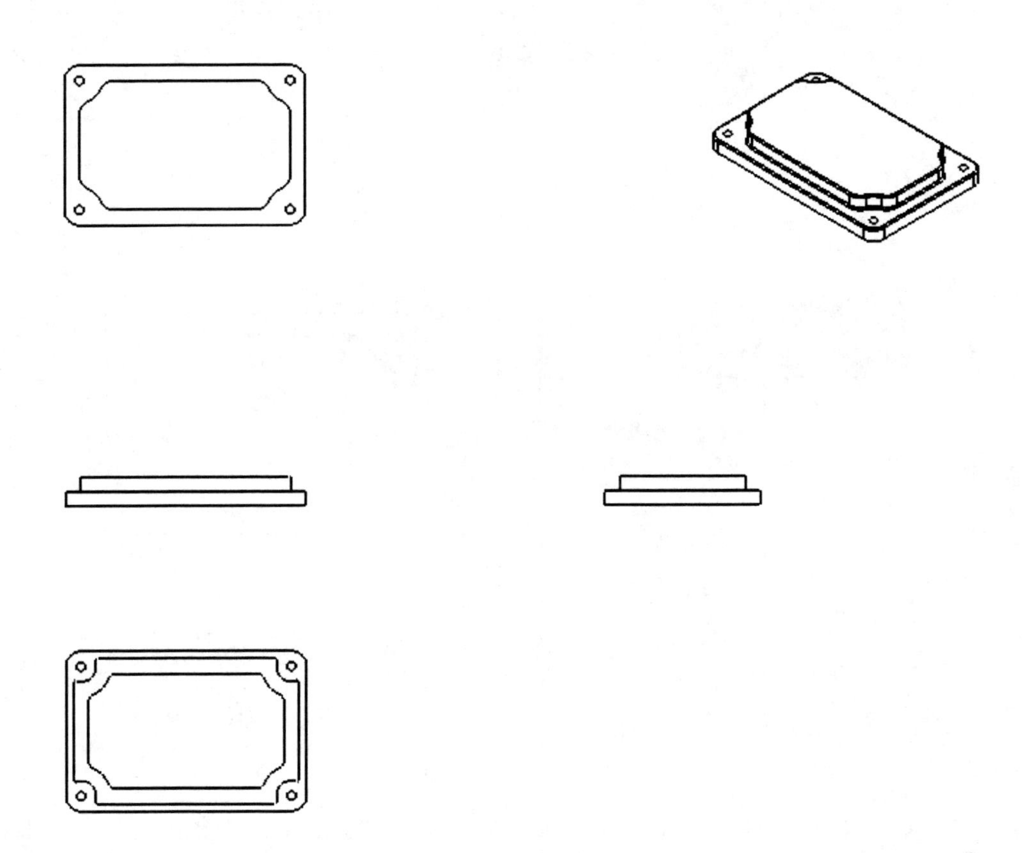

181. - Now we need to make a section through the Top view to get more information about the cross section of the cover. Select the Section View icon and draw a horizontal line as indicated; then locate the section just below the Top View.

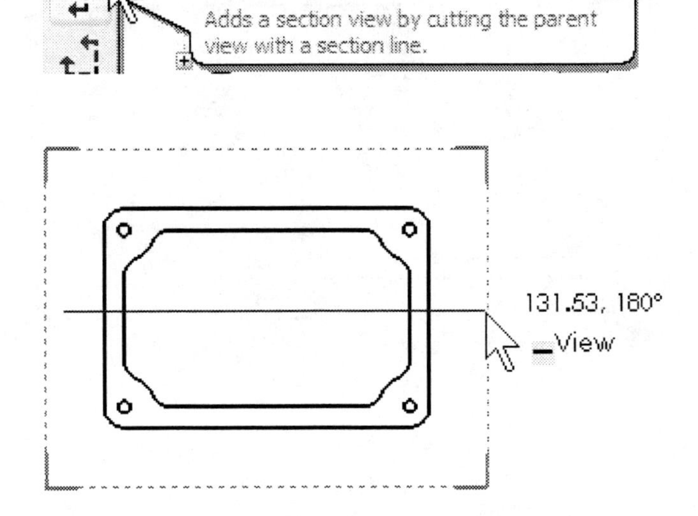

182. - After locating the dimension, you may have to activate the "Flip Direction" checkbox to reverse the direction of the section view. This can also be done by selecting the section view from the Property Manager. We also want to check the "**Display only surface**" option; this option will show only the section's surface ignoring the rest of the model behind it, as if we had only taken a thin slice of the part.

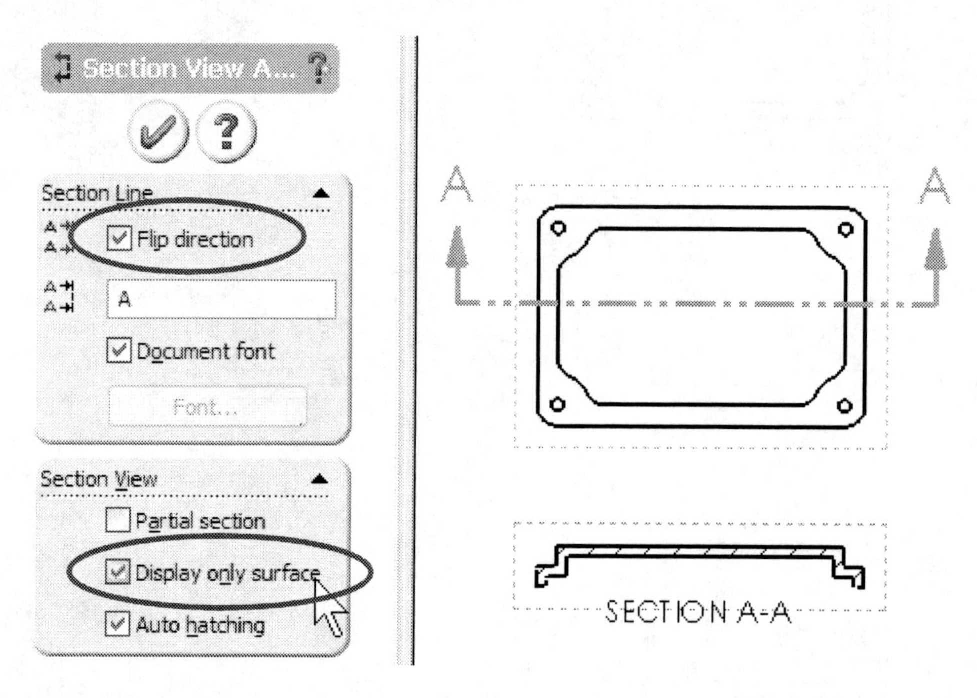

183. - To get an even better view of the cross section, we'll make a detail of the right side of the Section View. Select the "Detail View" icon and draw a circle as shown; locate the detail under the Right view to distribute the drawing evenly in the sheet.

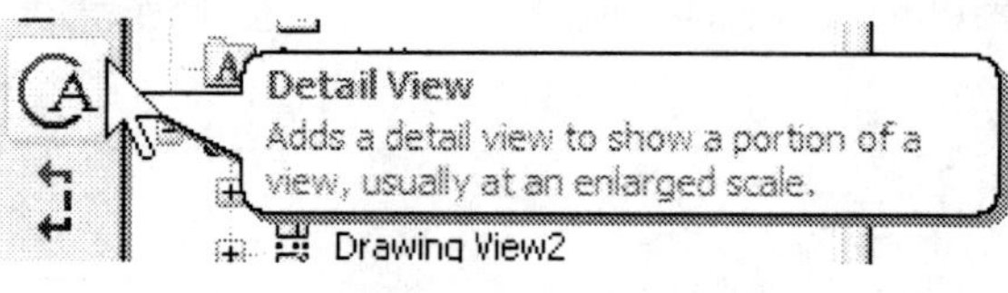

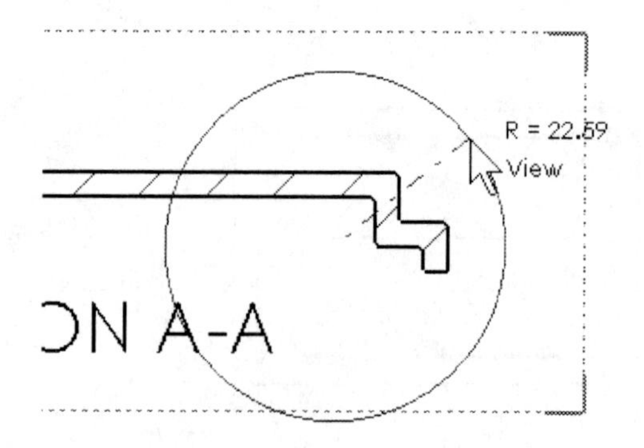

184. - When we make detail and section views, and make a mistake or we are simply not happy with the result and delete it, SolidWorks increases the new view label to the next available letter. Then, if we make a new view, the new view will have this new label. The good news is that the label can be changed in the view's Properties by selecting the view and then changing the label.

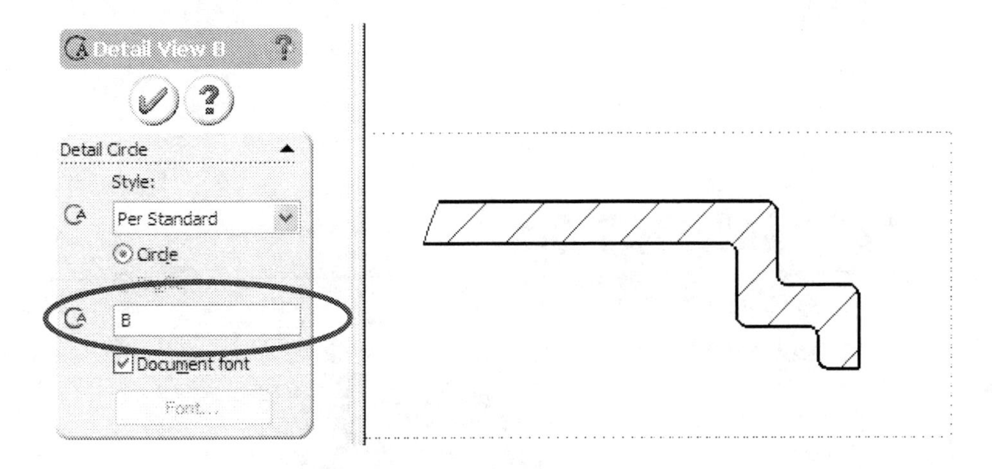

185. - The next step is to import the model dimensions from the part. Go to the "Insert" menu and select "Model Items"; make sure to select the "Import items into all views" and "Entire Model" options. Delete and arrange the dimensions as needed to match the next image.

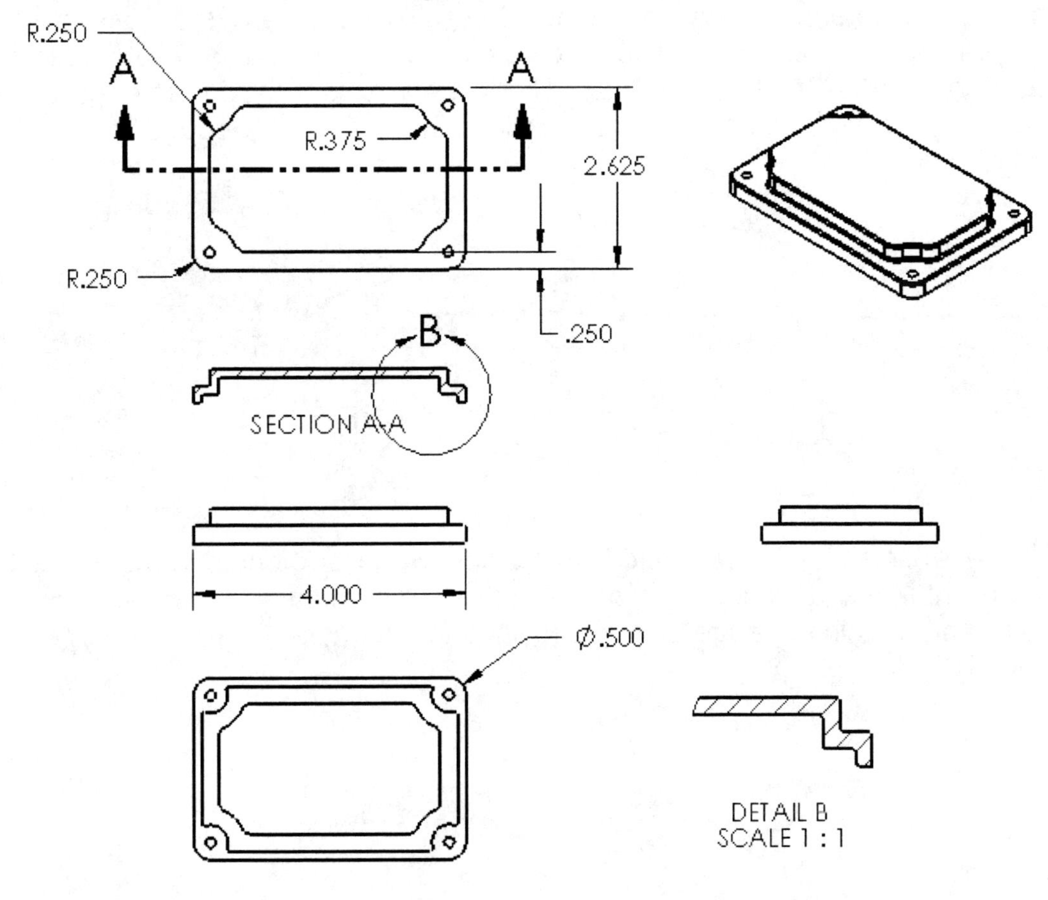

186. - Sometimes it's convenient to add **ordinate dimensions** to a drawing. The ordinate dimensions have to be added manually. Click in the graphics area with the Right Mouse Button, and from the pop-up menu select the "More Dimensions, Vertical Ordinate".

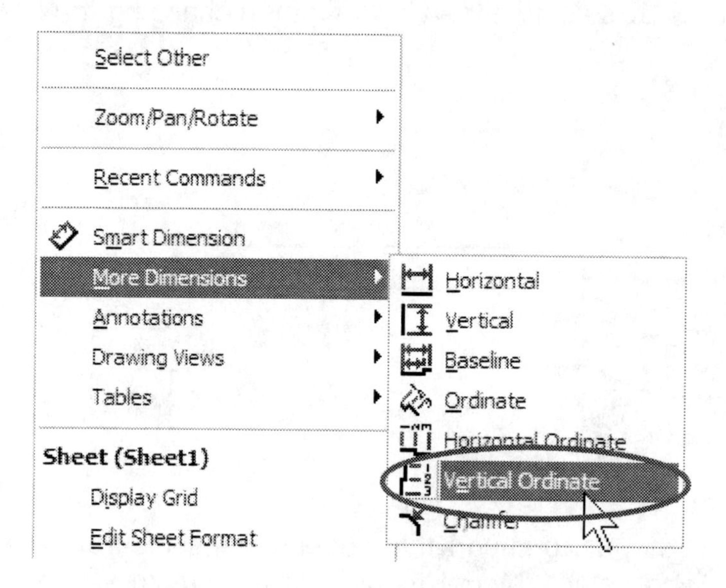

187. - Adding Ordinate dimensions is very simple: first select the edge that will be the zero reference, click to locate the "0" and select the rest of the edges/vertices to dimension. The Ordinate Dimensions will be automatically aligned and jogged if needed. For the Detail view, select the lower edge to be zero, click to the right to locate it, and click in the rest of the edges to add the dimensions.

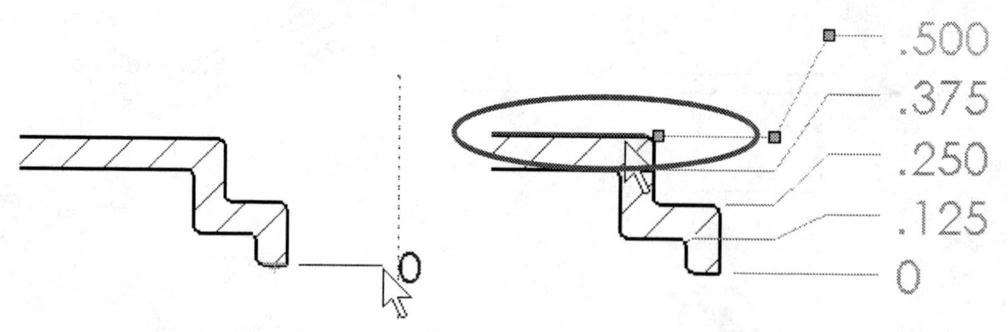

NOTE: If after finishing the Ordinate Dimensions we missed one dimension, click with the Right Mouse Button in one of the ordinate dimensions, select "Add to Ordinate" and click in the edge that was missed to dimension it.

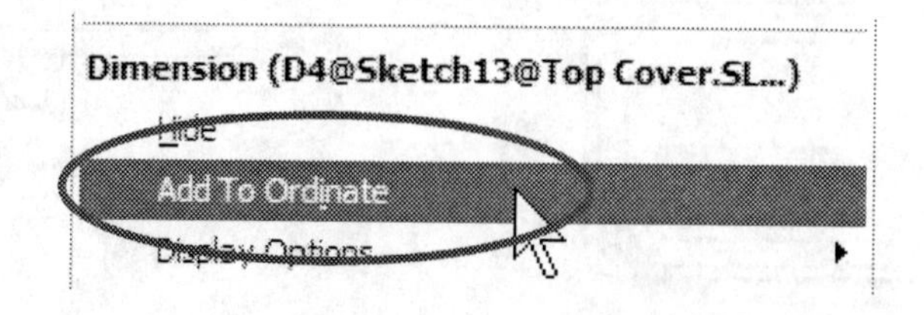

188. - The last thing we are going to do to complete the drawing is to add a note. Notes are a common element in drawings to communicate important information, such as material, finish, dates, etc. To add a note to the drawing, select the "**Note**" icon from the Annotations Toolbar...

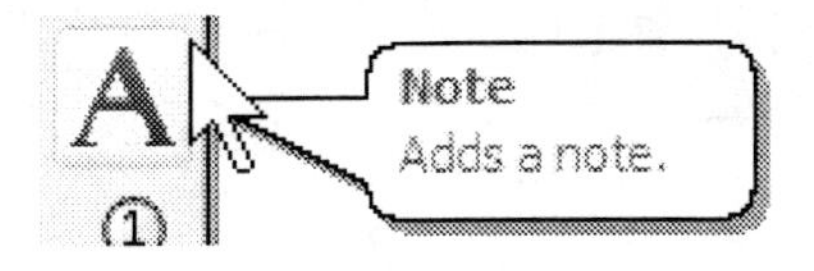

...and click in the drawing to locate the note; the **Formatting Toolbar** is automatically displayed next to the note if not already visible. This toolbar is similar to the one found in many other Windows applications to change the text font and style.

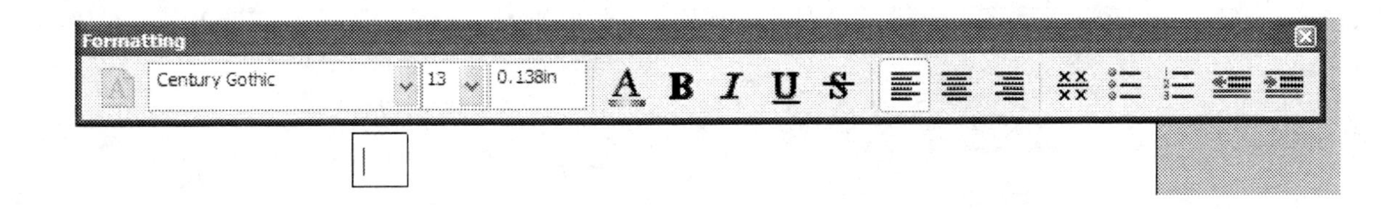

189. - Once the note's location is placed in the drawing, we are free to type anything we want and format it as well. Click OK when finished. If multiple notes need to be added, instead of OK click where the next note will be and repeat the process.

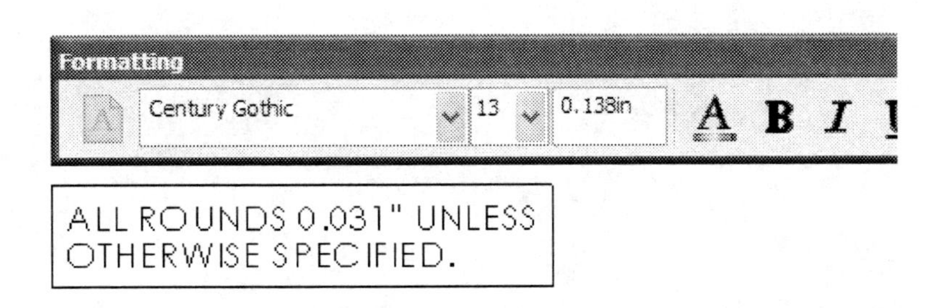

190. - Save and close the drawing file.

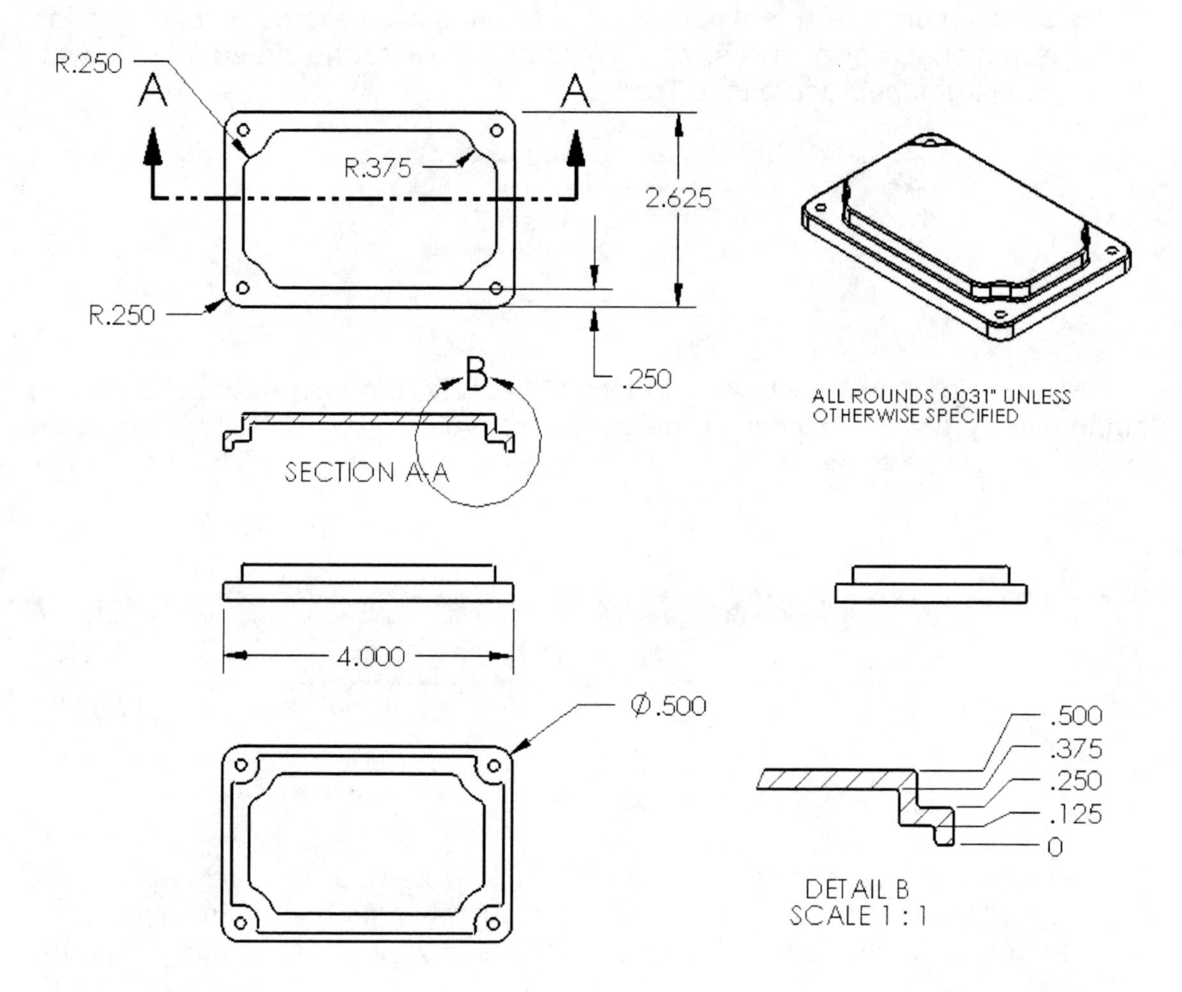

Drawing Four: The Offset Shaft

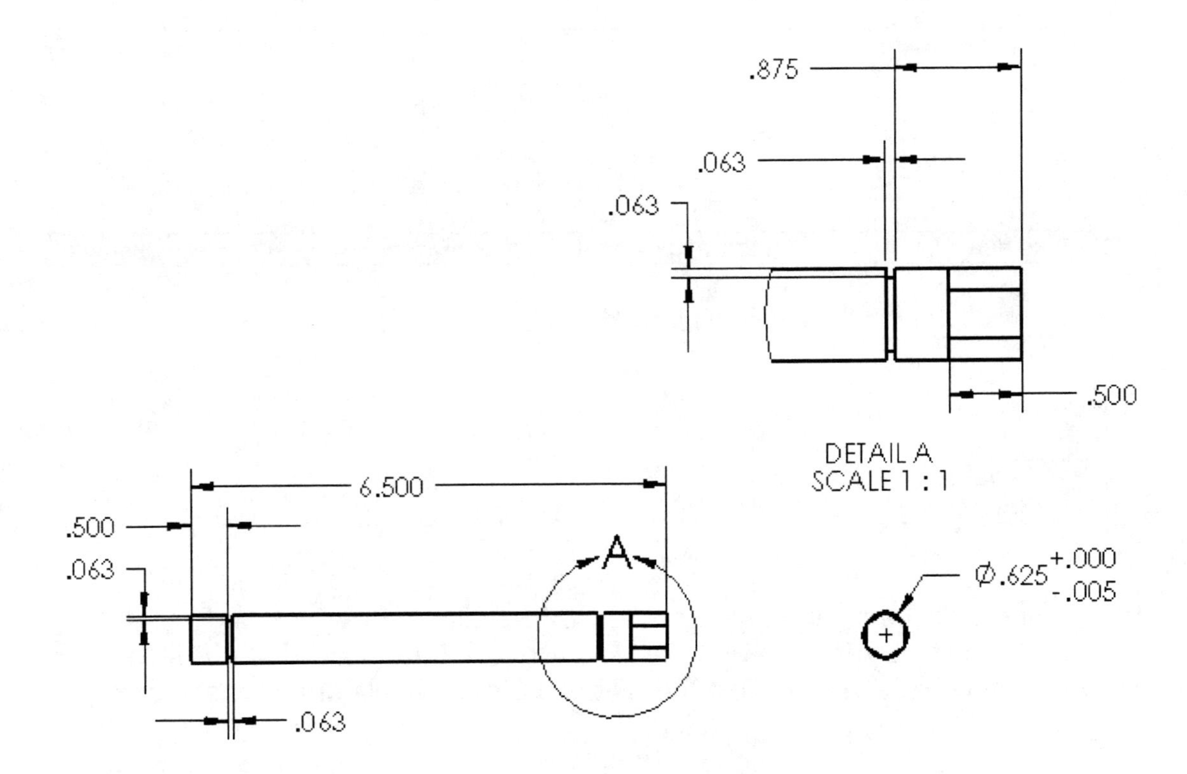

The Offset Shaft drawing, although a simple drawing, will help us reinforce previously covered commands including standard and detail views, importing dimensions, manipulating and modifying dimension appearance. In this exercise we will make Front, Right and detail views, import the model dimensions and modify their appearance following the next sequence:

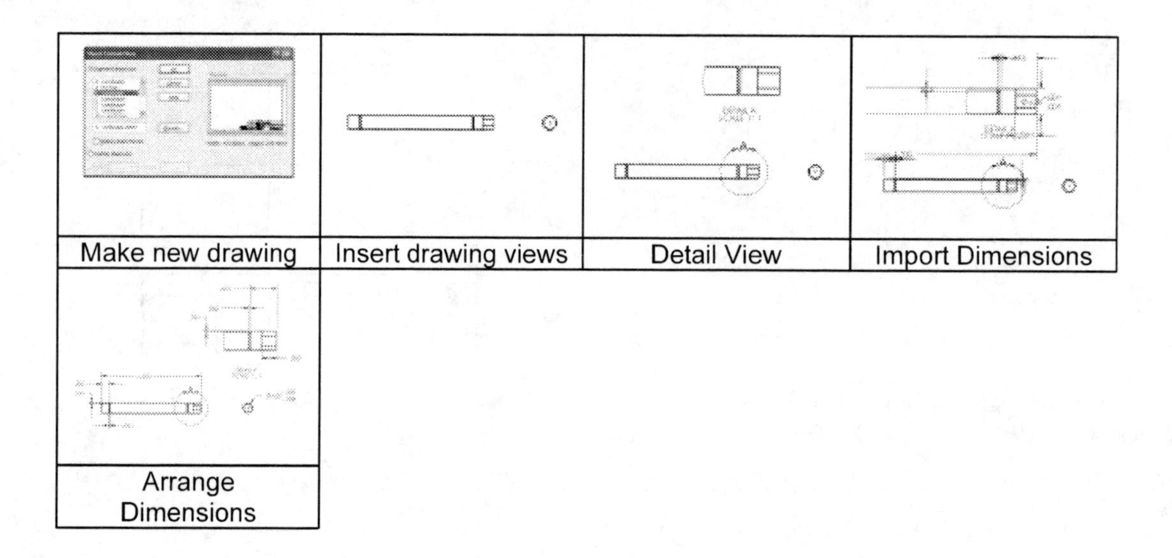

191. - Since we have already shown how to make a new drawing, we'll simply ask you to make a new drawing file using the "A-Landscape" sheet size. Then add Front and Right views using the "Hidden Lines" display mode as shown.

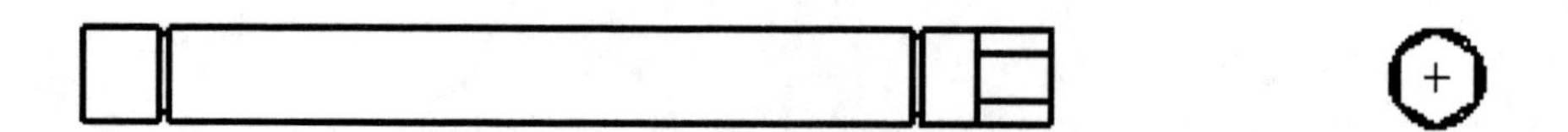

192. - We also covered how to make a detail view in previous exercises. Select the "Detail View" icon from the Drawing Toolbar, draw the circle at the right end of the shaft, and locate the view just above the drawing.

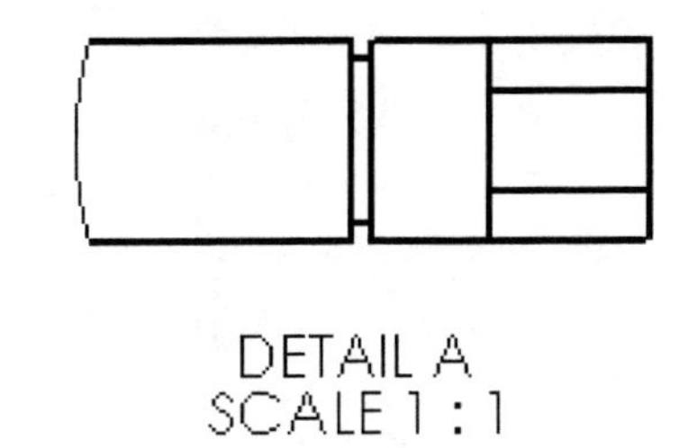

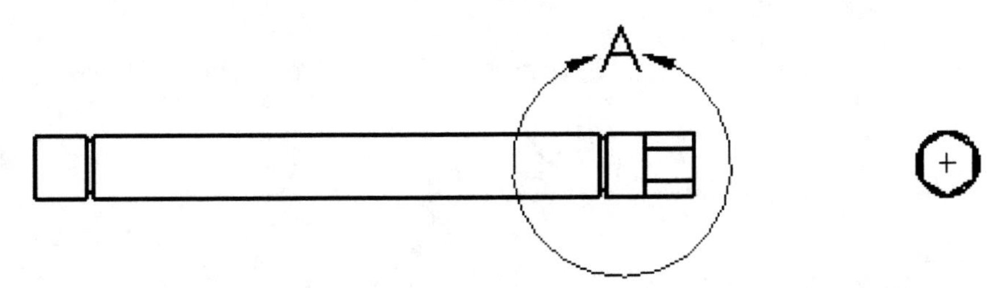

193. - You probably know what we are going to do now... Yes, import the dimensions from the part. Go to the "Insert" menu, "Model Items"; remember to select the "Import items into all views" and "Entire Model" options. In this case, SolidWorks added most of the dimensions to the detail drawing, but we'll fix that in the next step.

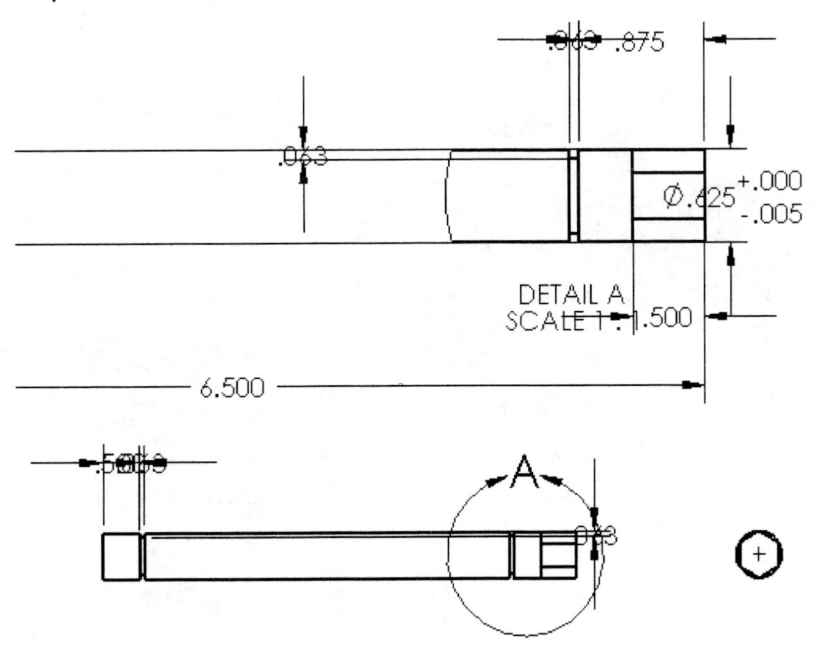

194. - Move the dimensions as shown in the next image; remember to hold down the "SHIFT" key while dragging them.

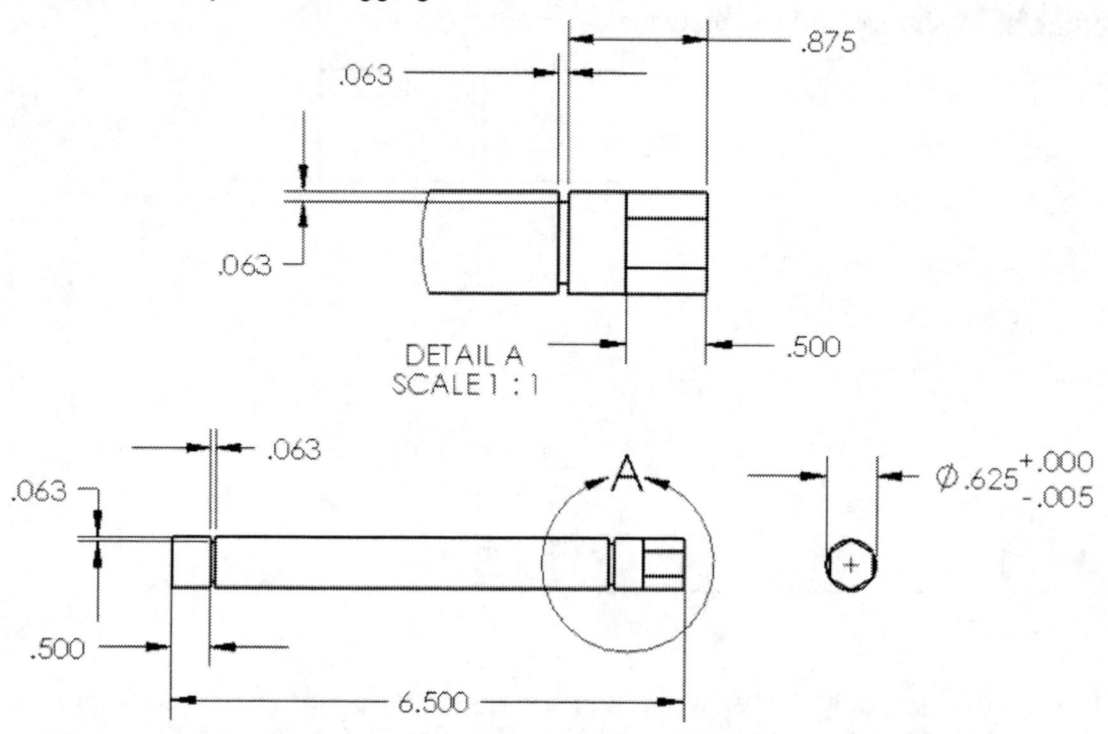

195. - For the diameter dimension, we'd like to see a radial dimension instead of a linear one. To change the dimension display, select it with the Right Mouse Button, and from the pop-up menu, select "Display Options, Display as Diameter".

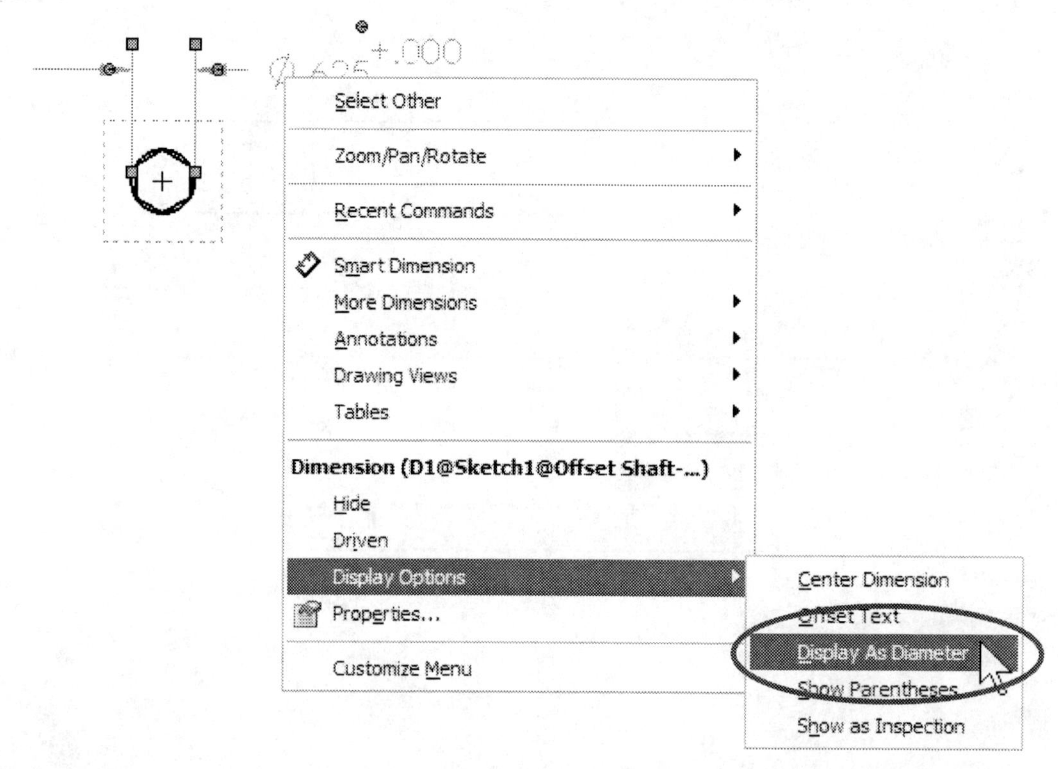

196. - Save and close the drawing file.

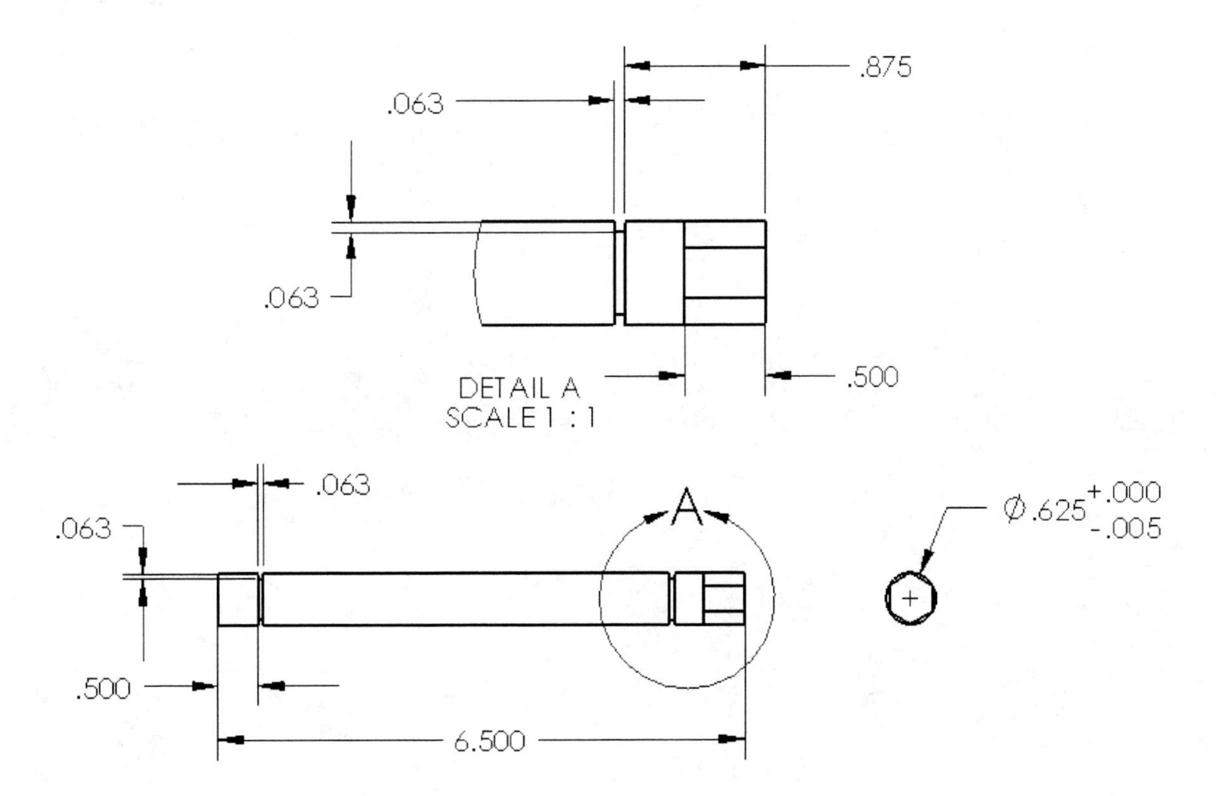

Notes:

Drawing Five: The Worm Gear

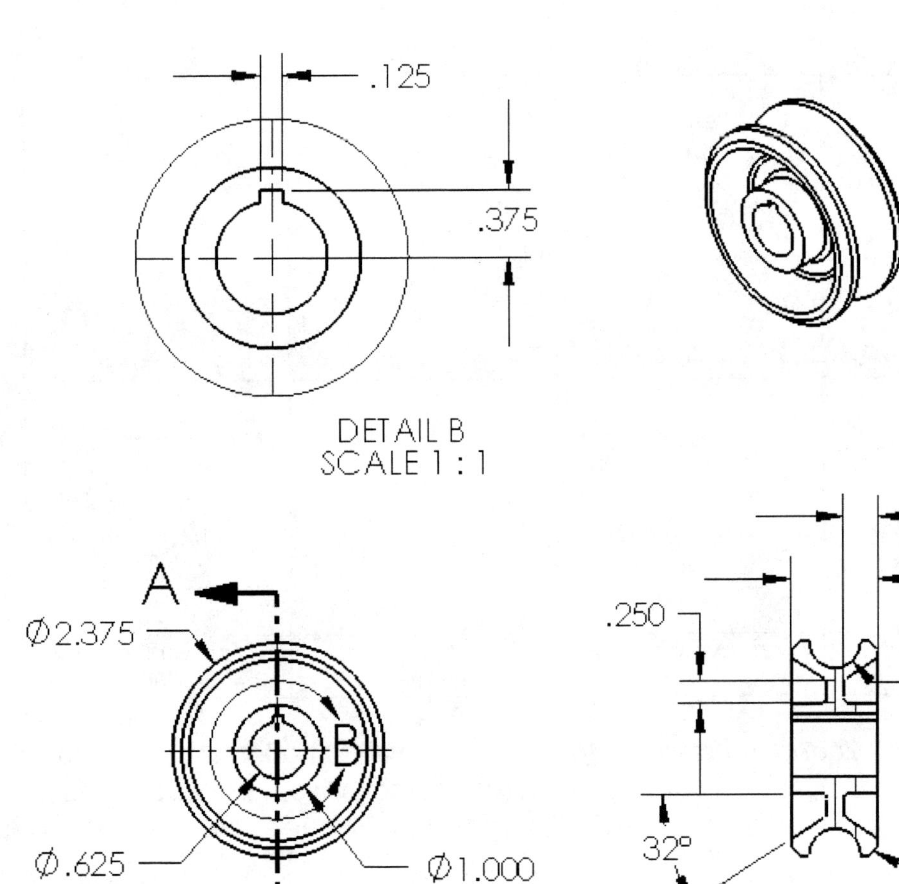

In this lesson we'll review previously covered material and a couple of new options, like adding angular dimensions, changing a dimension's precision and adding chamfer dimensions. The detail drawing of the Worm Gear will follow the next sequence.

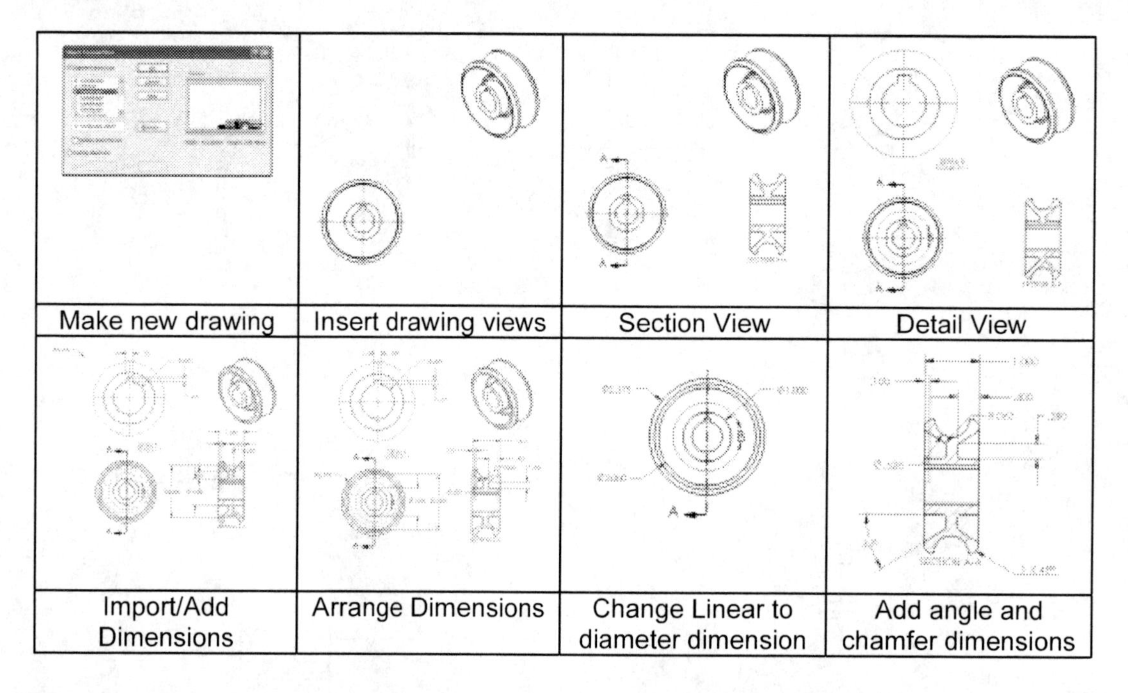

197. - As we have done in previous drawings, we will first open the "Worm Gear" part and select the "Make Drawing from Part/Assembly" icon. For this component use the Drawing Template with an "A-Landscape" sheet size and no sheet format.

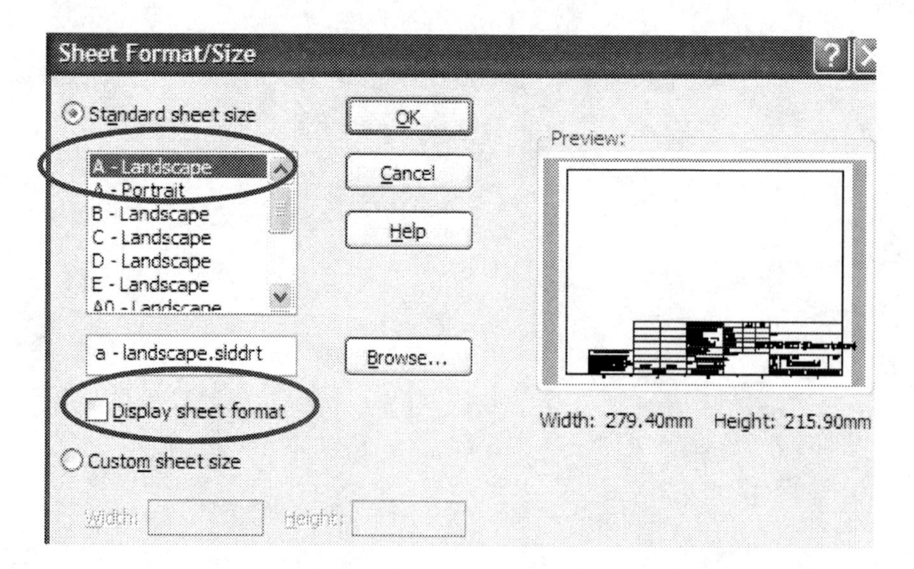

198. - Just as we've done in previous drawings, the first step is to add the main views for the model. Start by locating the Front and Isometric Views. Immediately after, make section and detail views as shown. This is a review of material previously covered. Notice that we are using "Hidden Line" display mode for all views and "Tangent Edge Removed" for all views except the Isometric.

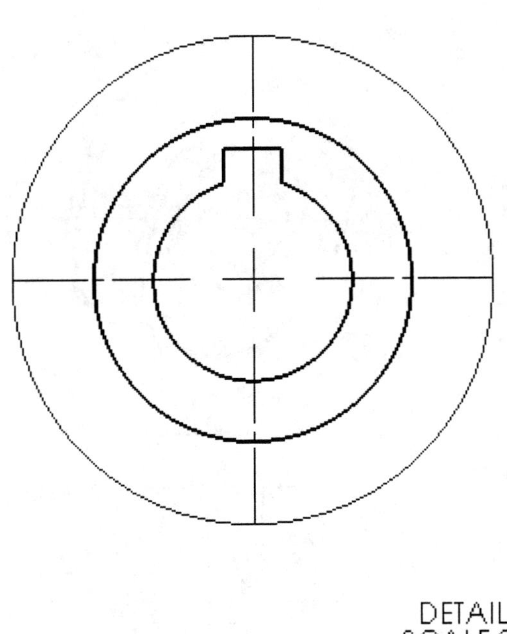

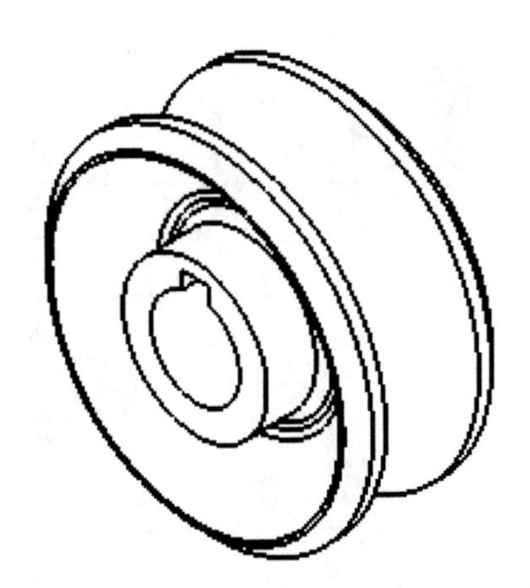

DETAIL B
SCALE 2 : 1

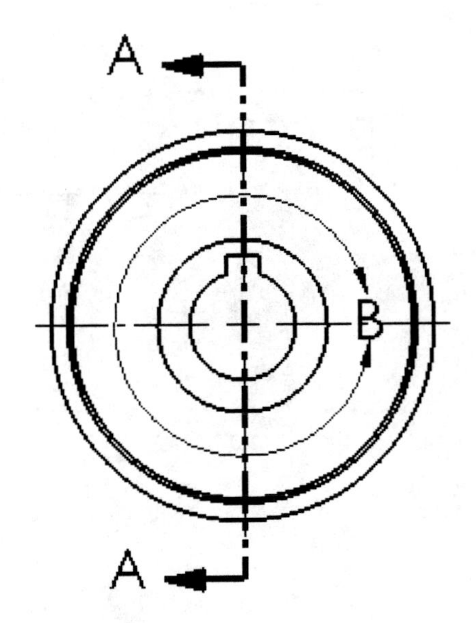

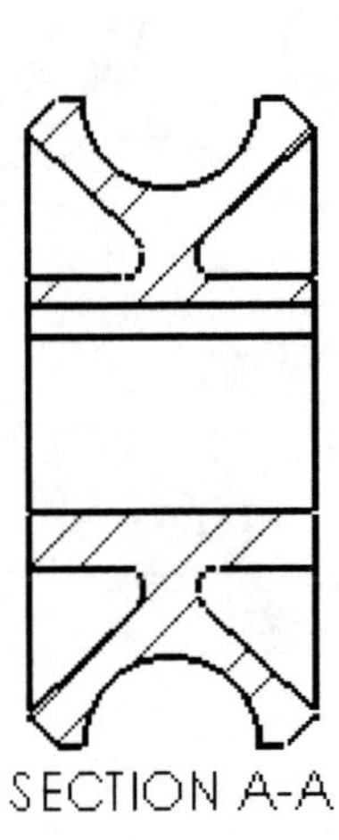

SECTION A-A

199. - Now import the dimensions from the model using the "Insert" menu, "Model Items". (Remember to select the options "Import items into all views" and "Entire Model".) After importing the dimensions, delete and arrange as needed to match the following picture; you may have to change a dimension to diameter display (see step 195).

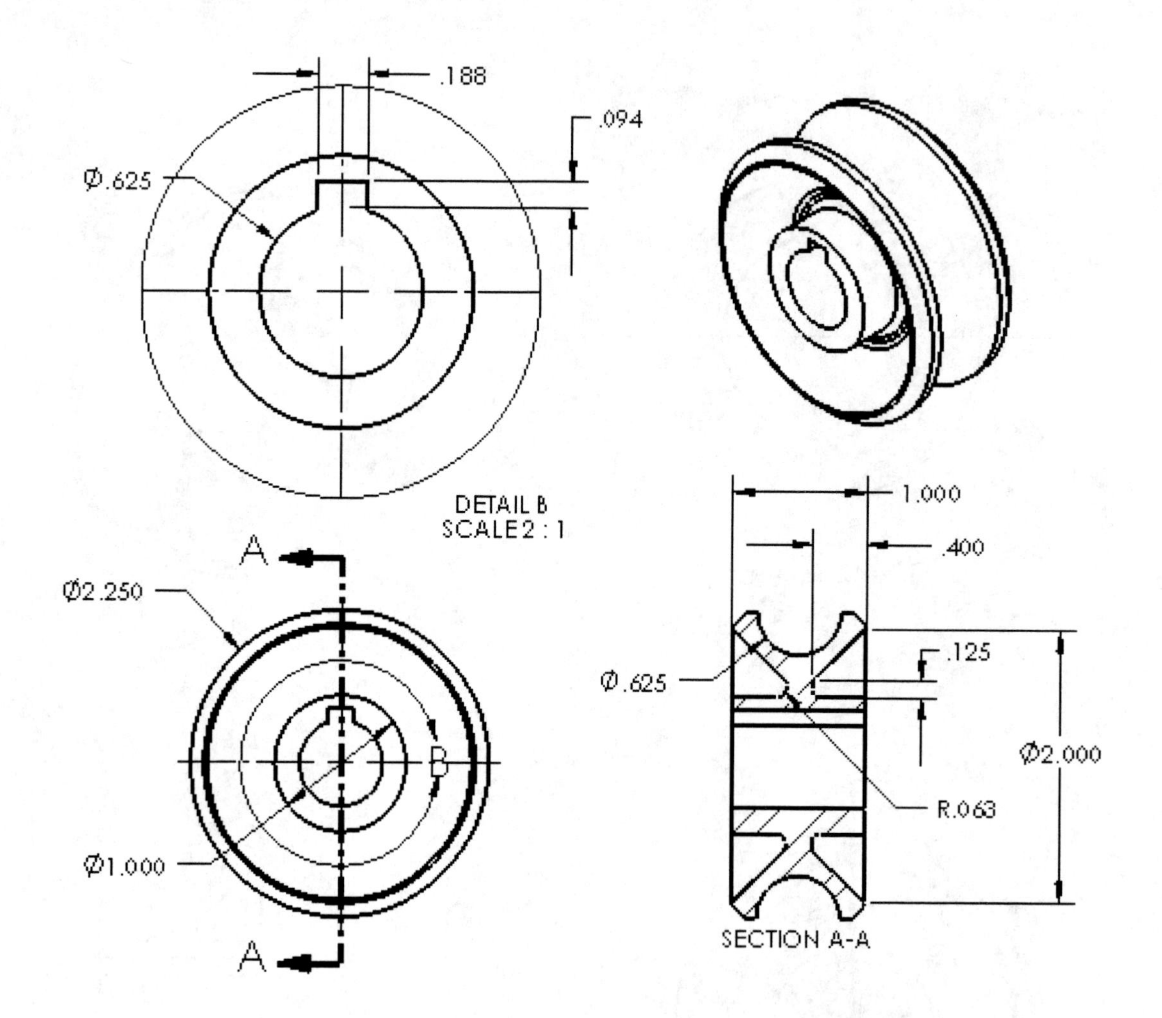

200. - Now add an angular dimension. To do this click on the "Smart Dimension" icon and select the two edges to add the 43.15° dimension. After locating the dimension, we see the dimension has a precision of two decimal places. This is the drawing's default settings for angular dimensions. We'll change the precision of this dimension to remove the decimal places.

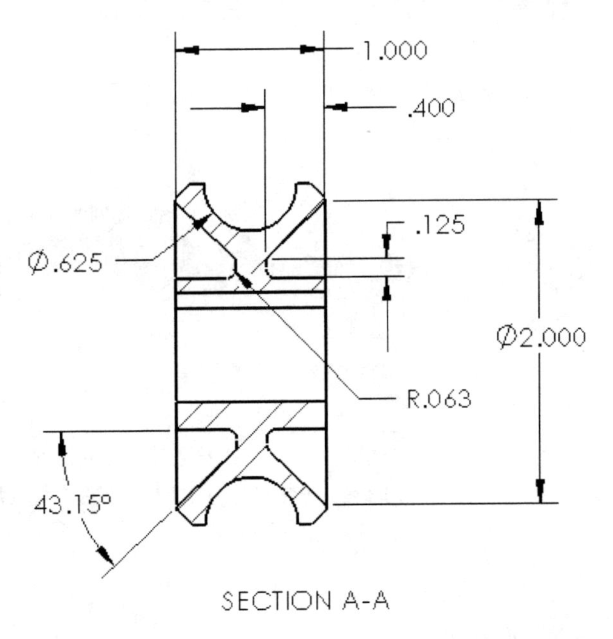

201. - Select the angular dimension, and from the dimension's properties select "None" from the drop down list under the "**Tolerance/Precision**" options. Multiple dimensions can be changed at the same time if needed by pre-selecting them using the "Ctrl" key.

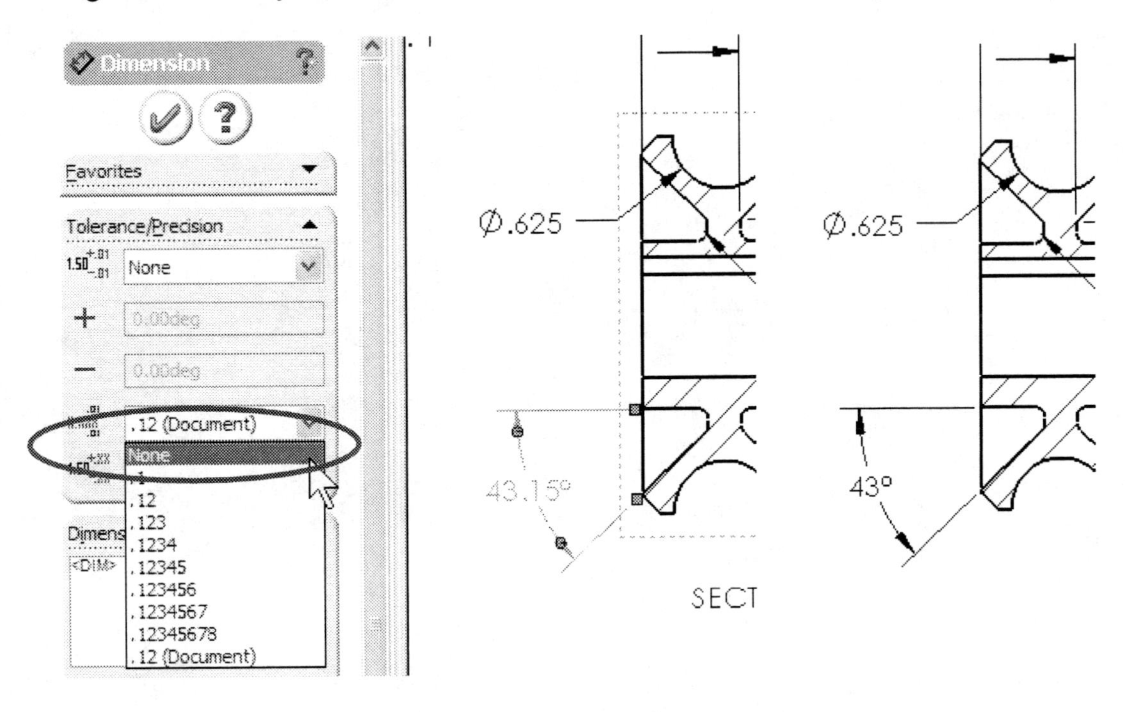

202. - The last step in this drawing is to add the **chamfer dimension**. Click in the graphics area with the Right Mouse Button, and from the pop-up menu select "More Dimensions, Chamfer".

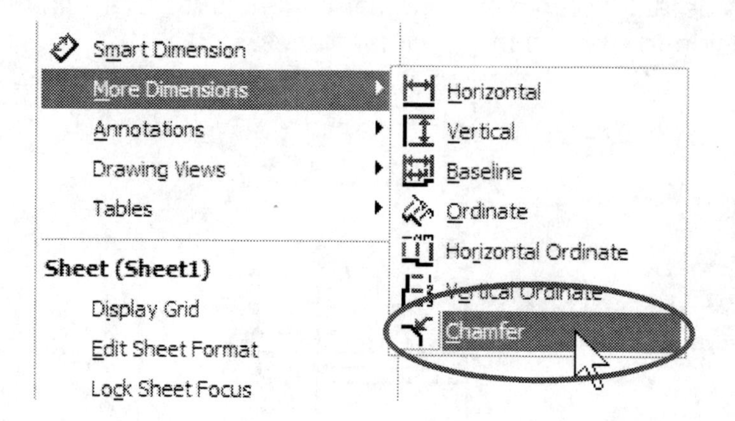

203. - To add the Chamfer dimension, select the Chamfered edge first, then the vertical edge to measure the angle against it and locate the dimension.

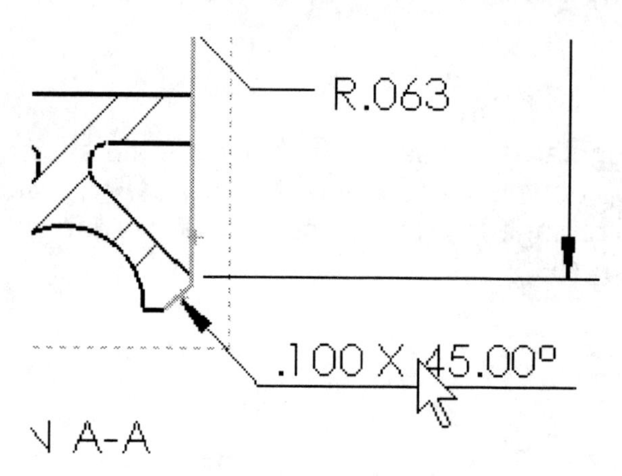

204. - Finish the Chamfer Dimension by changing its precision for the distance and angle. Select the Chamfered dimension, and from the dimension's properties change the "Tolerance/Precision" to ".1", this will change the dimension to one decimal place, and then change the "2nd Tolerance/Precision" to "None" to remove the decimal places from the angular value.

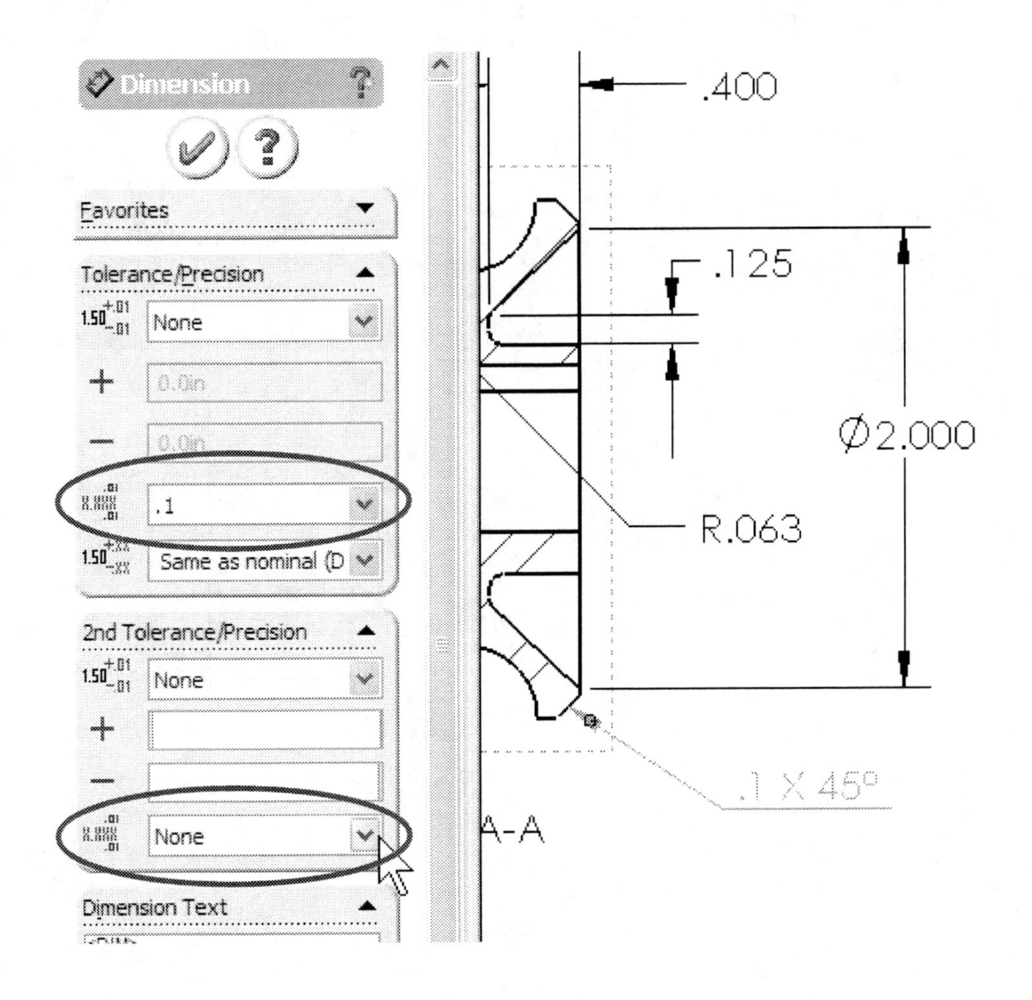

205. - Save and close the drawing file.

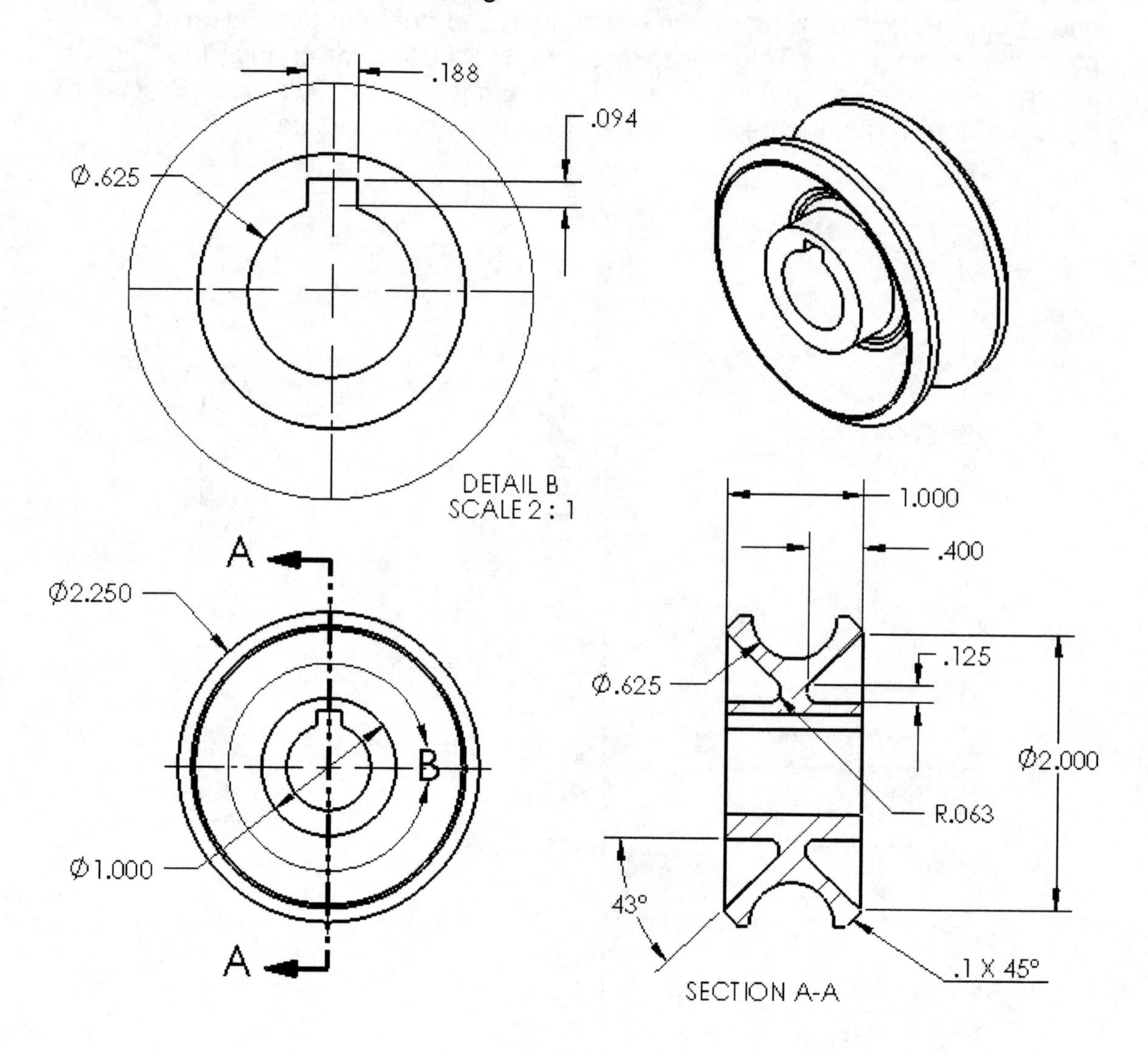

Drawing Six: The Worm Gear Shaft

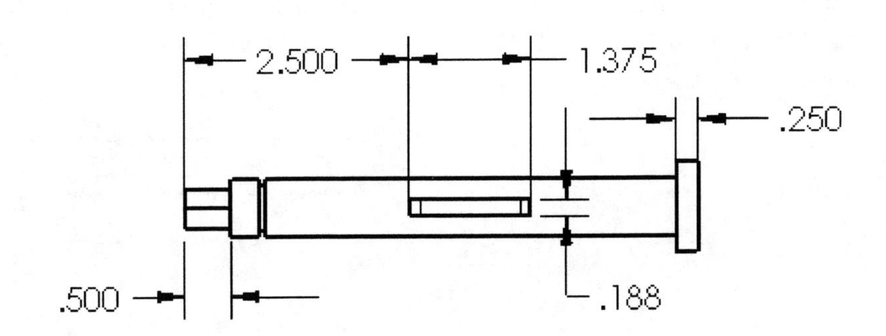

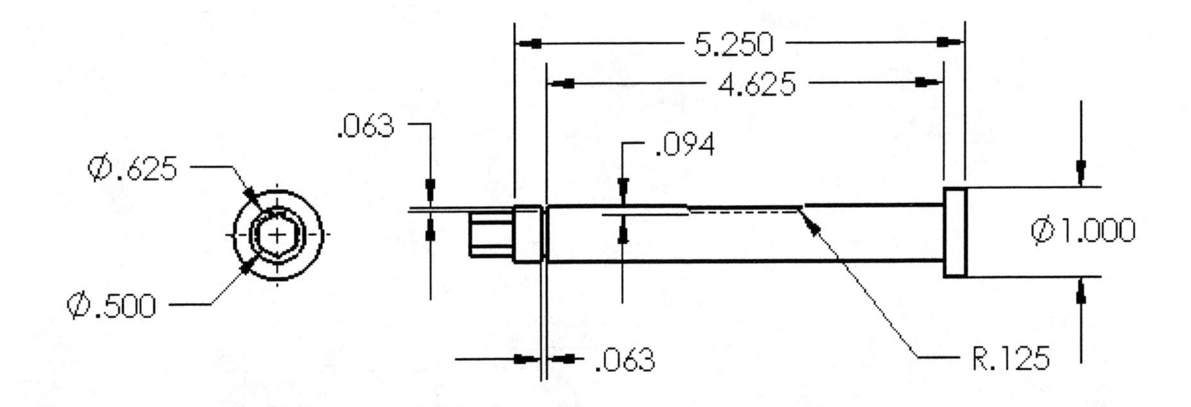

Just as we did with the Offset Shaft drawing, we'll reinforce making new drawings, adding views, importing dimensions, moving dimensions from one view to another and changing a diameter's display style. In this exercise we will make Left, Front and Top views, import and arrange the model dimensions to complete the drawing.

206. - We'll start by making a new drawing using the "A-Landscape" sheet drawing template, and add Front, Top and Left Front views using "Hidden Lines Visible" mode.

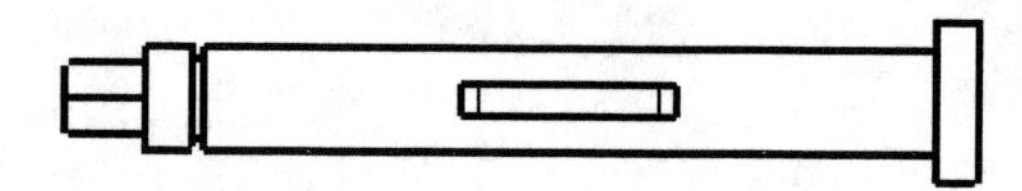

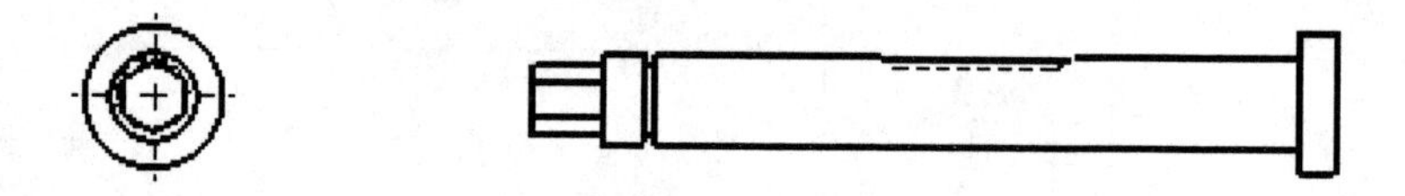

207. - We'll now Import the dimensions from the model...

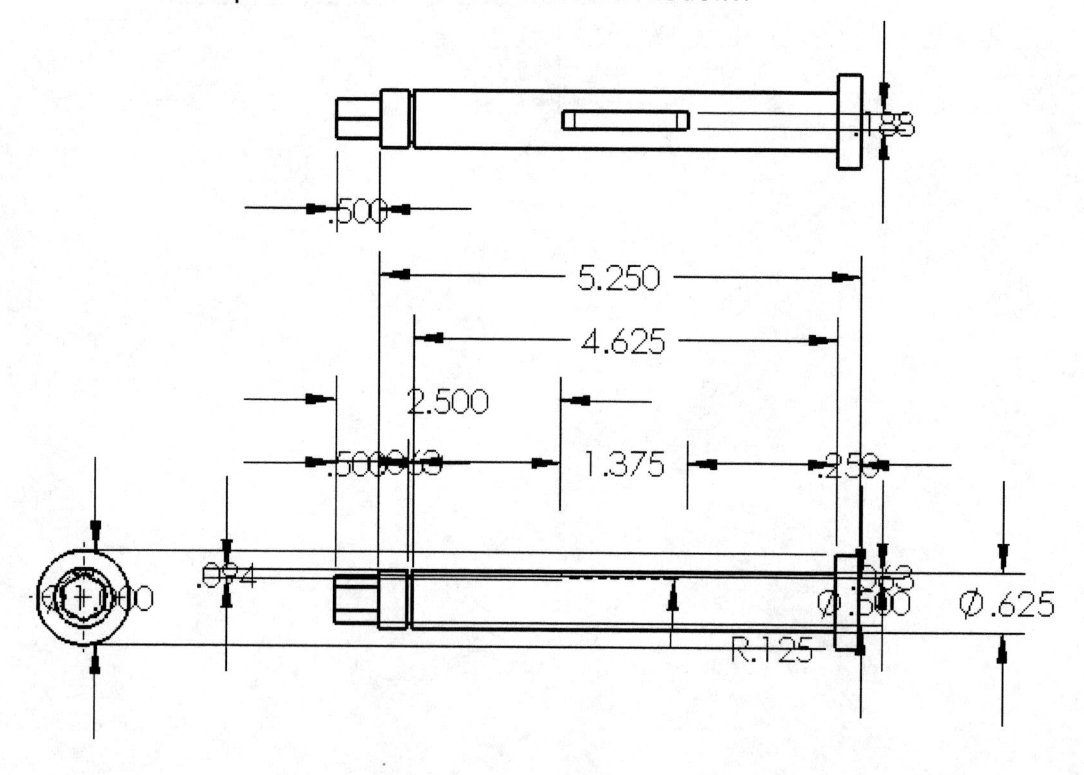

208. -...and arrange them as indicated.

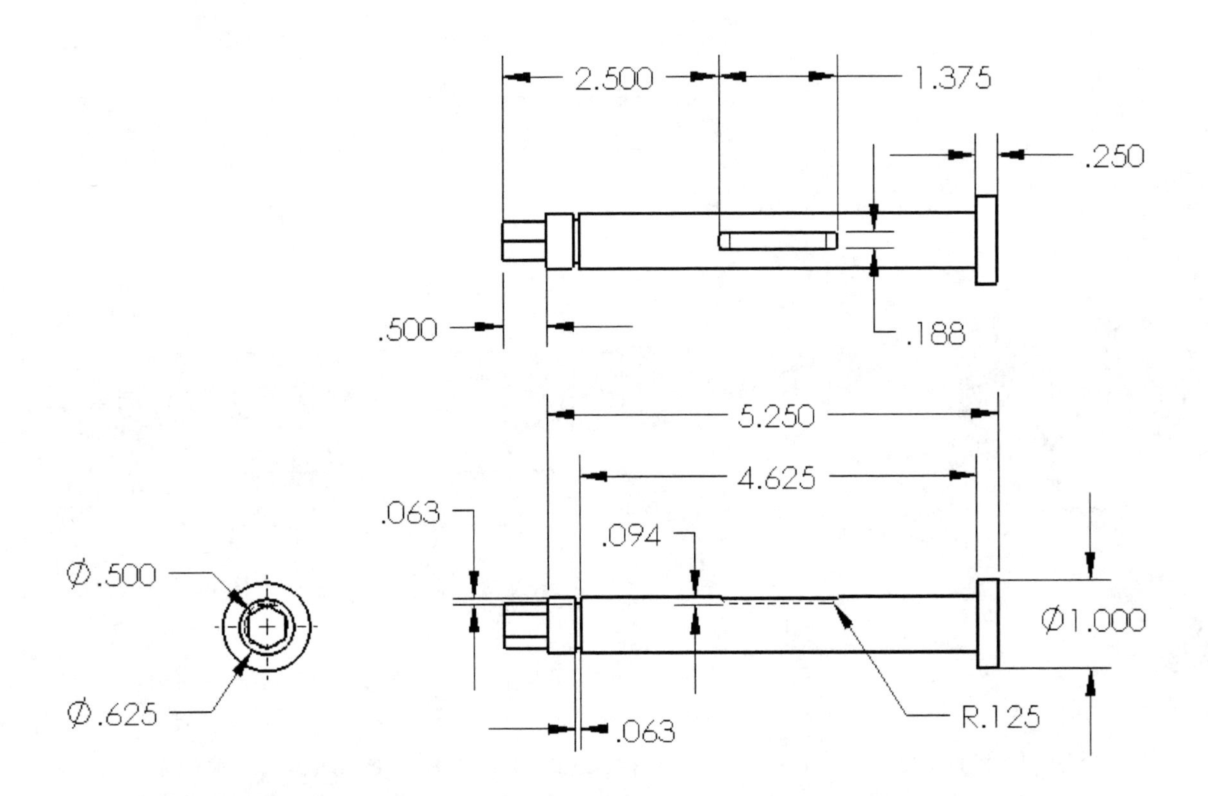

One more thing we need to add to the drawing is a centerline. SolidWorks allows us to automatically insert a centerline from every cylindrical face of the model in as little as two clicks.

209. - To add centerlines to our drawing views, select the "Centerlines" icon from the Annotations Toolbar, and click in the view that we want to add centerlines to. When finished adding centerlines click on OK to finish.

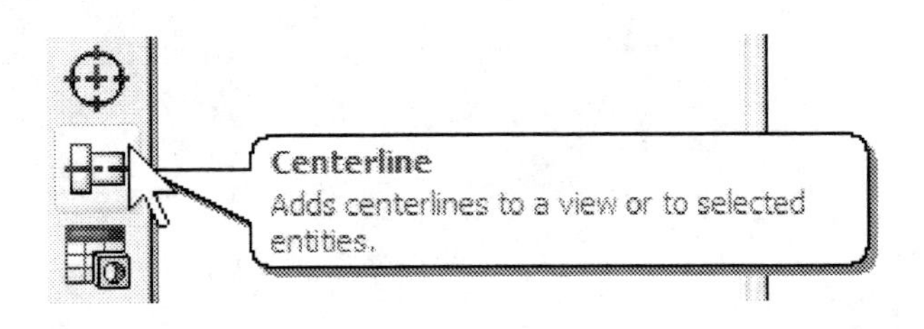

210. - Save the drawing and close the file.

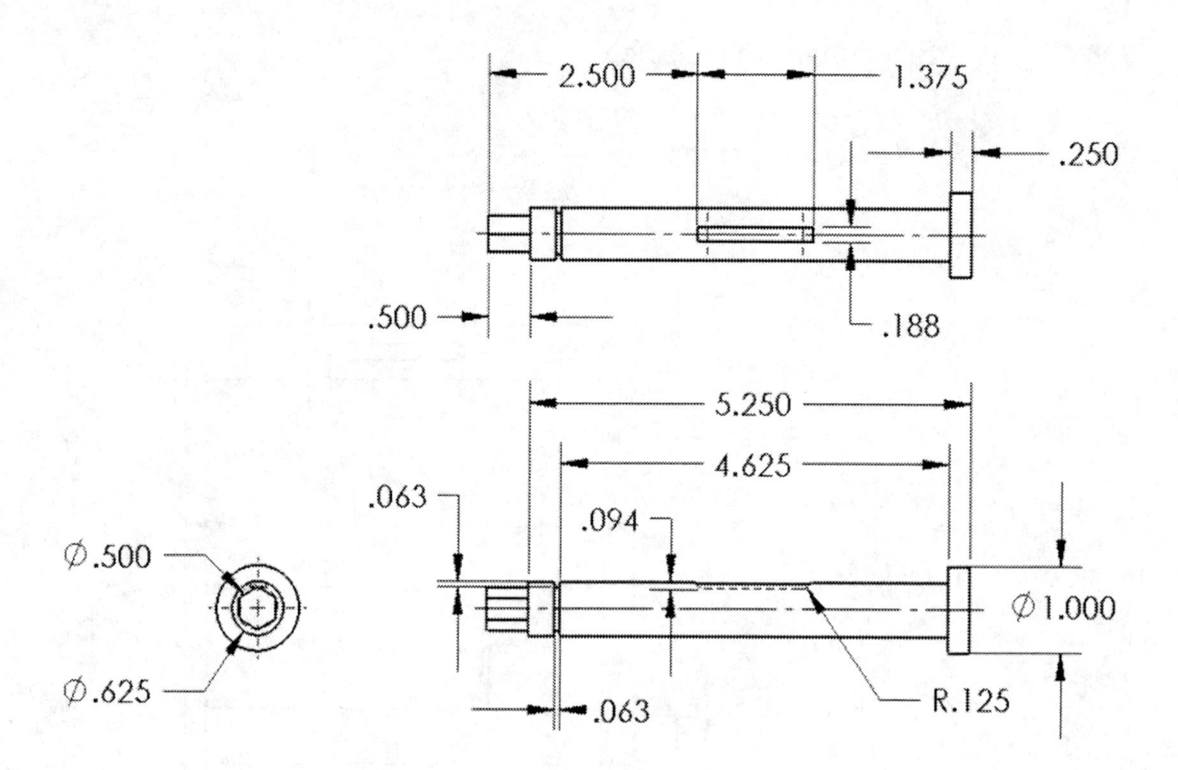

EXTRA: In this drawing, it may be a good idea to make a detail view of the Left view to show the detail of the hexagonal cut and add the dimensions manually.

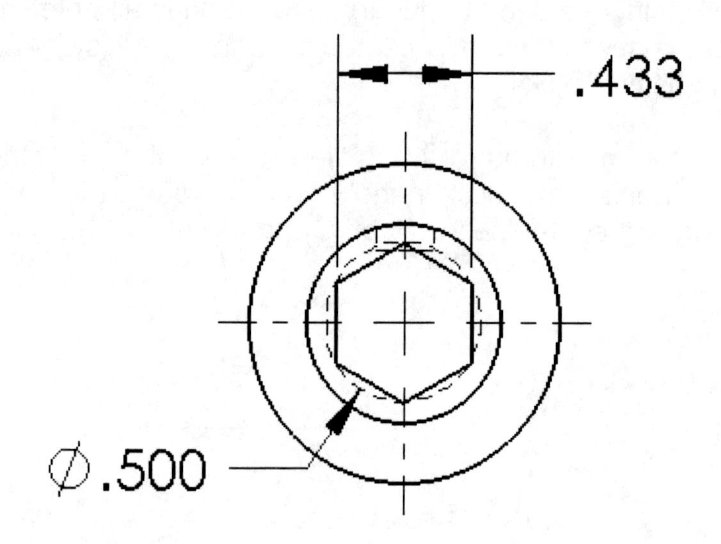

Assembly Modeling

The third part of the design after making the parts and drawings is to assemble all the components. Designing the parts first and then assembling them is known as "Bottom-Up Design". This is similar to buying a bicycle; when you open the box you get all the pieces needed ready for assembly. A different approach, usually known as "Top-Down Design" is where the parts *are designed while working in the assembly*; this is a very powerful tool that allows us to match parts to each other and make one part change if another one is modified. In this book we'll cover the Bottom-Up Design technique, since it is easier to understand and is also the basis for the advanced Top-Down Design. Most of the time it's preferable to design the parts, make the full assembly to make sure everything fits and works as expected (Form, Fit and Function), AND THEN make the drawings of the parts and assemblies; this way the drawings are done at the end, when you are sure everything works correctly.

So far, we've been working on parts, single components and the building blocks of an assembly. In the assembly we have multiple parts, and maybe other assemblies (in this case called sub-assemblies). The way we tell SolidWorks how to relate components (parts and/or sub-assemblies) together is by using Mates or relations between them. Mates in the assembly are similar to the geometric relations in the sketch, but in the assembly we reference faces, planes, edges, axes, temporary axes, vertices and even sketch geometry from different components.

In an assembly, the components are added one at a time until we complete the design. In a SolidWorks assembly every component has six degrees of freedom, meaning that they can move and/or rotate six different ways: three translations in X, Y and Z, and three rotations about the X, Y and Z axes. Once a component is mated to another component, we restrict how it moves, or which degrees of freedom are unconstrained. This is the basis for assembly motion and simulation.

The first component added to the assembly has all six degrees of freedom fixed by default. Therefore, it's a good idea to make this fixed component the one that will serve as a reference for the rest of the components. For example, if we make a bicycle assembly, the first component added to the assembly would be the frame. For the gear housing we are designing, the first component added to the assembly will be the Housing, as the rest of the components will be attached to it.

Notes:

The Gear Box Assembly

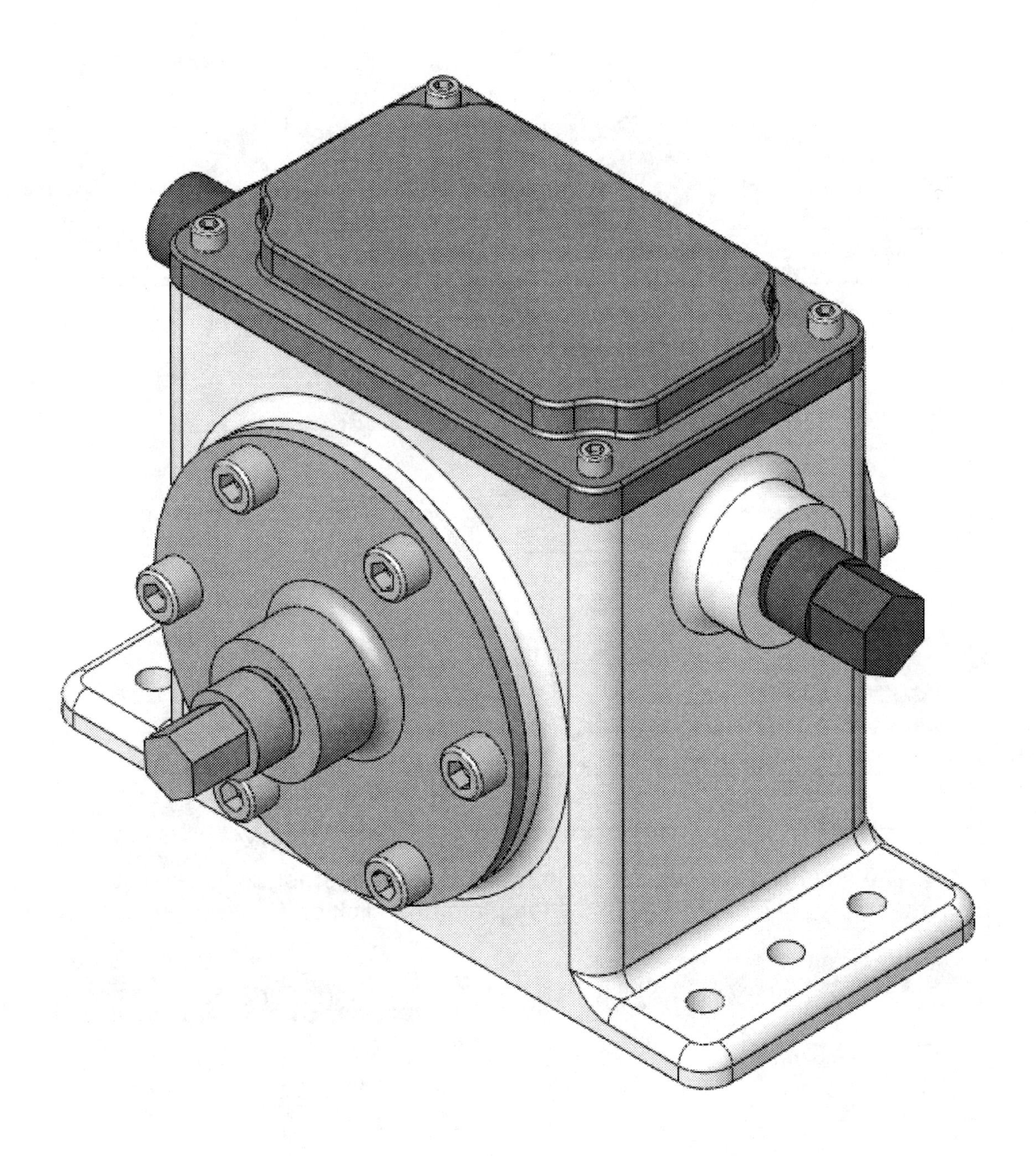

In making the Gear Housing assembly we will learn many assembly tools and operations, including making new assemblies, how to add components to it, add Mating relations between them using model faces and planes, add fasteners using the Toolbox library, and an exploded view of the assembly. The sequence that we will follow to complete the Gear Housing assembly is the following:

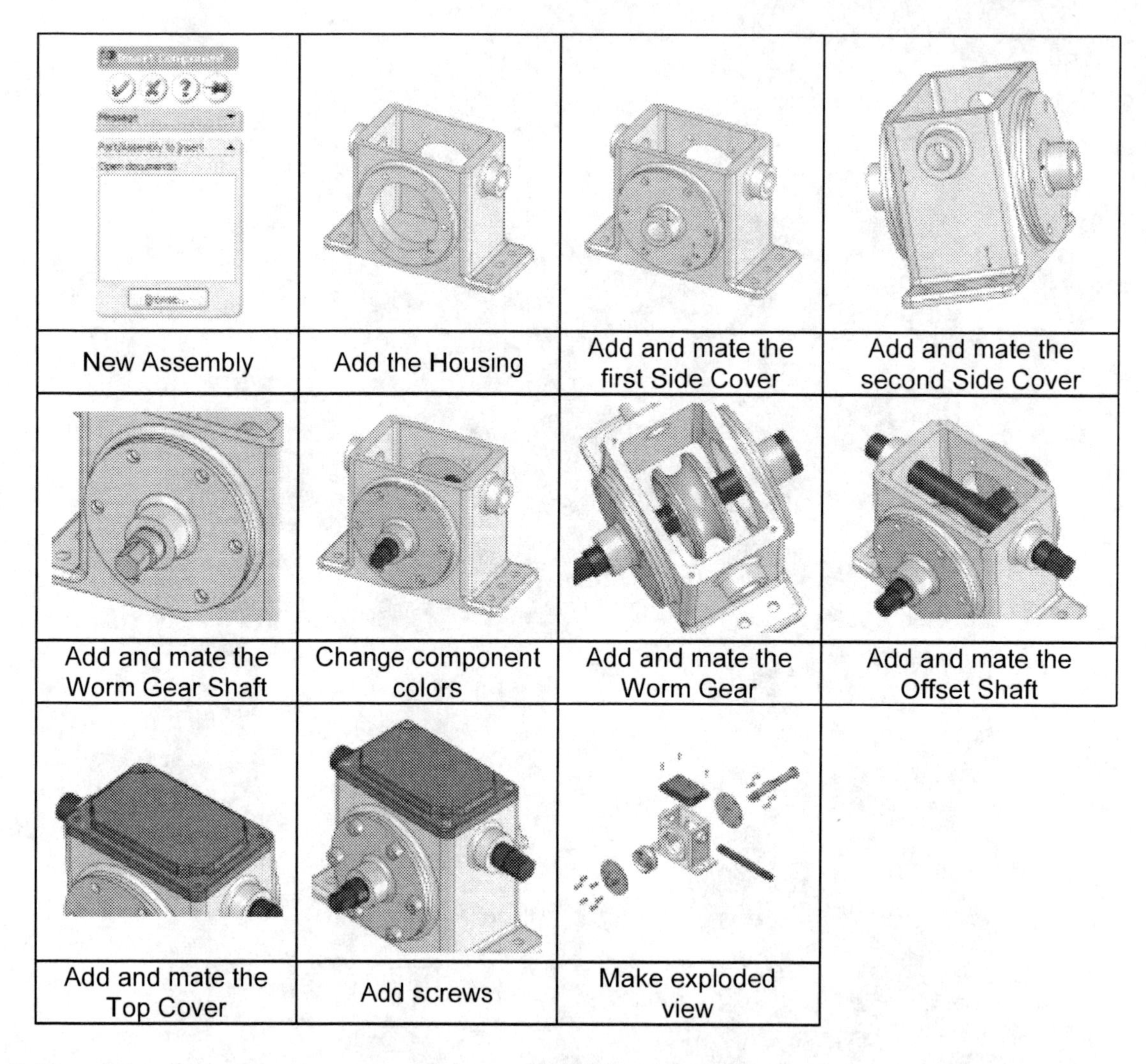

New Assembly	Add the Housing	Add and mate the first Side Cover	Add and mate the second Side Cover
Add and mate the Worm Gear Shaft	Change component colors	Add and mate the Worm Gear	Add and mate the Offset Shaft
Add and mate the Top Cover	Add screws	Make exploded view	

211. - The first thing we need to do to make a new assembly is to select the New Document icon, select the Assembly template and click on OK.

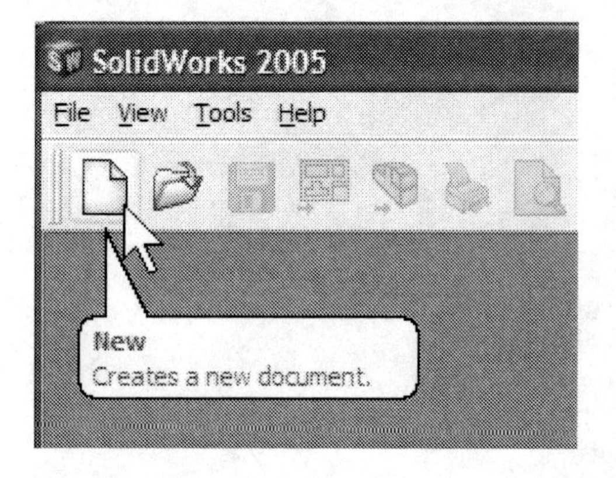

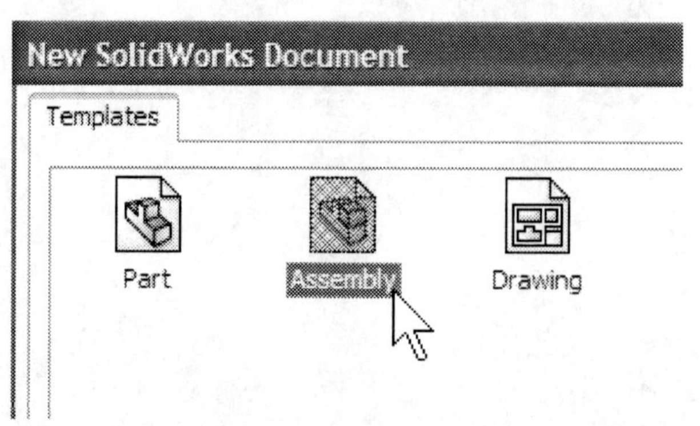

212. - The first thing that we see when we make the new assembly is the "**Insert Component**" dialog to start adding components. As we discussed previously, the Housing will be the first component. Click on "Browse" to locate the Housing, select it in the open dialog box and click "Open". If you have the "Graphics Preview" option box checked, you will see a transparent preview of the component being inserted. If you cannot see the component that you want, make sure you are looking in the correct folder and have the "Files of Type" set to "Part" (it can be either "Part" or "Assembly").

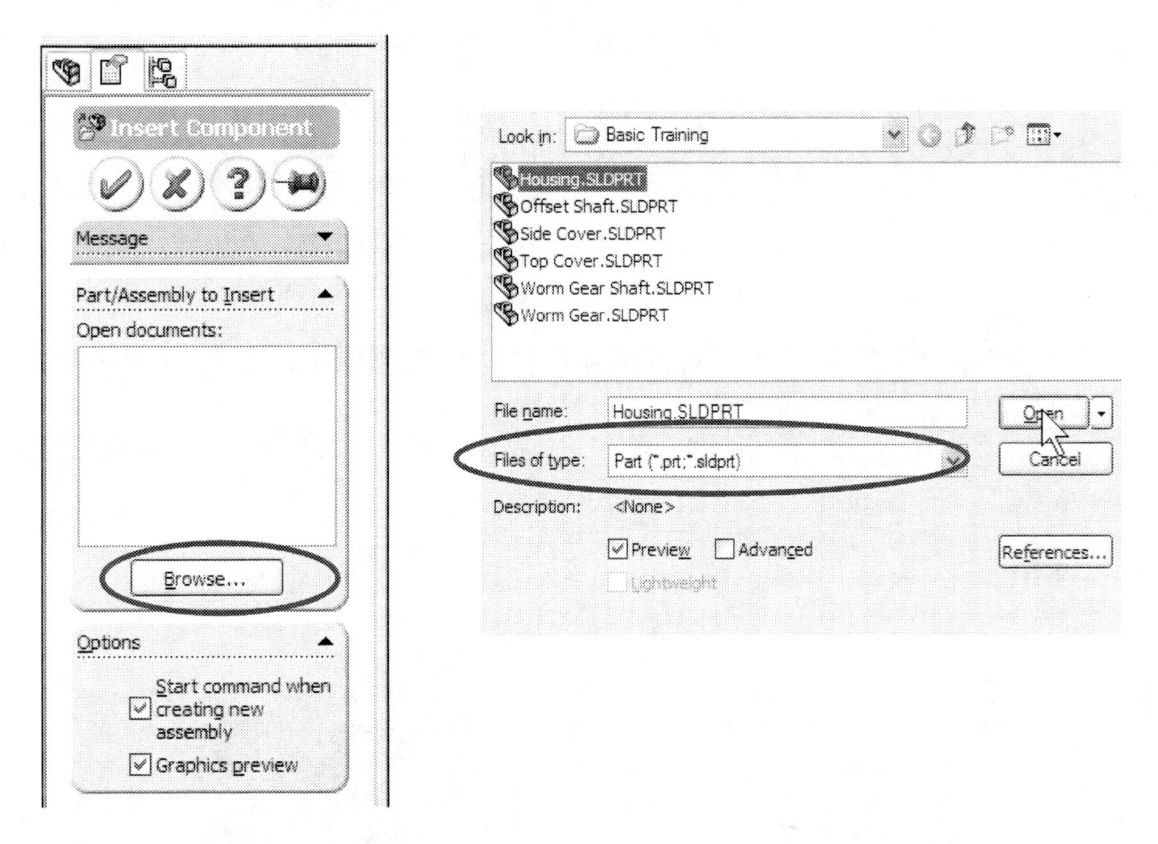

213. - We now have to locate the Housing in the assembly; as you can see, the preview follows the mouse. What we want to do is to locate the Housing at the assembly origin. If you cannot see the assembly origin, turn it on by selecting the "View" menu, "Origins" (this can be done while you are inserting a component). The reason to locate the Housing at the assembly origin is that the Housing's planes and origin will be aligned with the assembly's planes and origin. Move the mouse pointer into the assembly origin and click to insert it. (You will see the housing "snap" in place, and the cursor will have a double arrow next to it.)

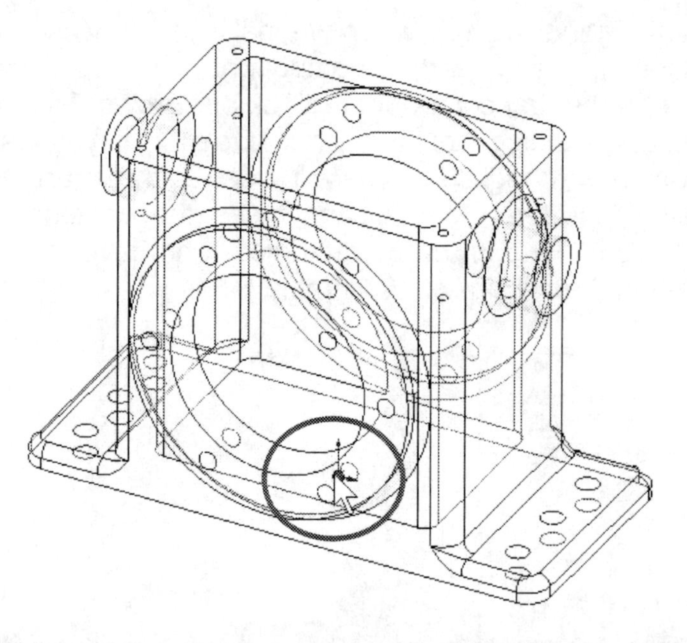

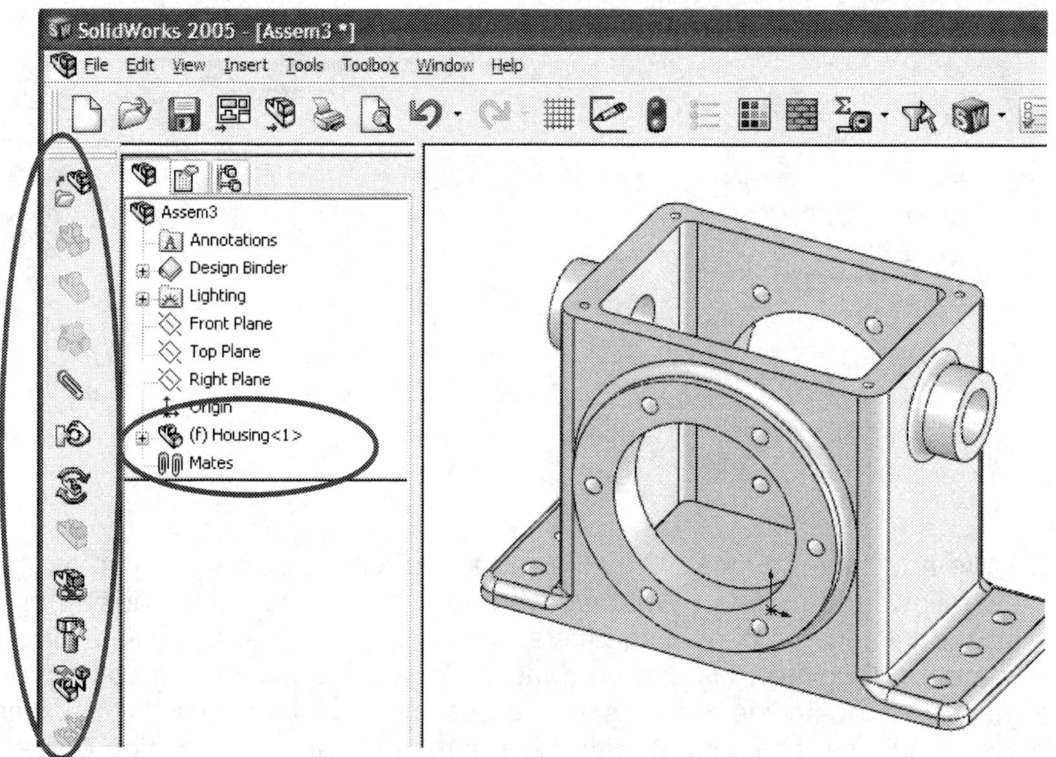

214. - Notice at the bottom of the Assembly Feature Manager a special folder called "Mates". This is where the relations between components (Mates) will be stored. Also, notice the assembly toolbar is activated when SolidWorks is in the assembly environment and the other toolbars are hidden.

215. - To the left of the "Housing" in the Feature Manager, we can see a letter "**f**"; this means that the part is "Fixed" and its six degrees of freedom are constrained. To add the second component to the assembly click the "**Insert Component**" icon, navigate to the folder where the Side Cover part was saved; select the "Side Cover" part and click "Open".

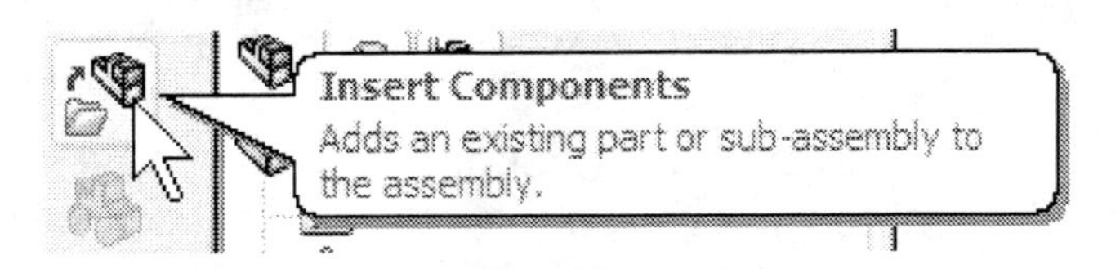

216. - Place the Side Cover next to the Housing as seen in the next image. Don't worry about the exact location; we'll place it accurately using mates.

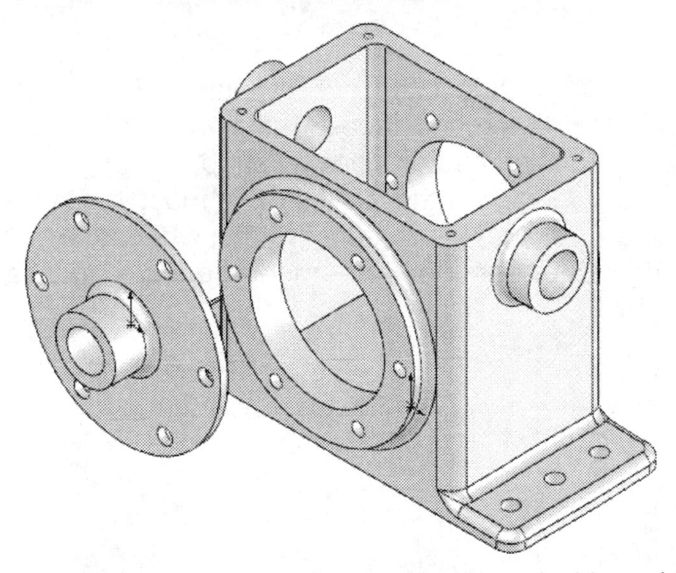

217. - The part name in the Feature Manager is preceded by a (-), this means that the part has at least one unconstrained degree of freedom. Since this part was just inserted, all six degrees of freedom are unconstrained and the part is free to move in any direction and rotate about all three axes.

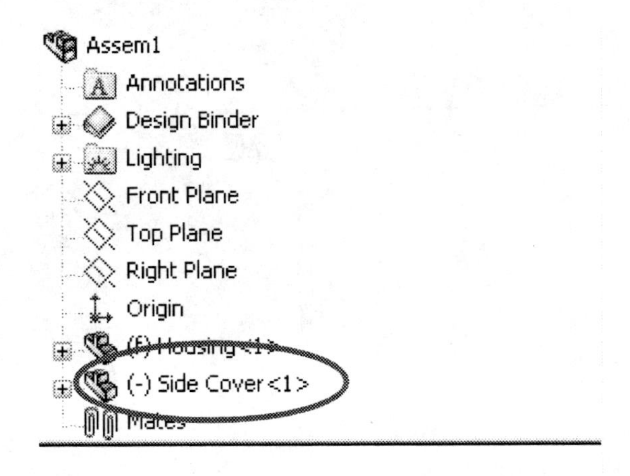

218. - Click on the "**Mate**" icon or select the "Insert" menu, "Mate" to mate the Side Cover to the Housing.

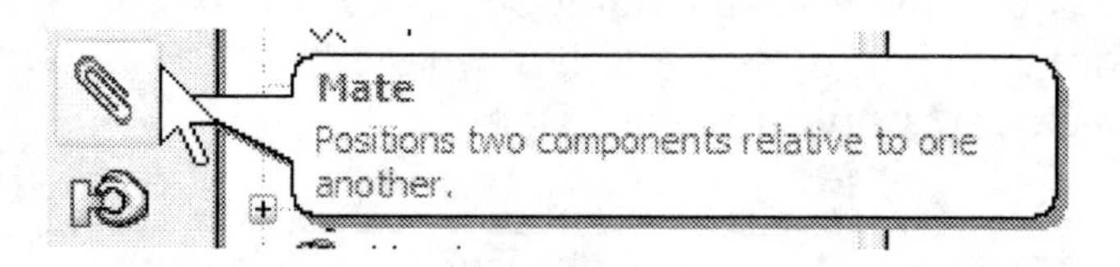

Now we are ready to add the relations (mates) between the Housing and the Side Cover. The mates will reference one component to another restricting the motion of the components. As explained earlier, mates can be added between faces, planes, edges, vertices, axes and even sketch geometry. When possible, select model faces as your first option; they are easier to select and visualize most of the time.

219. - For the first mate, select the two cylindrical faces indicated in the next picture. As soon as the second face is selected, SolidWorks recognizes that both faces are cylindrical and automatically "snaps" them with a Concentric Mate, (SolidWorks defaults to concentric as it is the most logical option). The Side Cover is the one that moves and/or rotates because it is the part with unconstrained degrees of freedom; remember the Housing was fixed when it was inserted.

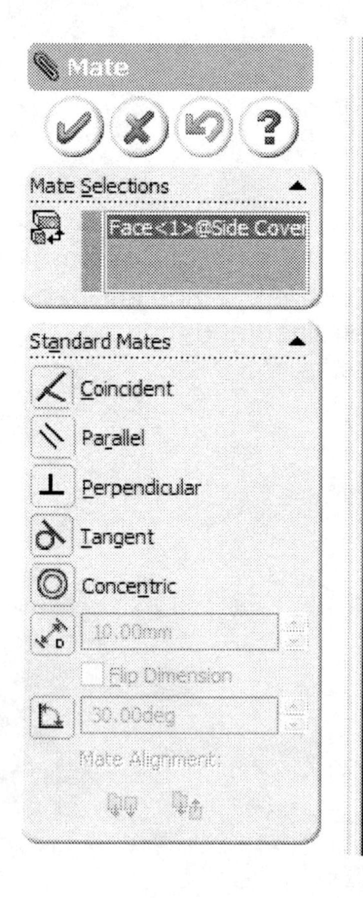

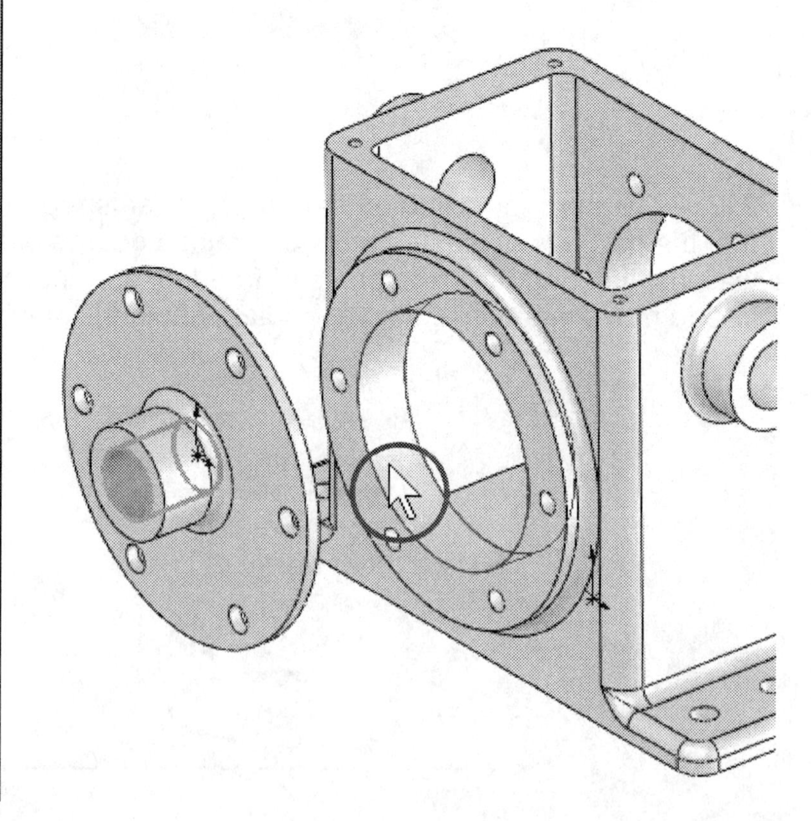

220. - The pop-up toolbar shows options to mate the selected entities. (These options are also listed in the Property Manager.) The pop-up toolbar is a way to help us be more productive and minimize mouse travel. The following picture shows the basic mate options available in SolidWorks. Depending on the type of entities selected, the options available will be different.

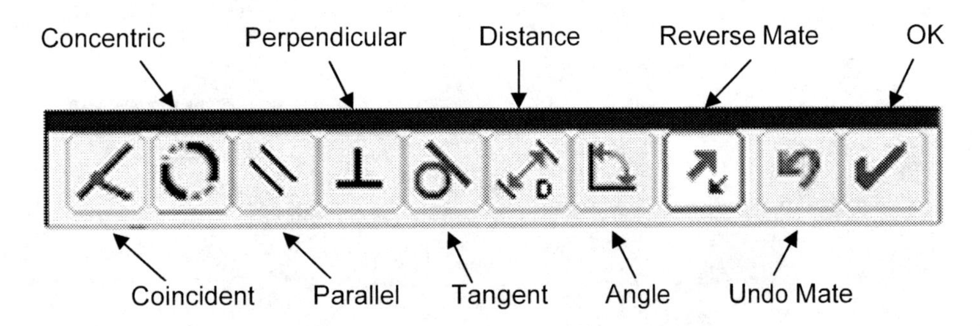

In this case the **concentric mate** is pre-selected; click the OK button to add this mate.

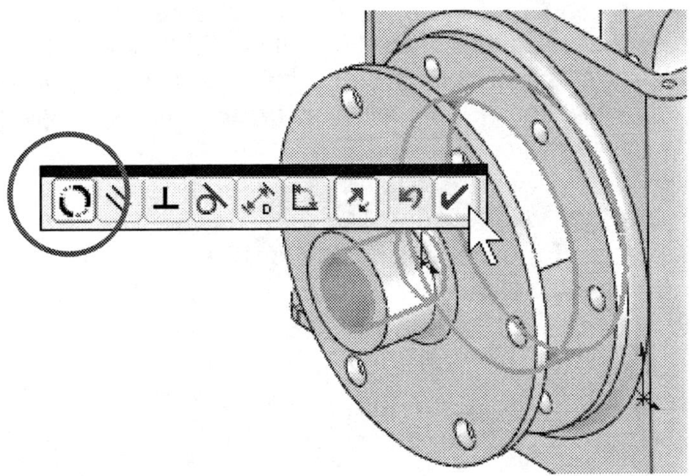

221. - After adding the mate the dialog remains visible and ready for us to add more mates. It will remain until we click in Cancel or hit "ESC" on the keyboard. Notice that the mate added is listed under the "Mates" box in the Property Manager.

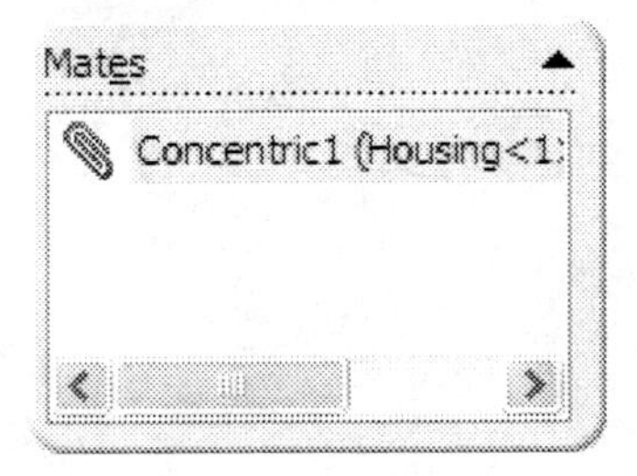

222. - For the second mate, select the two cylindrical faces indicated (one from the Side Cover, and one from the Housing); SolidWorks defaults again to a Concentric mate and rotates the Side Cover to align the holes. This will prevent the Side Cover from rotating. Click OK to add the mate.

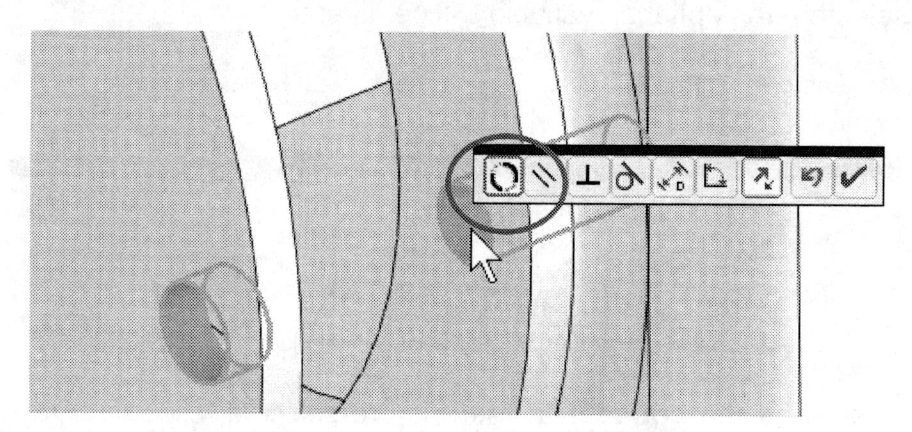

NOTE: In this case the concentric mate works as expected and aligns the Side Cover to the Housing because the holes are located exactly at the same distance from the center in both parts; in reality, it is generally a better idea to align the two components using planes and/or faces, with either a parallel or coincident mate, as will be shown later in the book.

223. - The last mate that will be added to this Side Cover is a **Coincident mate** between the back face of the cover and the front face of the Housing; if needed, rotate the view to select the faces. The cover will move to make the selected faces coincident and show the pop-up toolbar with the Coincident mate pre-selected. Click OK on the pop-up toolbar to add the mate and hit the Escape (ESC) key to cancel the mate command. (If the Side Cover is *inside* the Housing, click and drag it to move it outside, if needed.)

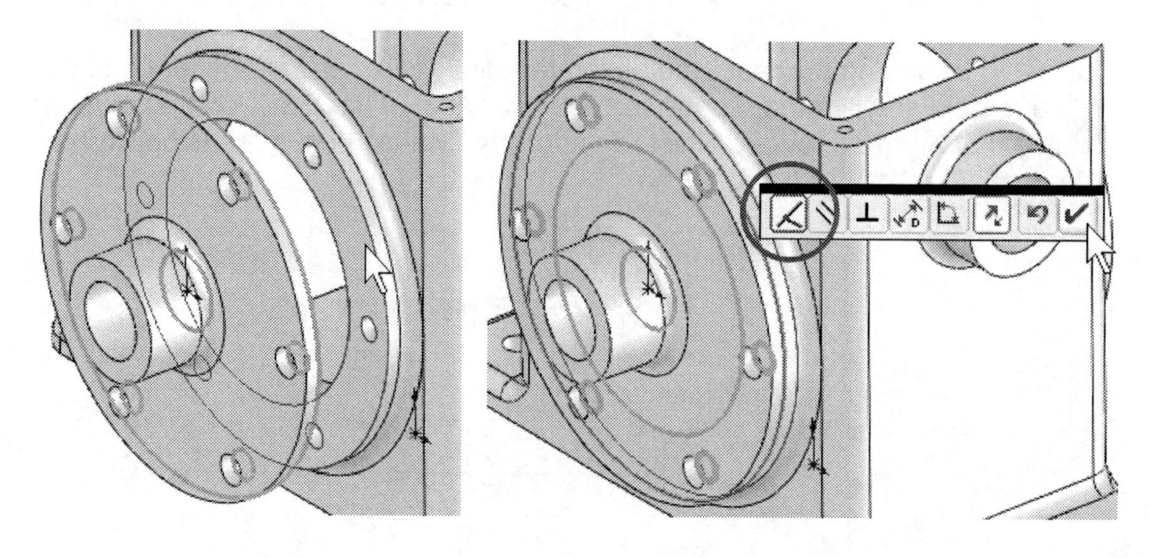

224. - All six degrees of freedom of the Side Cover have now been constrained using mates; this can be seen in the Feature Manager where the "Side Cover" is no longer preceded by a (-) sign. Notice that the "Mates" folder now includes the Concentric and Coincident mates just added; SolidWorks automatically adds the names of the components related in each mate.

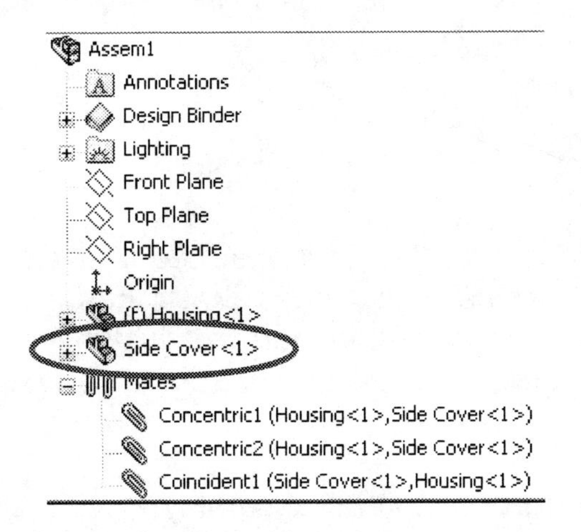

IMPORTANT NOTE: If a (+) sign precedes the part name, you probably also received an error message telling you that the assembly has been over defined. If this is the case, it means you added conflicting mates that SolidWorks cannot solve, or inadvertently selected the wrong faces or edges when adding mates.

The easiest way to correct this error is to either hit the Undo button or delete the last mate in the "Mates" folder; you will be able to identify the conflicting mate because it will have an error icon next to it. If multiple mates have errors, start deleting the mate at the bottom (this was the last mate added) and chances are this is the cause of the problem. If you still have errors, keep deleting mates with errors from the bottom up until we clear all the errors. It's not a good idea to proceed with errors, as it will only get worse.

225. - We are now ready to insert the second Side Cover. Repeat the "Insert Component" command from step 215 to add a second Side Cover and locate it on the other side of the Housing approximately as shown. Don't worry too much about the exact location; we'll move and rotate the part in the next step.

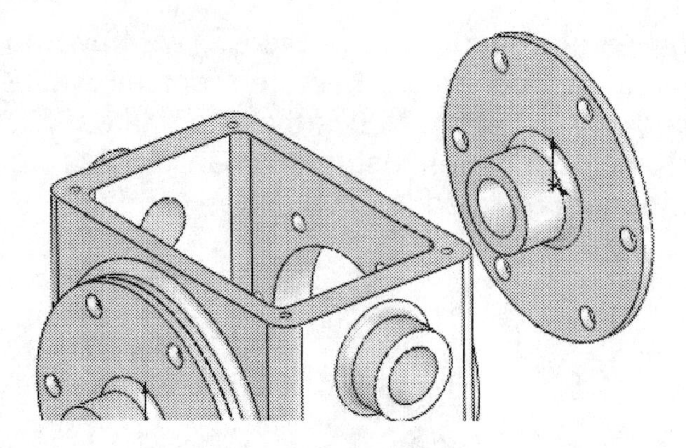

226. - When the second Side Cover is inserted it has a (-) sign next to it in the Feature Manager; remember this means that it has at least one unconstrained Degree of Freedom (DOF); since this part was just inserted in the assembly, ALL six DOF are unconstrained, and an unconstrained component can be moved and rotated.

227. - To **move a component** in an assembly, simply click and drag it with the Left Mouse Button; to **rotate** it click and drag it with the Right Mouse Button. Move and rotate the second Side Cover as needed to align it *approximately* as shown. Remember, we'll add more mates to locate it precisely.

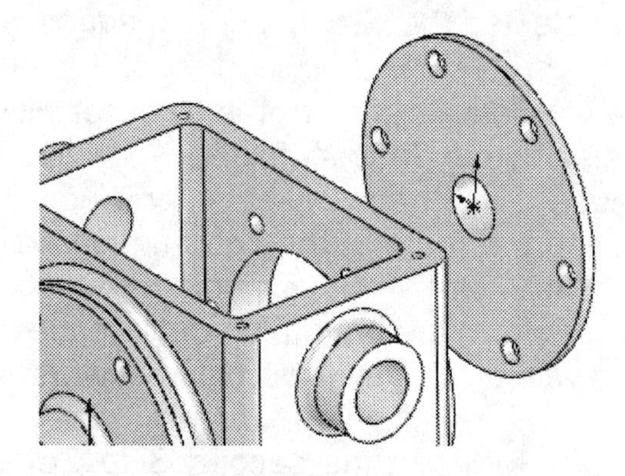

228. - We will now add three mates to the second cover as we did to the first one in steps 218 to 223. Select the "Mate" icon and add the Concentric mate selecting the faces indicated next. Rotate the view (not the part) to get a better view of the faces to select.

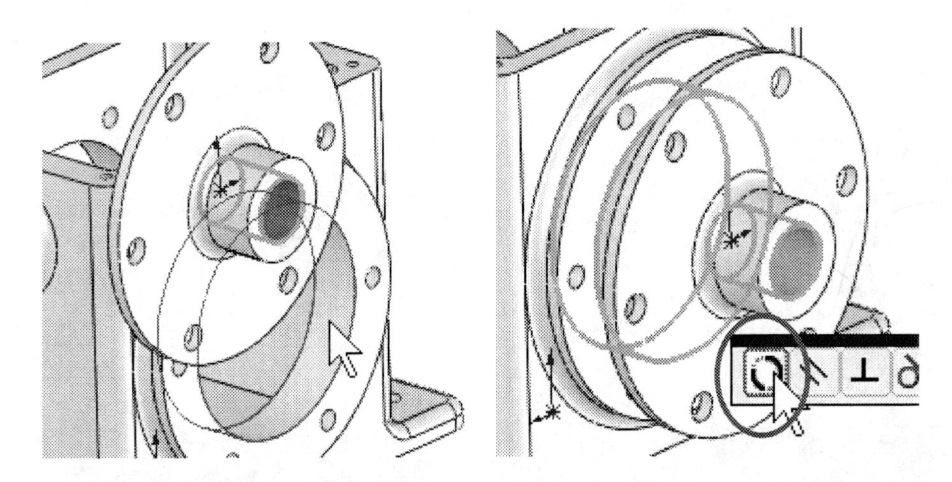

229. - Before adding the second Concentric mate, click and drag the cover and notice that it can rotate about its center and also move along the axis; under constrained components are the basis for SolidWorks to simulate motion.

230. - Now let's add the concentric mate to align the screw holes between the Cover and the Housing. Since the part is symmetrical, it doesn't really matter which two holes are selected. If the cover had an extra feature that needed to be aligned, then the orientation would be important.

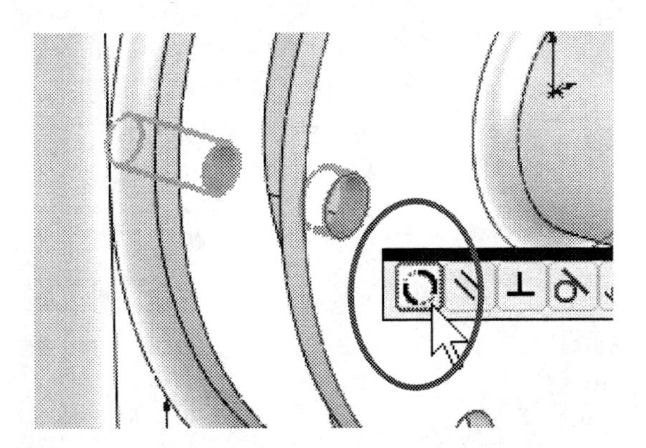

231. - Finally add a Coincident mate between the flat face of the Housing and the second Side Cover as we did with the first cover and click OK to continue.

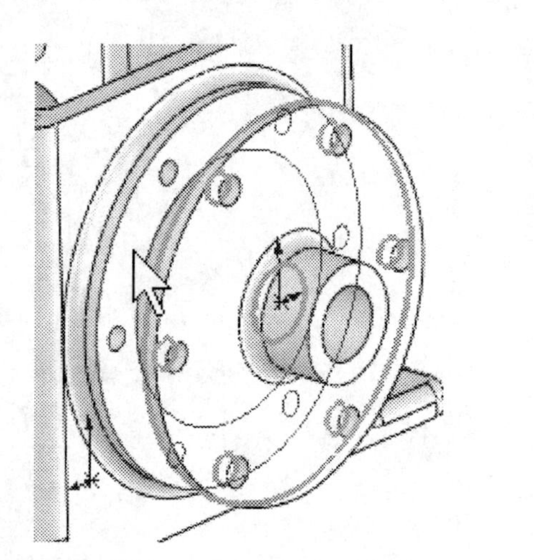

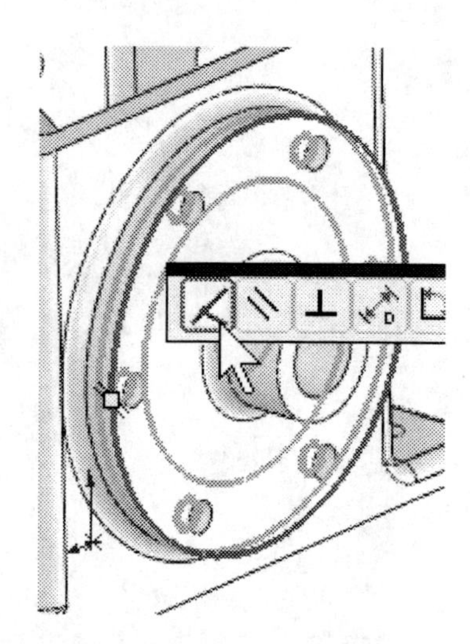

232. - It is very important to notice that these mates could have been added in any order; we chose this order to make it easier for the reader to see the effect of each mate. Whichever order you select, you will end up with both Side Covers fully defined (no free DOF) and six mates in the "Mates" folder.

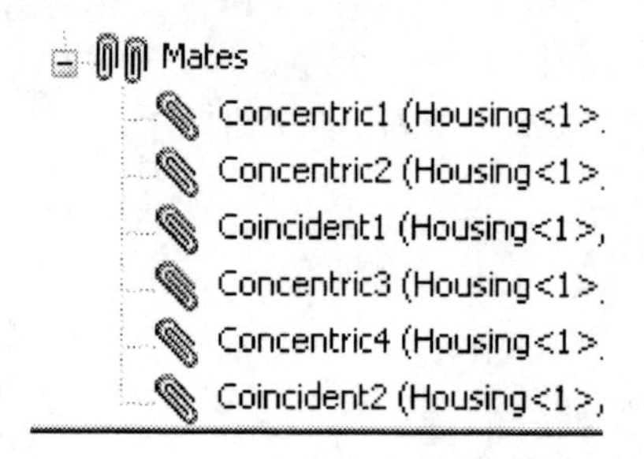

233. - Now that we have correctly mated both Side Covers, we'll insert the next part, the "Worm Gear Shaft" using the same procedure as before and locating it approximately as shown.

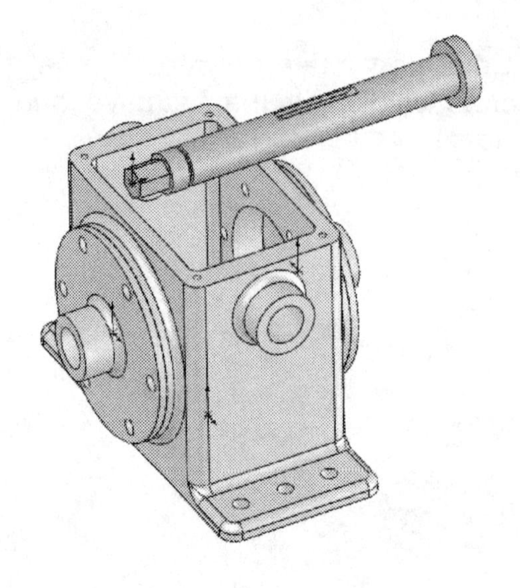

234. - With four different components in the assembly, all of them the same color, it's a good idea to change their color to easily identify them. To change a component's color select the **Edit Color** icon from the main toolbar.

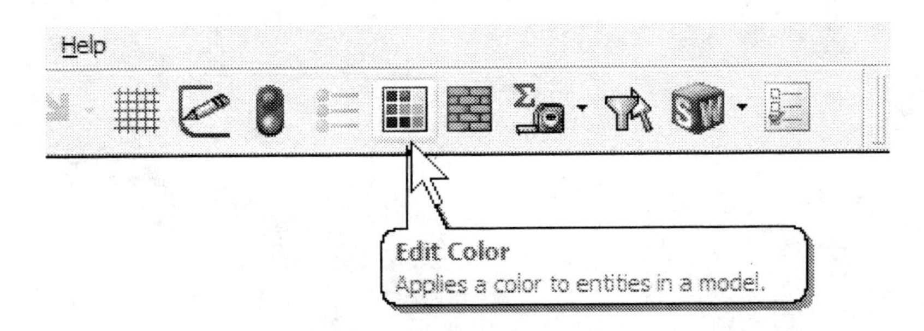

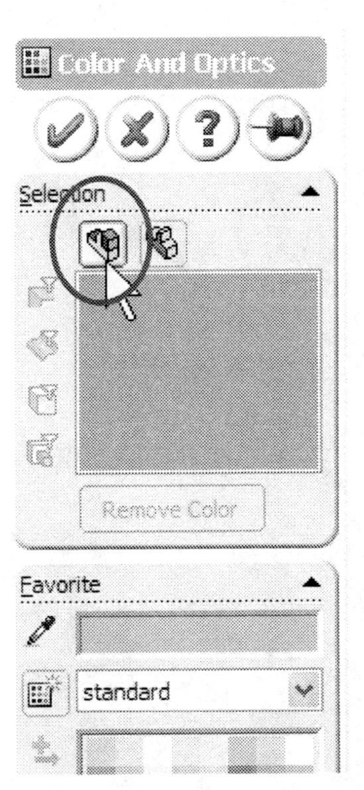

In SolidWorks we can change the color of the part in the assembly, or in the part as well. Picture it like this: we paint the part before we assemble it, or we paint it after we assemble it. In the second case, the part color changes only in the assembly, and is not changed in the part. For this example we want to change the color at the assembly level, so make sure the "Apply Changes at Assembly Component Level" option is selected.

235. - From the graphics area, click to select the components that we want to change; they will be listed inside the "Selection" box. Pick a new color from the "Favorite" color selection box and click OK when done.

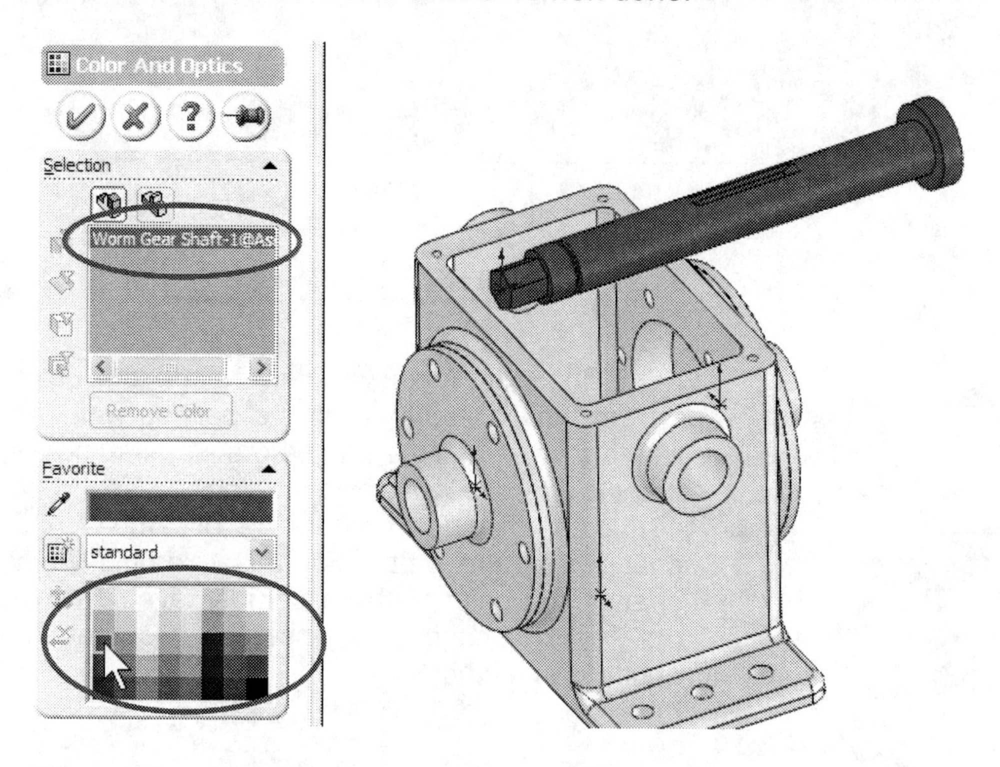

236. - Change the color of both covers to your liking and continue adding mates.

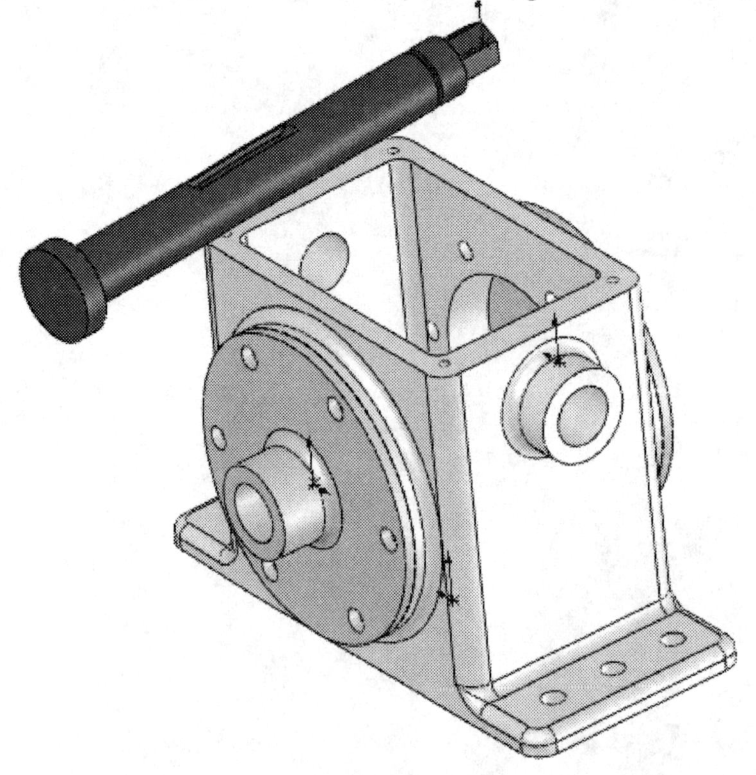

237. - Since we covered Concentric and Coincident mates and practiced them with the side covers, we'll simply ask the reader to add a Concentric and a Coincident mate using the faces indicated in the next image.

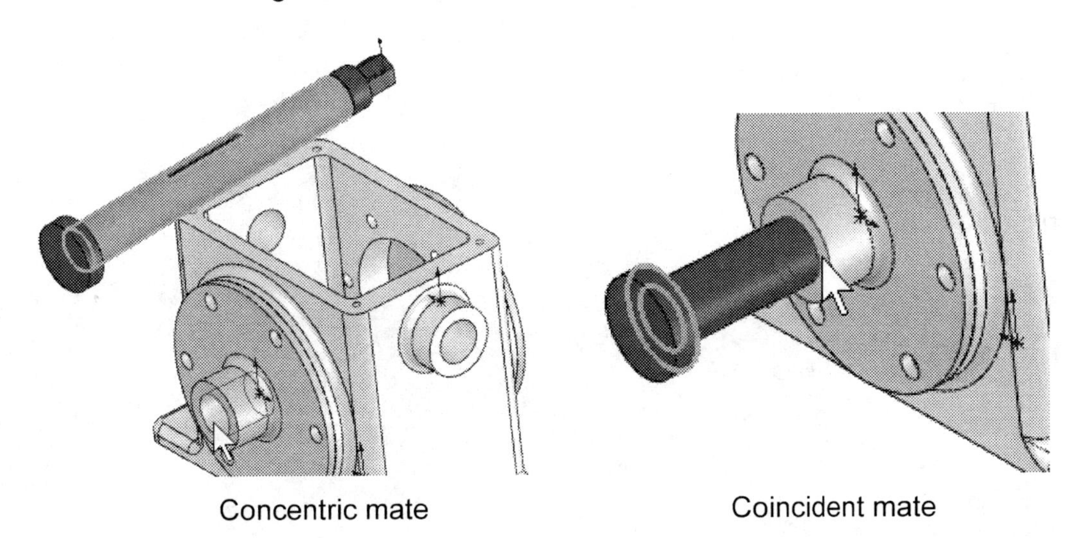

Concentric mate Coincident mate

238. - Notice that the Shaft is not yet fully defined, it still has a (-) sign in the Feature Manager. In this case the only DOF left unconstrained is to rotate about the axis of the shaft, and that is exactly what we want. The shaft is supposed to rotate. If we click and drag the shaft, we'll see the keyway rotate.

239. - The next component will be the Worm Gear. Insert it with the "Add Component" command as before and place it in the assembly as shown. And now that you know how, change its color too.

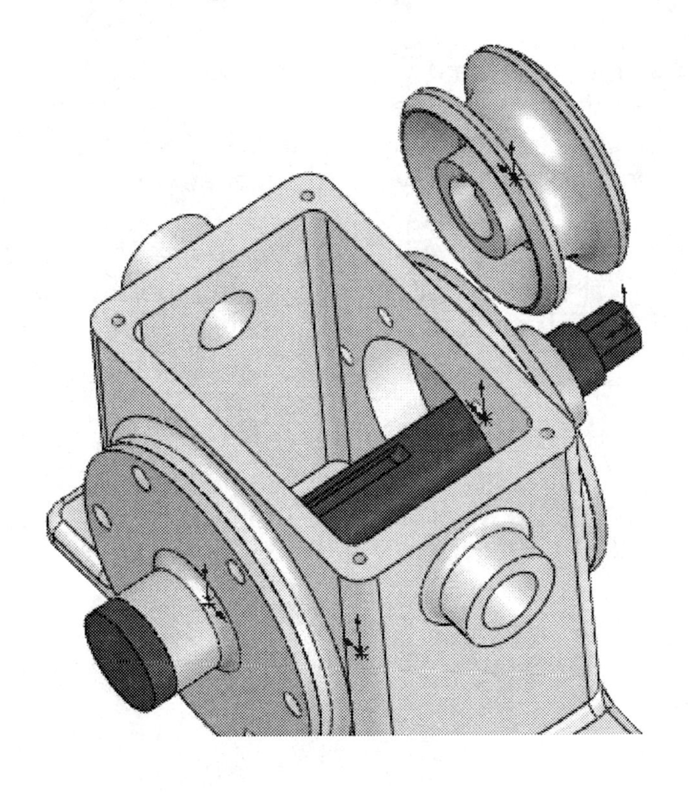

240. - To mate the Worm Gear in place, select the "Mate" icon and add a Concentric mate with the Worm Gear Shaft using the inside face of the Worm Gear and the outside face of the shaft.

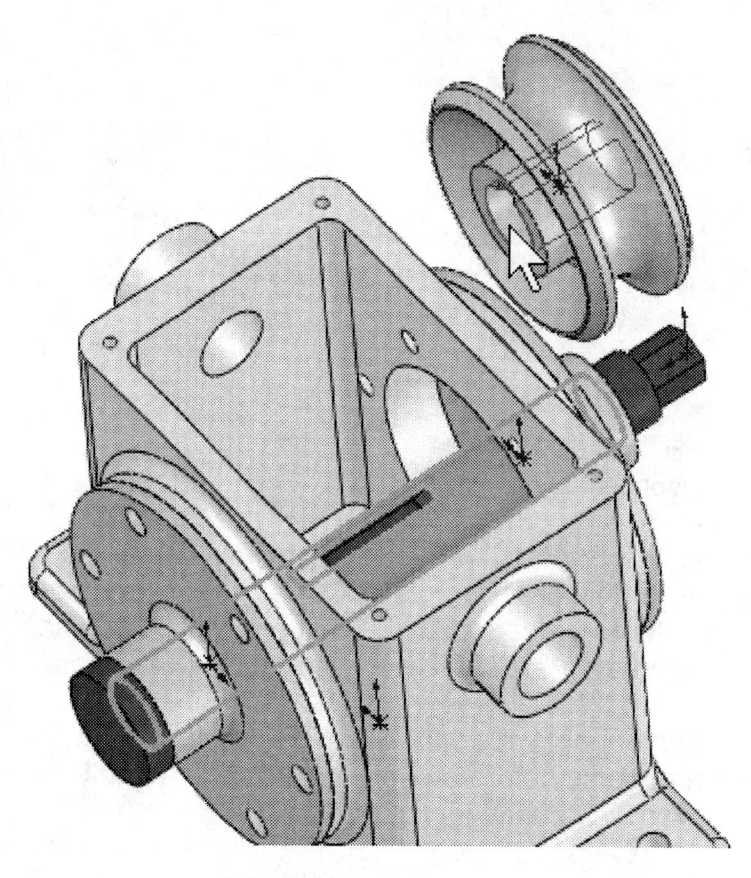

241. - Now we need to center the Worm Gear with the Housing. In order to do this we have to use the Worm Gear's Front plane and the Front plane of the Housing; this last one conveniently located in the middle of the Housing. (Remember we made the housing symmetrical about the origin?)

Select the Mate command (if it's not open already) and add a Coincident mate selecting the Front Plane of the Housing and the Front Plane of the Worm Gear from the fly-out Feature Manager as indicated.

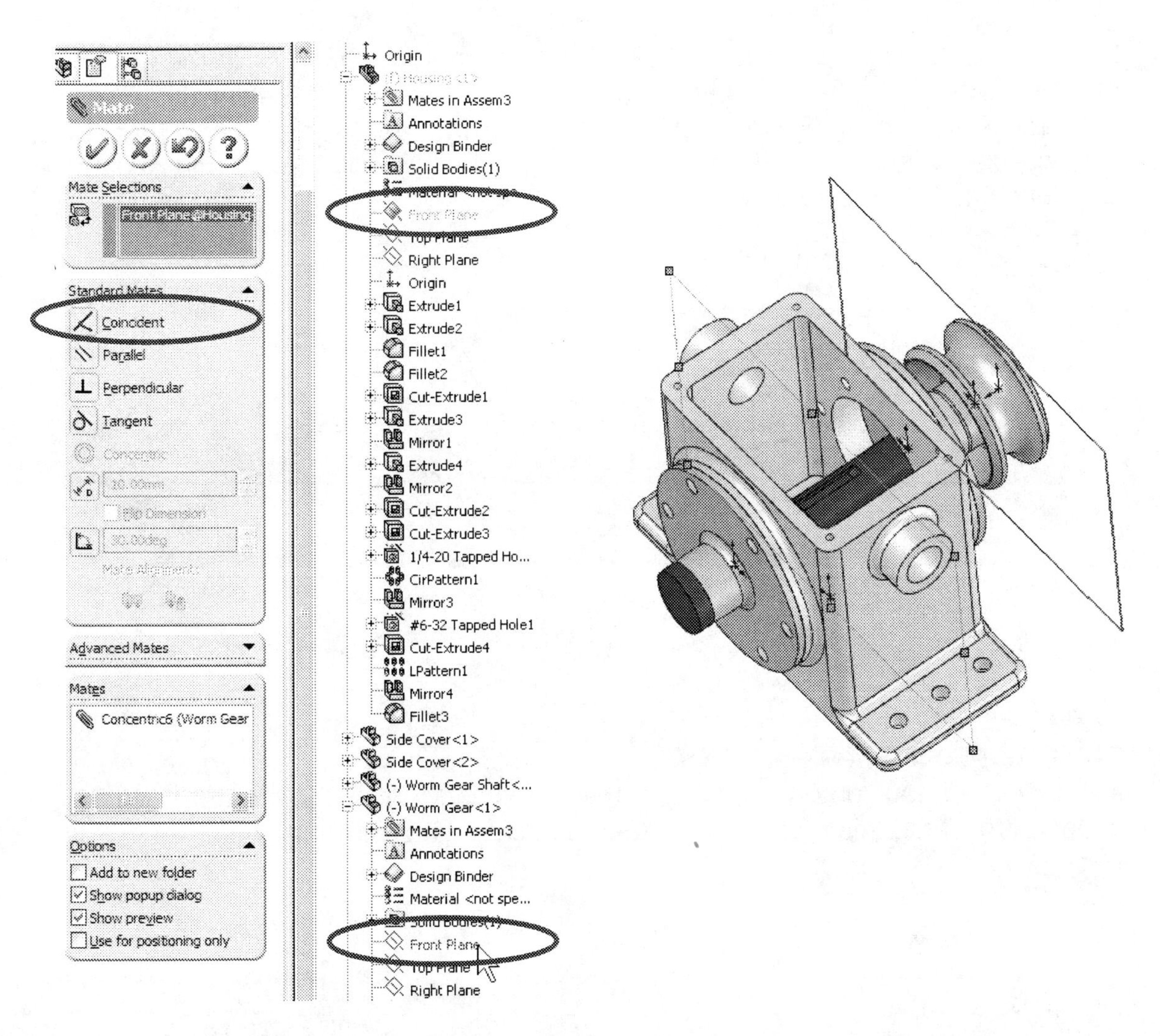

242. - The Worm Gear is still under defined, and we want the shaft and the gear to move together. To accomplish this we'll add either a Coincident mate using the corresponding planes from the parts, or a Parallel mate between the faces of the keyways. We'll do the Parallel mate here and let the reader explore the second option.

243. - As luck would have it, the Housing is obstructing the view to the keyways, and we'll have to hide it. Cancel the "Mate" command and select the Housing with the Right Mouse Button from the Feature Manger, and select "**Hide**" from the pop-up menu. This will hide the part from view. We'll show it again after we are done with this mate.

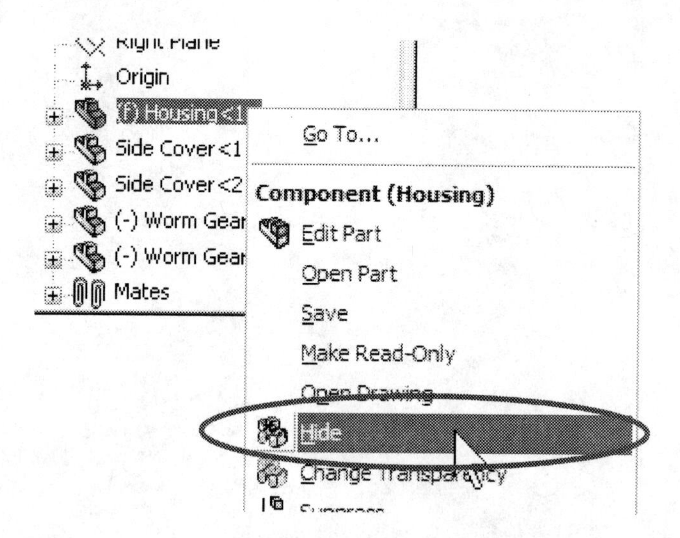

244. - As soon as we select "Hide", the part disappears from the screen. We can see inside without obstructions and the Housing icon in the Feature Manager changes to white; this is how SolidWorks indicates that the part is hidden.

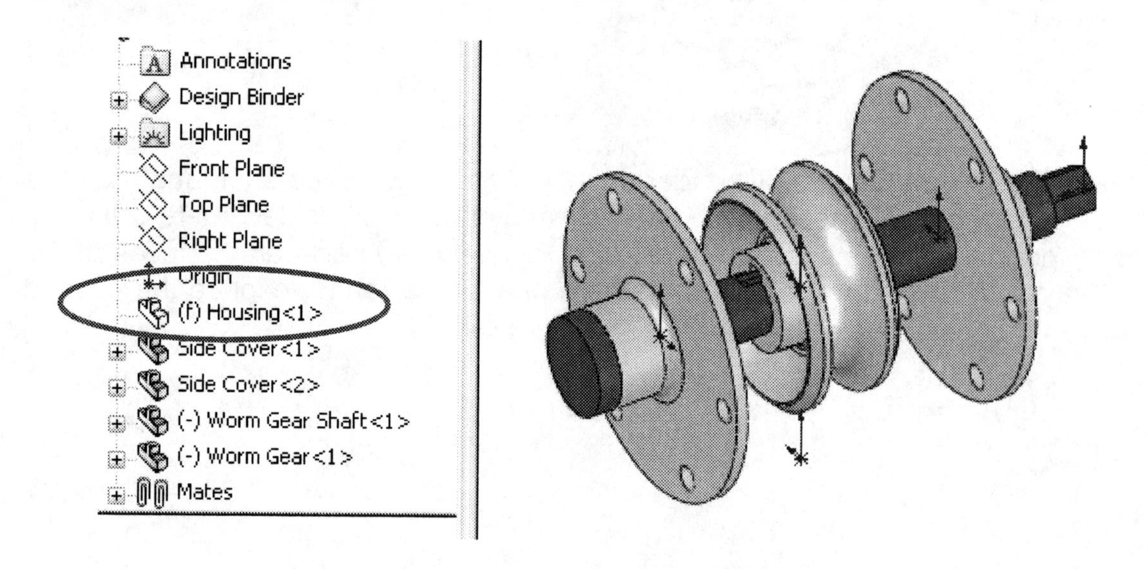

245. - Select the "Mate" icon again and select the flat faces of the shaft and gear keyways. In this case, since both keyways are the exact same size, we can use either a "Coincident" or "Parallel" mate. The **Parallel mate** is a good option, as it can help us absorb any small differences there may be between the parts. This is a more "forgiving" option.

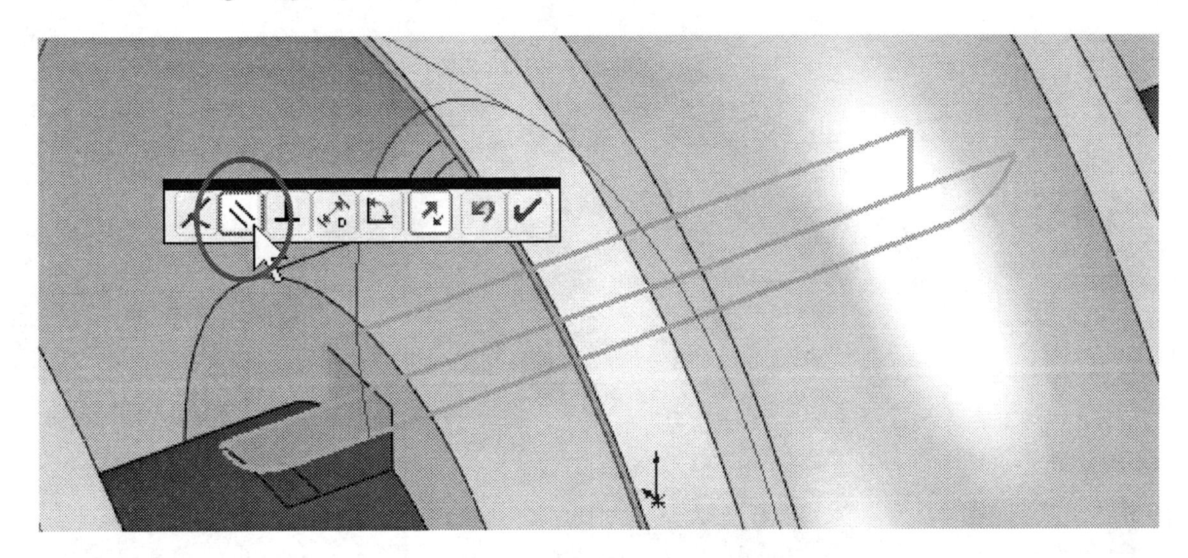

If we click and drag the gear or the shaft, we'll see both moving at the same time, as if they had the keyway inserted.

Extra credit: Make a keyway and add it to the assembly, mating the keyway to the shaft, and the Worm Gear to the keyway.

246. - To show the Housing again, we'll reverse the procedure that we used to hide it. Select the Housing from the Feature Manager with the Right Mouse Button, and select "Show". If a component is hidden, you'll get "**Show**"; if the component is visible, you'll get "Hide", like a toggle switch.

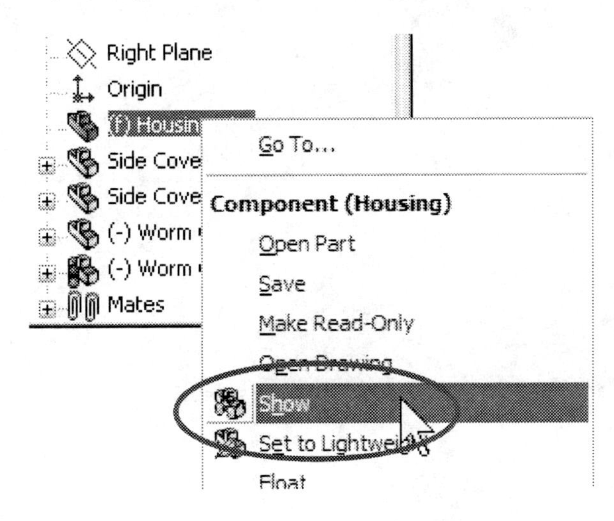

247. - Now it's time to add the Offset Shaft to the assembly and mate it in place. Insert the Offset Shaft using the "Insert Component" command. (By now adding components to your assembly should be easy). As you can see, we also changed its color for visibility.

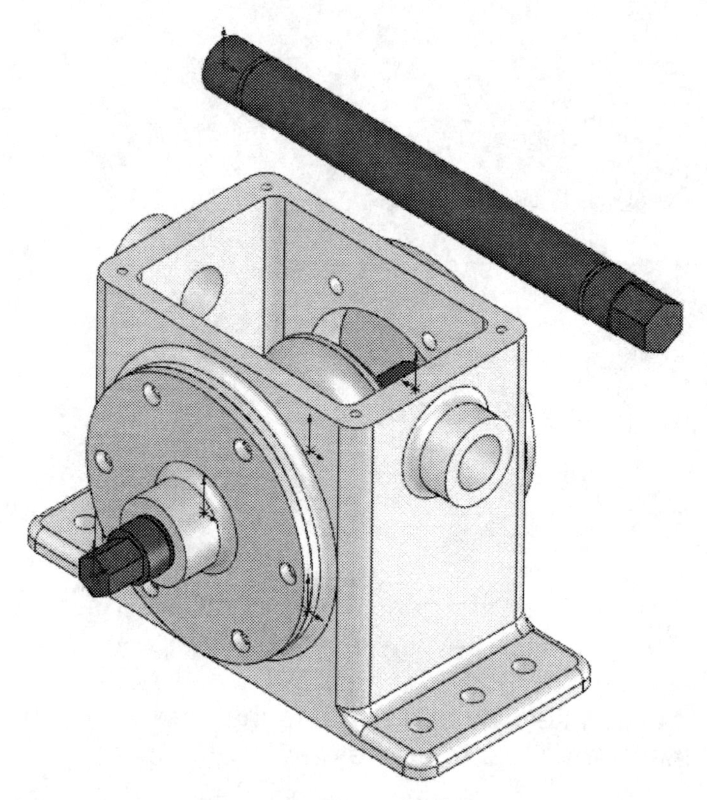

248. - To mate the shaft in place, select the "Mate" command and add a Concentric mate using the faces indicated.

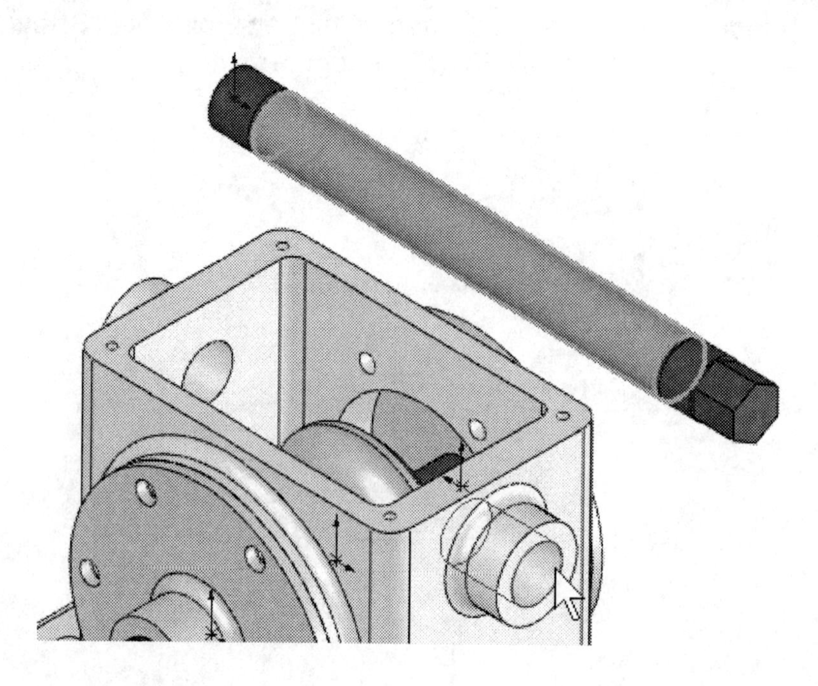

249. - Now we need a Coincident mate to prevent the shaft from moving along its axis. We'll use the groove in the shaft for this mate. In this case we can use either the flat face or the edge of the groove; selecting the edge may be a bit easier than the face. To select the face we may need to zoom in a little bit.

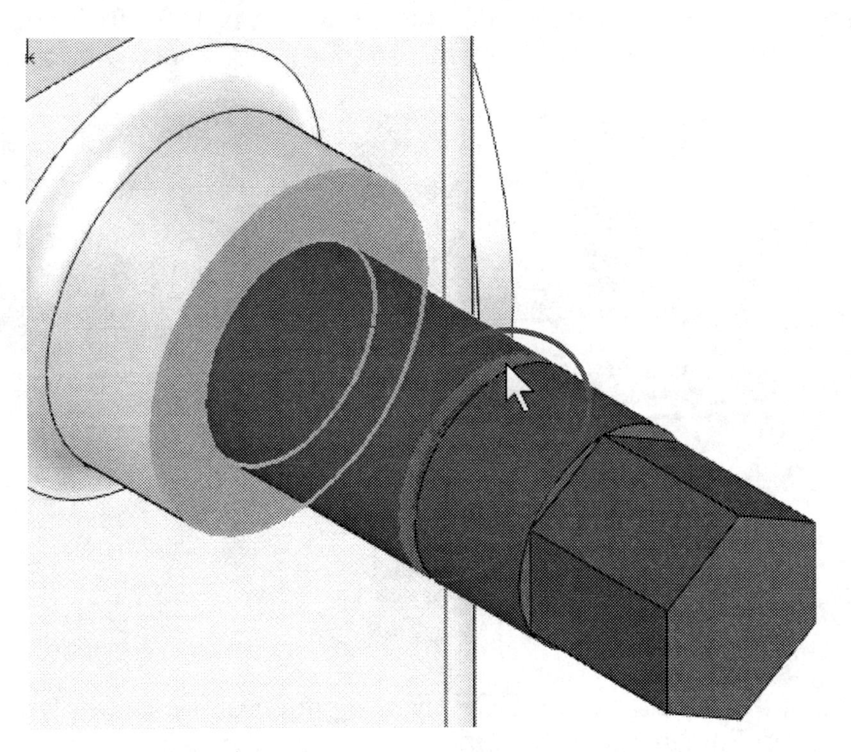

250. - Offset shaft assembled.

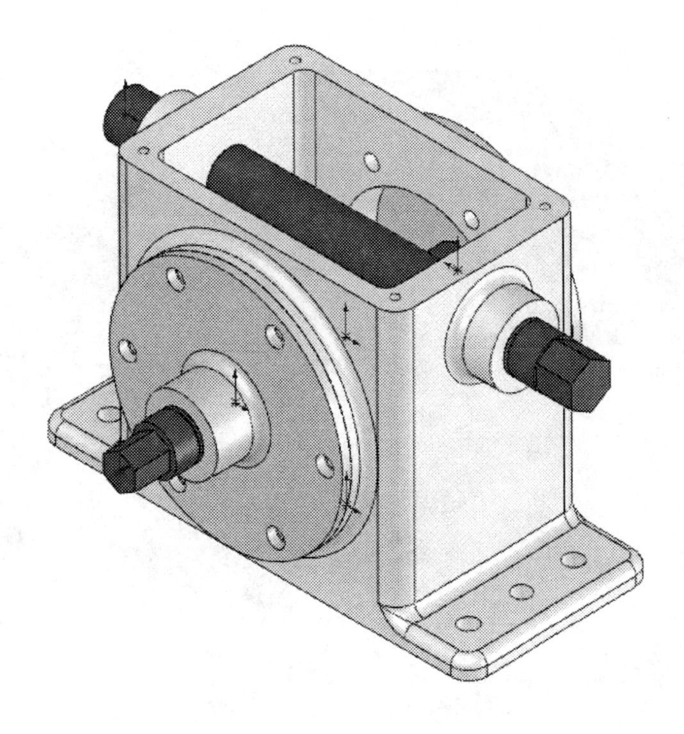

251. - The last component we're adding is the Top Cover. Add the Top Cover and if you wish, change its color. First add a Concentric mate to align one of the holes of the cover to a hole of the housing. For this mate you can use either faces or circular edges; it will work the same. An interesting detail to notice is that when we add a Concentric mate, it removes four DOF from the component, two translations and two rotations (if mated to a fixed or fully defined reference).

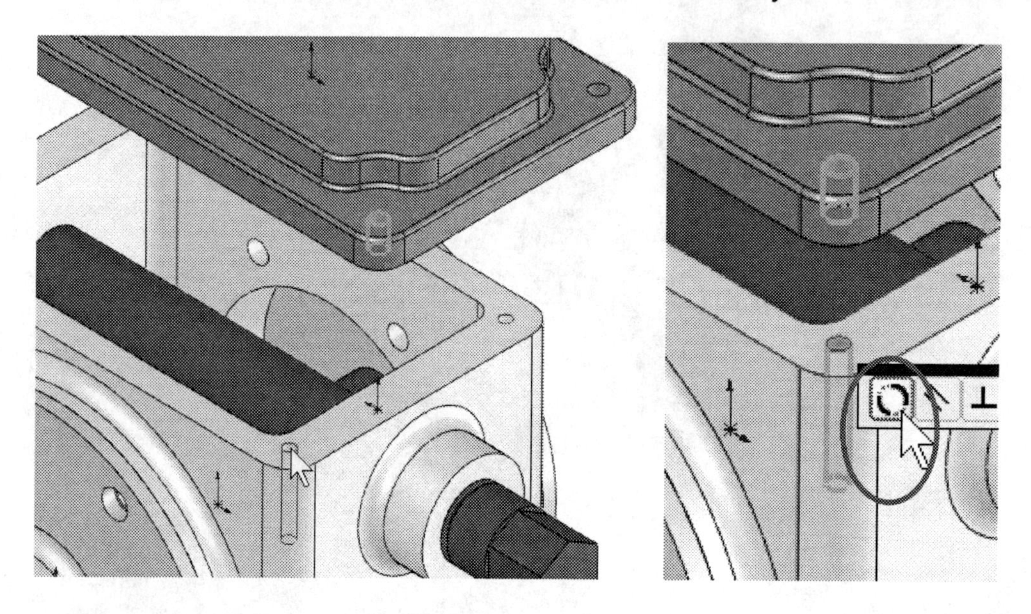

252. - Check the remaining two DOF by dragging the Top Cover with the mouse; it will rotate and move up and down. By adding a Coincident mate between the bottom face of the Top Cover and the top face of the Housing we'll remove one more DOF.

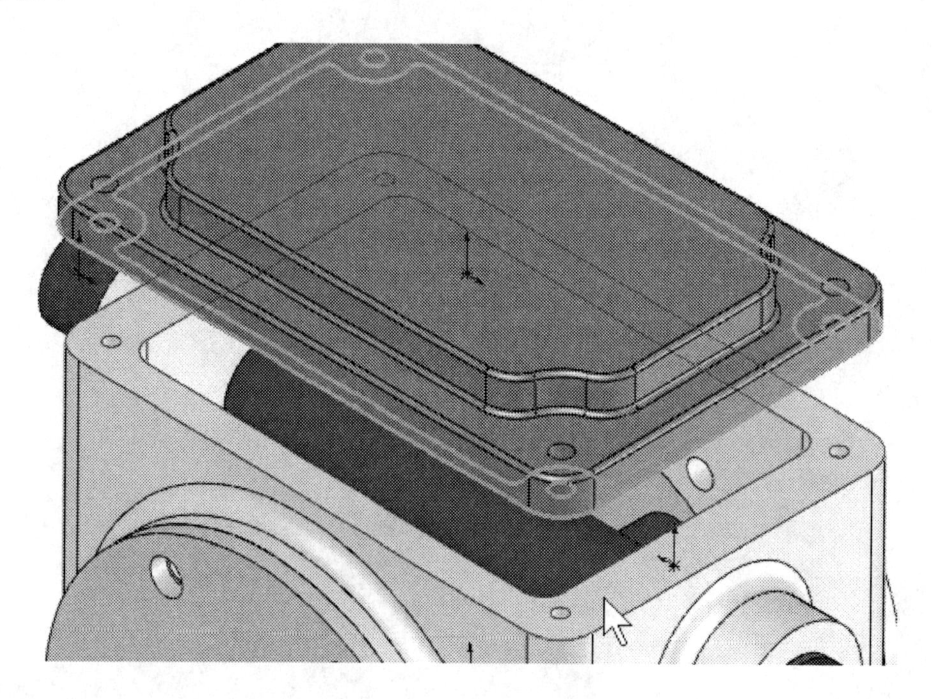

253. - If we now click and drag the Top Cover it will turn; now we have only one DOF left. To finish constraining the cover we'll add a Parallel mate between two faces of the Top Cover and Housing. As we explained earlier, the reason for the Parallel mate is that sometimes components don't match exactly, and if we add a Coincident mate, we may be forcing a condition that cannot be met, over defining the assembly and getting an error message. The parallel mate can be added between faces, planes and edges. In this particular case we can use either Coincident or Parallel mate, since the parts were designed to match exactly. However, in real life they may not; that's why we chose the Parallel mate.

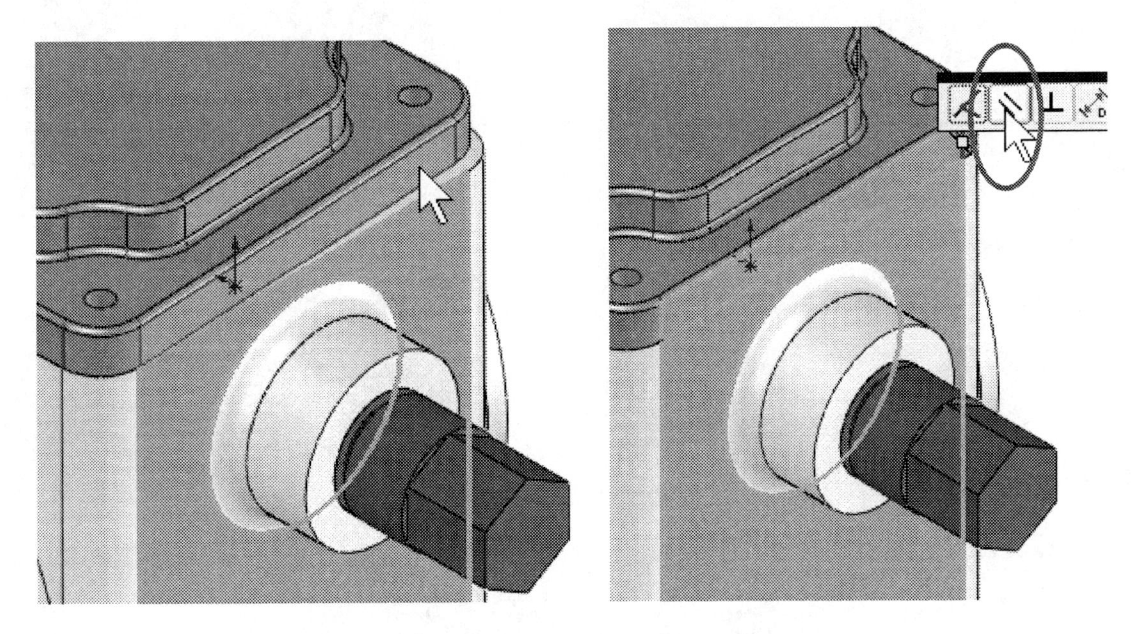

Your assembly should now look like this.

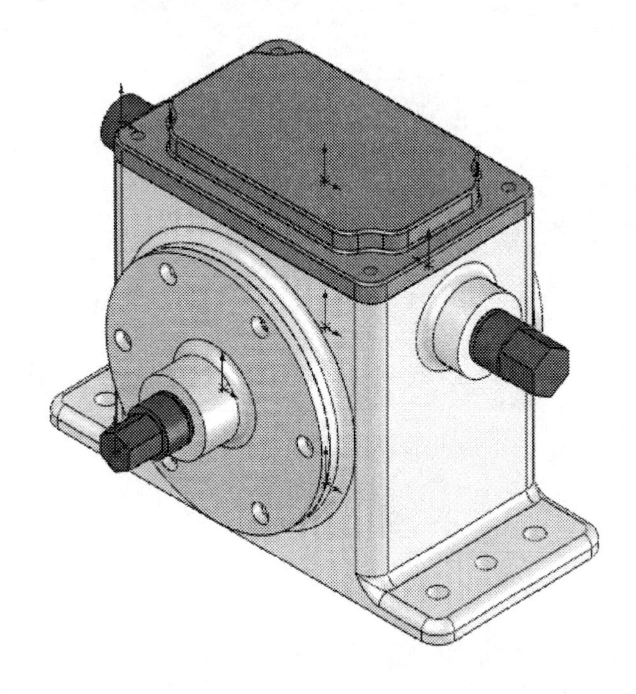

Notes:

Fasteners

The commercial version of SolidWorks Office, as well as the educational editions, includes a library of hardware called "**SolidWorks Toolbox**" that includes nuts, bolts, screws, pins, washers, bearings, bushings, structural steel, gears, etc. in metric and inch standards. The Toolbox is an Add-In that has to be loaded through the "Tools" menu, "Add-ins". For Toolbox to work correctly we have to load "SolidWorks Toolbox" and "SolidWorks Toolbox Browser".

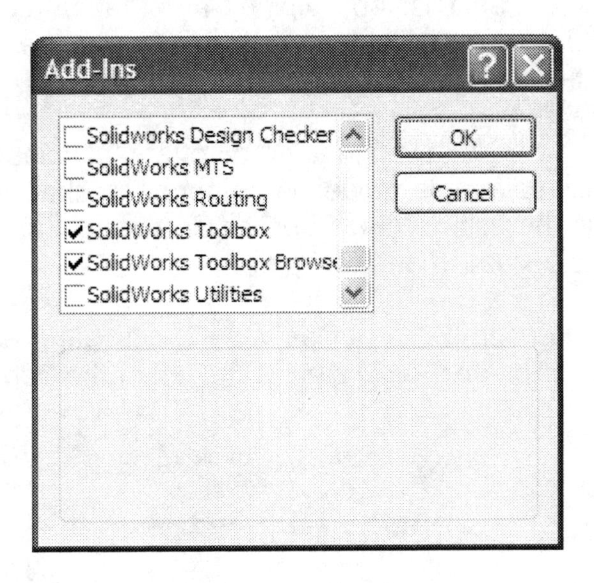

With the SolidWorks Toolbox we can add hardware to our assemblies by simply dragging and dropping components, and SolidWorks will automatically add the necessary mates saving us time. We can also add our own hardware to it, making it more versatile.

254. - To access the SolidWorks Toolbox we use the **Design Library**, which is located in the middle tab in the Task Pane toolbar. If the Task Pane is not visible, go to the "View" menu, "Toolbars, Task Pane" to show it.

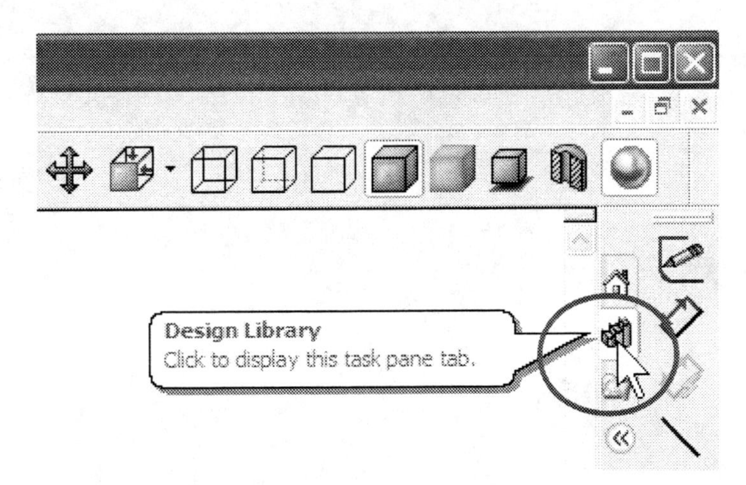

Clicking on the Design Library icon opens a fly-out pane that reveals the library. The Design Library contains three main areas:

- **Design Library**, which includes built in libraries of annotations, features and parts that can be dragged and dropped into parts, drawings and assemblies.
- **SolidWorks Toolbox**, which we just described, and
- **3D Content Central**, an Internet based library of user uploaded and supplier certified components, including nuts, bolts and screws, pneumatics, mold and die components, conveyors, bearings, electronic components, industrial hardware, power transmission, piping, automation components, furniture, human models, etc., all available for drag and drop use. All that is needed to access it is an Internet connection.

As we can see, the Design Library offers a valuable resource for the designer, helping us save time modeling components that are usually purchased or standard, and in the case of the Supplier Certified library, all components are accurately modeled for use in our designs.

255. - To start adding screws to our assembly, activate the Design Library from the Task Pane, and click on the (+) sign to the left of the Toolbox to expand it.

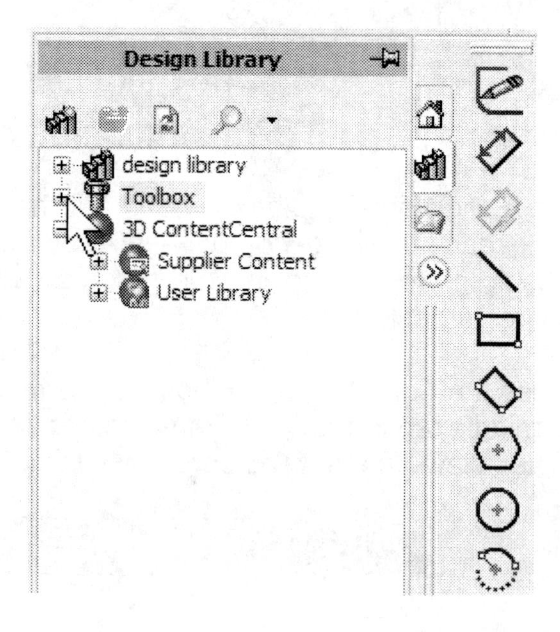

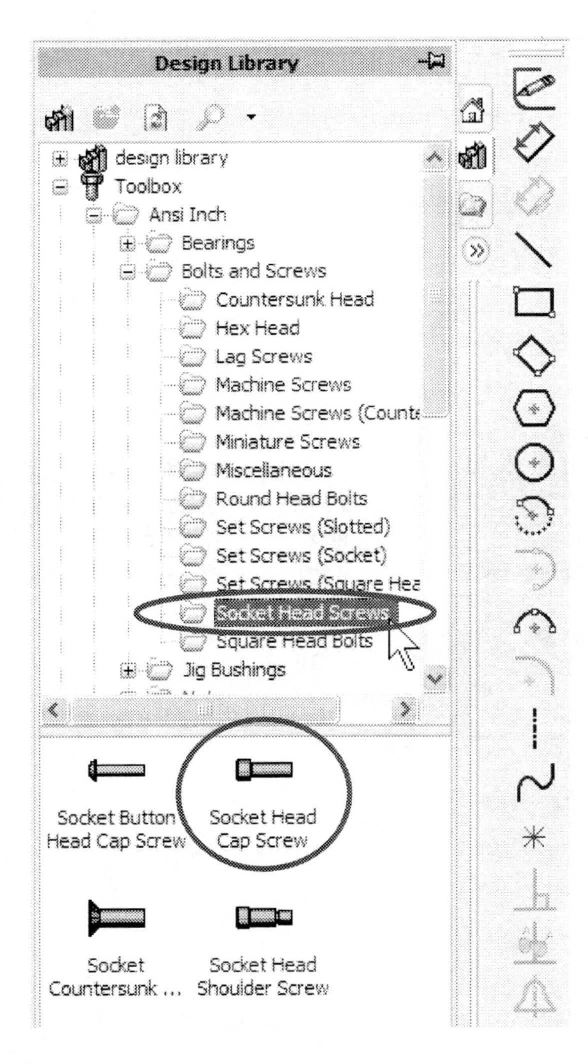

256. - After expanding the Toolbox we can see the many options available. In this exercise we'll go to "ANSI Inch", "Bolts and Screws" and select "Socket Head Screws".

When we click the "**Socket Head Screws**" folder we see in the lower half of the Design Library pane the available styles, including Button Head, Socket Head, Countersunk and Shoulder Screws. For our assembly we'll use Socket Head Cap Screws (SHCS).

257. - First we want to add the #6-32 screws to hold the Top Cover. From the bottom pane of the Toolbox, click and drag the Socket Head Cap Screw into one of the holes in the Top Cover. You will notice a yellow transparent preview of the screw, and when we get close to the edge of a hole, SolidWorks will automatically snap the screw in place. When we get the preview of the screw assembled where we want it, release the left mouse button. If we release the mouse button before, the screw will still be created, but it will not add any mates.

Do not worry if the preview in your screen is a big screw; it's just the default size in your computer. When we drop the screw in place we will select the correct size.

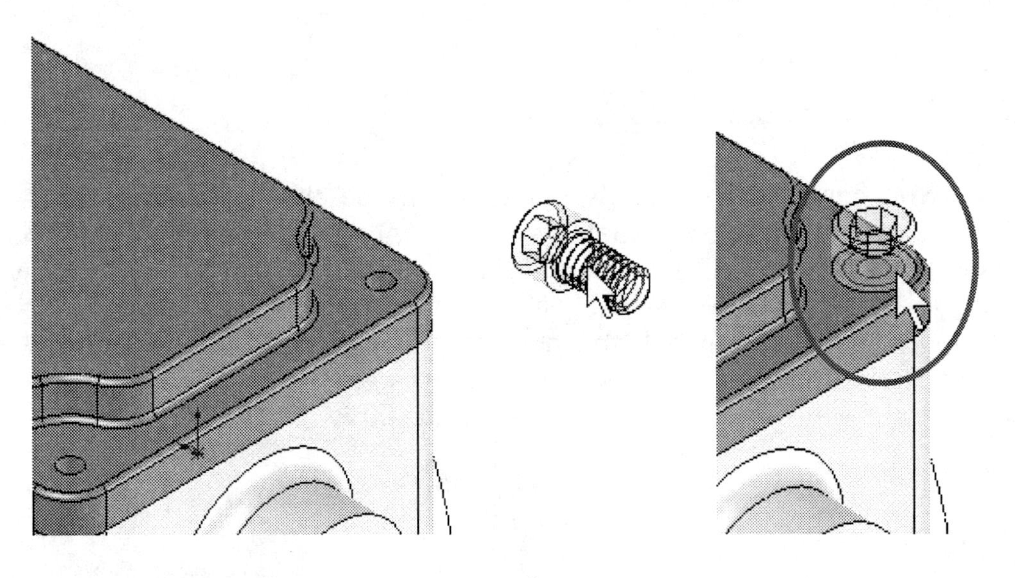

Notice that the Design Library hides away as soon as we drag the screw in the assembly; this is an automatic feature, unless the user pressed the "stay visible" thumb tack on the upper right of the Task Pane.

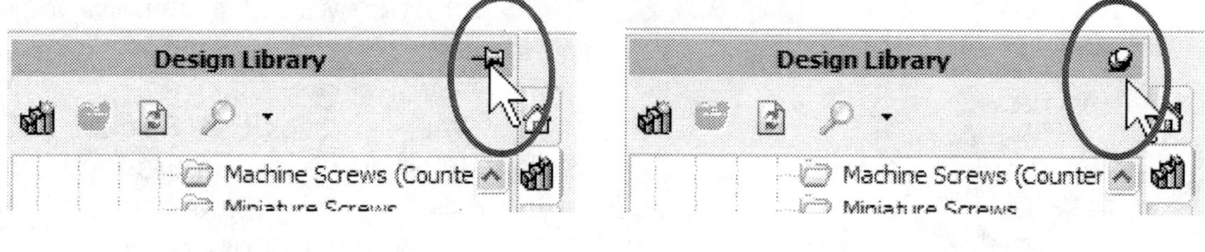

Auto-hide On Always visible

258. - As soon as we drop the screw in the hole, we are presented with a dialog box asking us to select the screw parameters. In this case, we need a # 6-32 screw, 0.5" long with Hex Drive and Schematic Thread Display. The "Add" button will add the screw size as a favorite for future use.

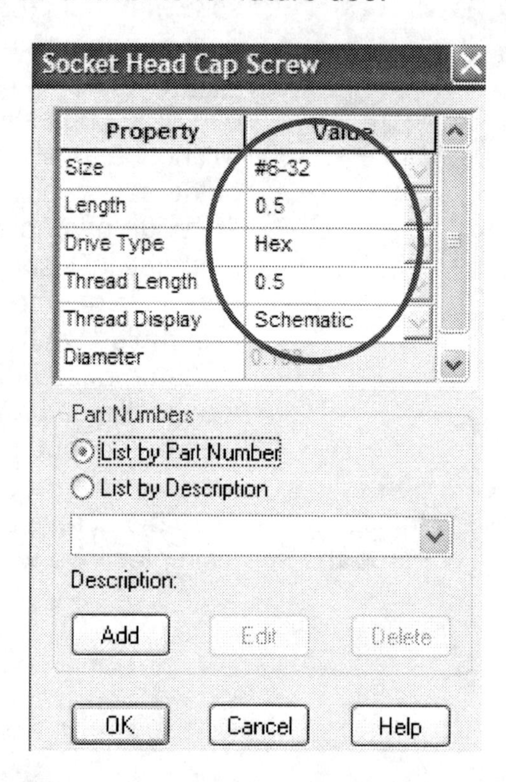

A word on Thread Display: The schematic thread display will add a revolved cut to the screw for visual effect. Helical threads can be added, but it's generally considered a waste of computer resources that doesn't really add value to a design in many instances. Helical threads are a resource intensive feature that is best left for times when the helix is a design requirement and not only cosmetic. The revolved cut gives a good appearance for all practical purposes and is a simple enough feature that doesn't noticeably affect performance.

259. - After selecting OK from the screw size options box, the screw is created with the selected parameters and mated in the hole where it was dropped. At this point we are ready to add more screws of the same size, if needed. In our example, we'll click in the other holes of the Top Cover to add the rest of the screws. Notice the graphic preview of the screw as we move the mouse and the snap when we get close enough to a hole. Click "Cancel" to finish adding screws.

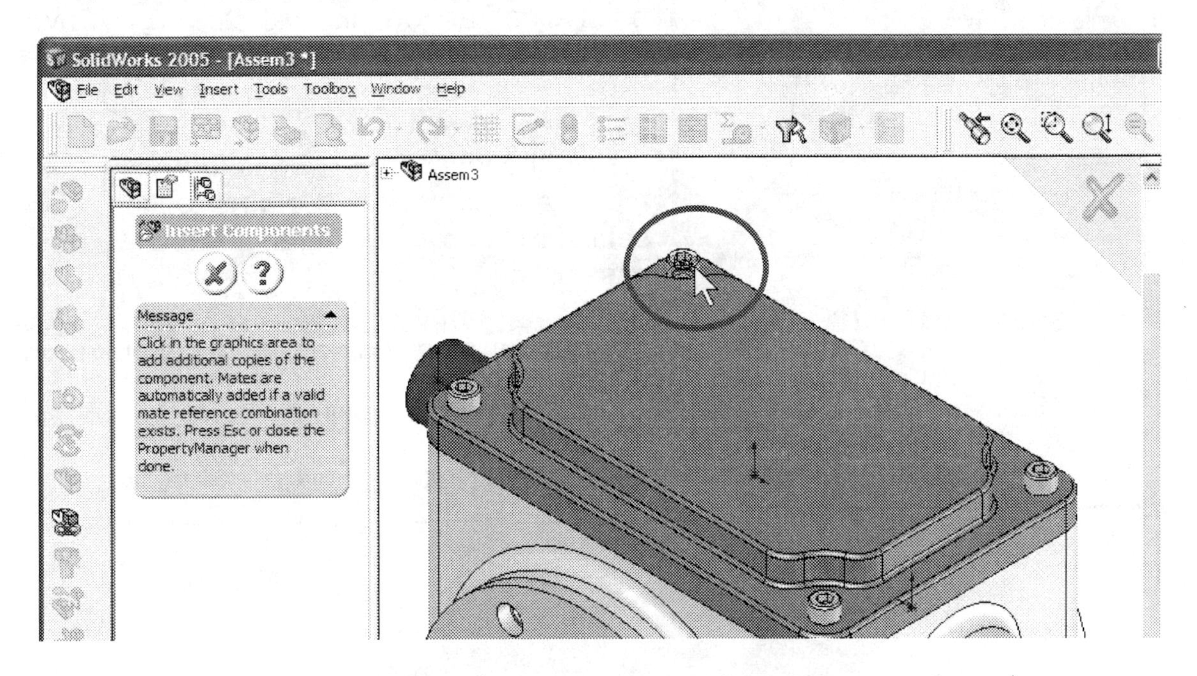

260. - We will now add the ¼-20 Socket Head Cap Screw to the side covers. Select the Design Library icon as we did before and drag and drop the Socket Head Cap Screw to one of the holes of the Side Cover, but be careful to "drop" the screw in the correct location. If you look closely, there are two different edges where you can insert the screw: one is in the Side Cover, and the other is in the Housing. The screw will be mated to the hole you "drop" the screw in. Use the snap preview to help you find the correct one.

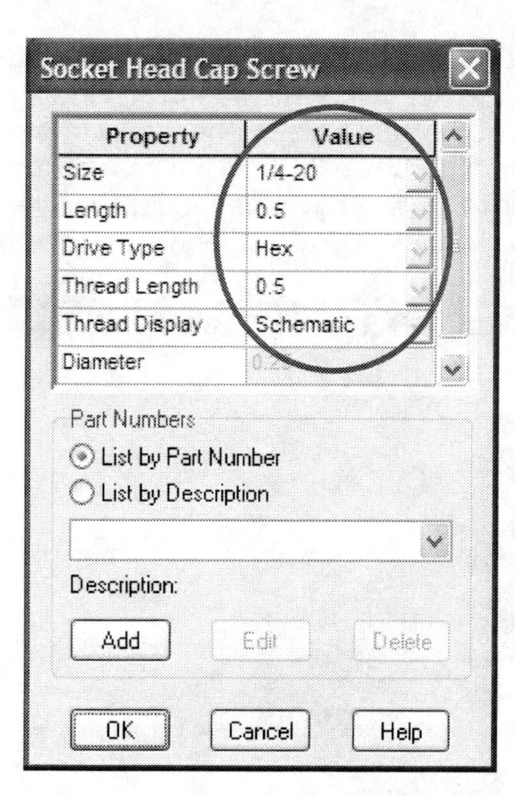

261. - After we drop the screw, the screw properties box is displayed. Select the ¼-20, 0.5" long screw with hex drive and schematic thread display. Click OK. SolidWorks makes the screw and mates it in the hole where we dropped it. Insert all twelve screws (six on each side) as we did with the Top Cover before.

262. - For display purposes, we can turn off the component origins. Go to the "View" menu, and turn off the "**Origins**" option. This setting will turn off all the origins in the assembly.

Your Gear Box assembly should now look like this.

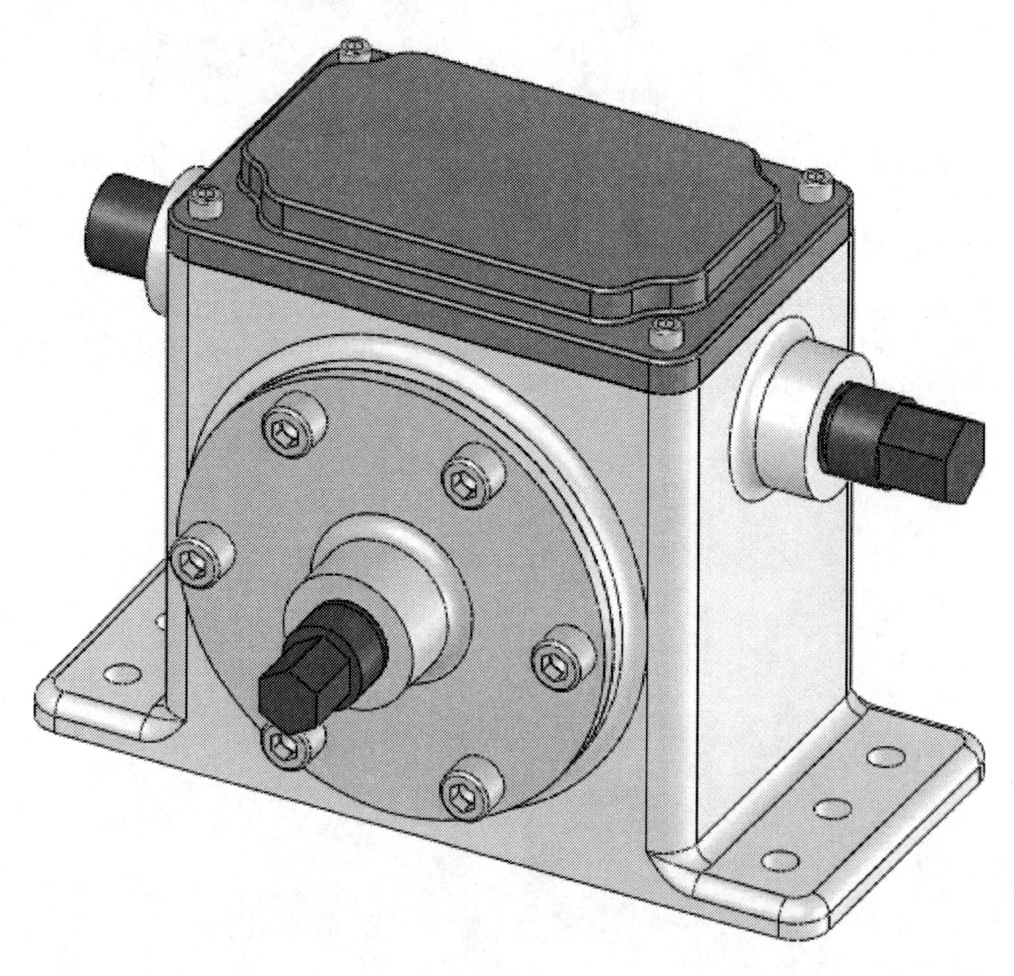

Exploded View

263. - The last step in the assembly will be to make an exploded view for documentation purposes. Exploded views are used to show how the components will be assembled together. Select the Exploded View icon from the Assembly Toolbar or from the "Insert" menu, "Exploded View" to display the Explode Property Manager.

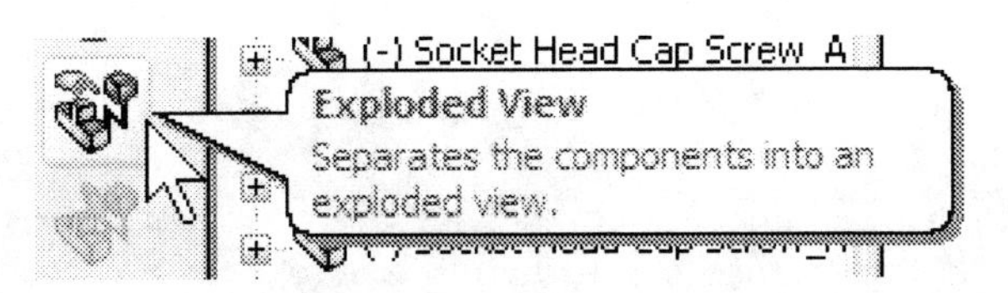

264. - In the Explode Property Manager we can see the "Settings" selection box is active and ready for us to select the component(s) that will be exploded. For this example we will leave the option "Auto-space components after drag" turned off.

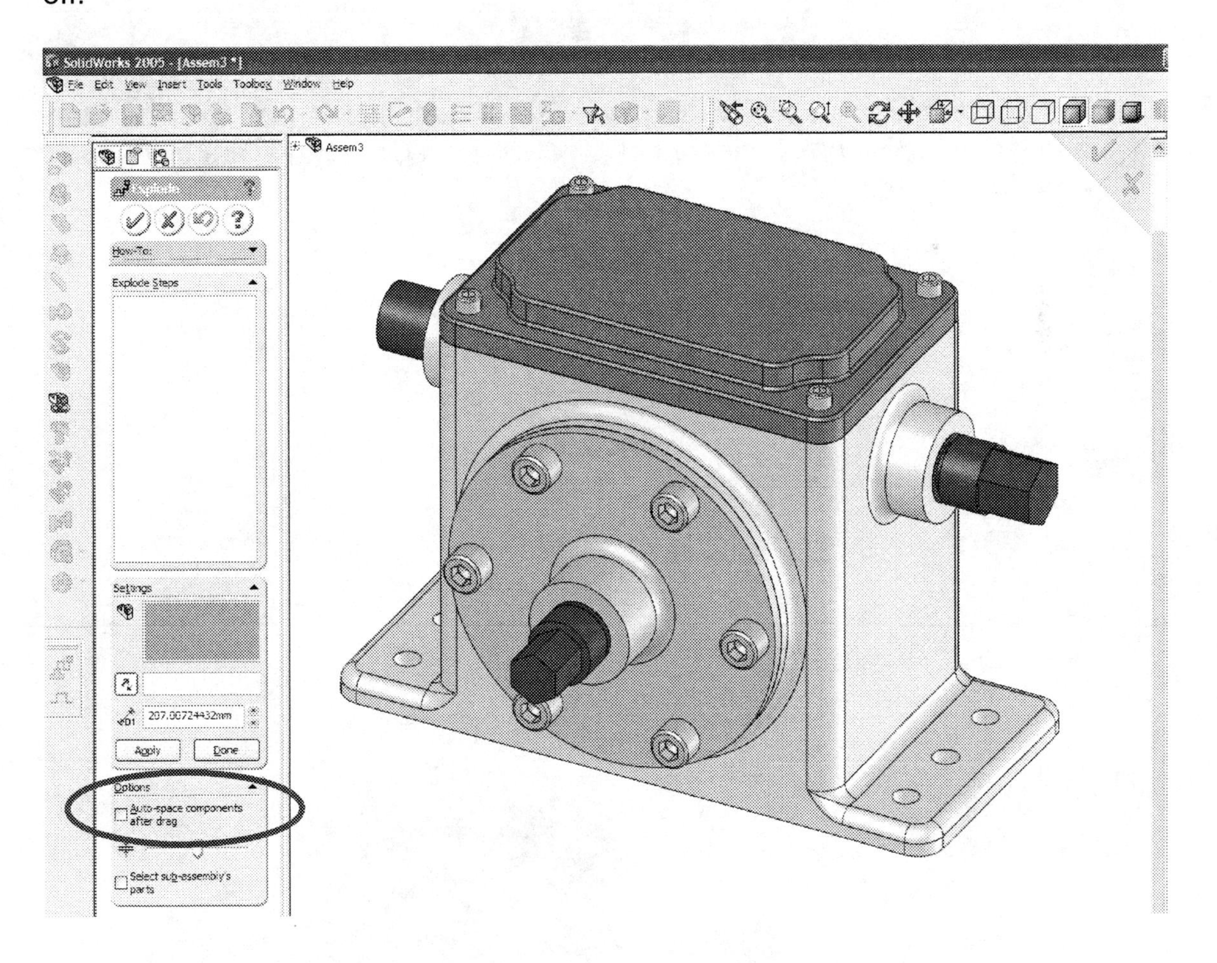

265. - Adding exploded view steps is very simple and straightforward. For the first explode step, select the four Screws on the Top Cover and drag the green tip of the manipulator arrow upwards as far as you want the screws exploded. Notice the four selected Screws are listed in the selection box and colored green.

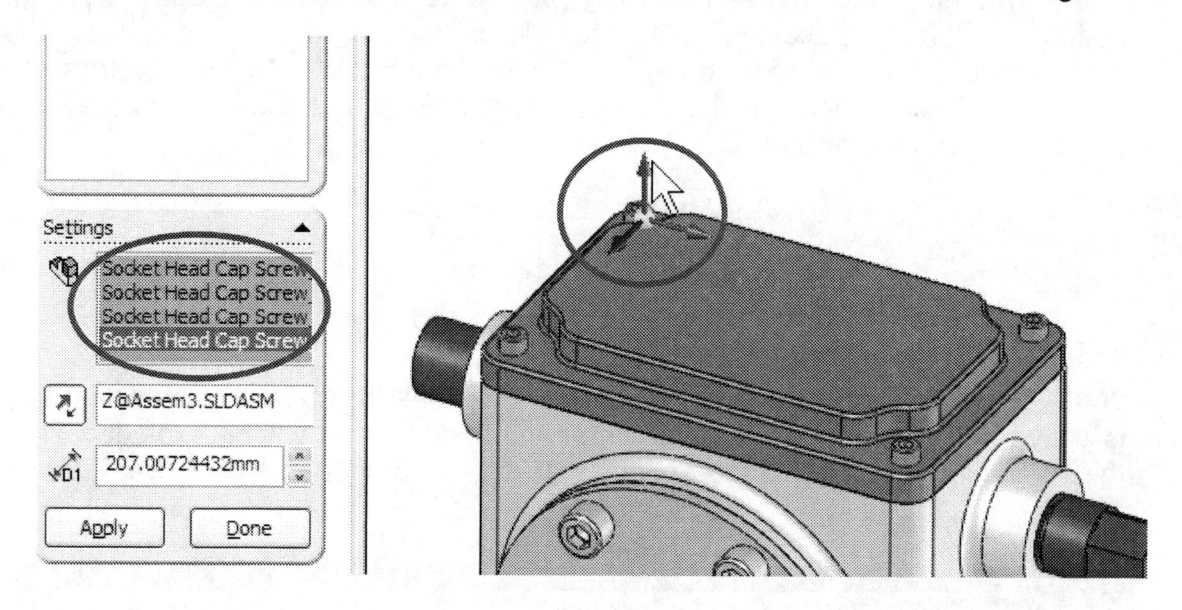

266. - After dragging the arrow, "Explode Step1" is added to the "Explode Steps" list, and the selection box is cleared. Do not click on OK yet; we are going to add more explode steps.

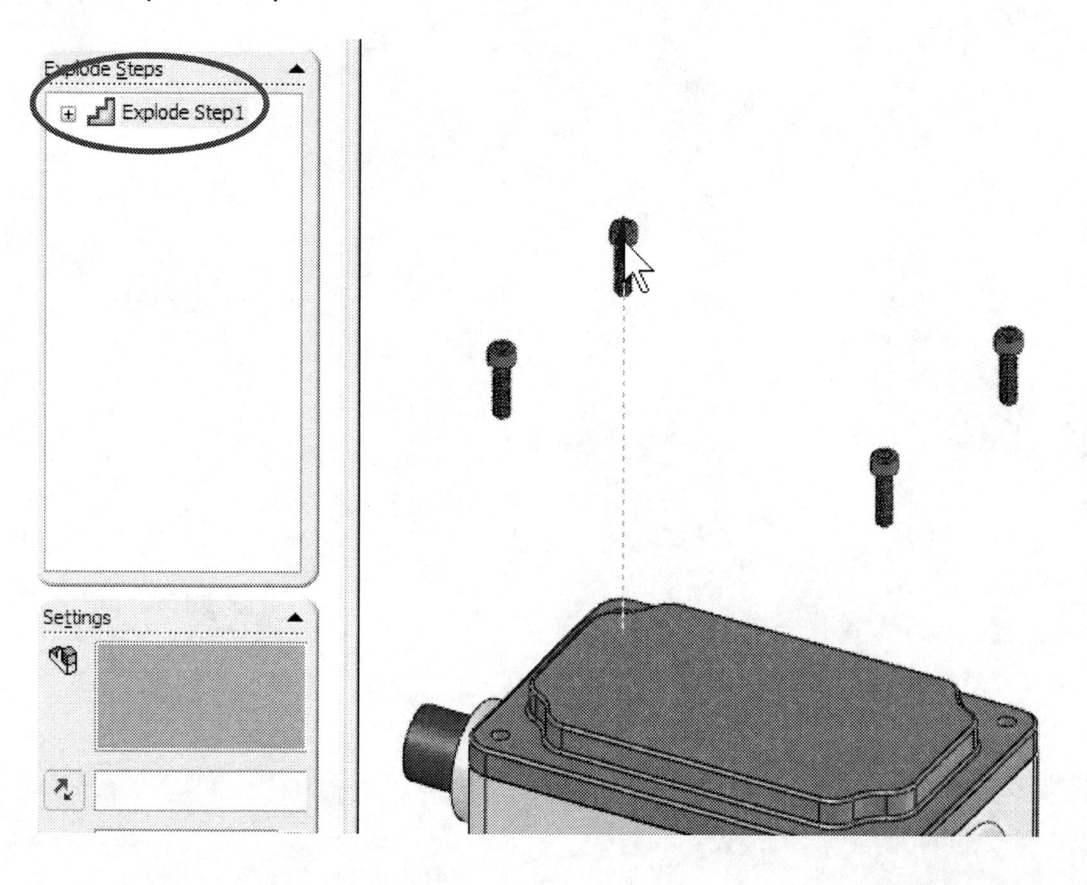

267. - For the second explode step, select the Top Cover and drag the green manipulator arrow upwards about halfway between the Screws and the Housing.

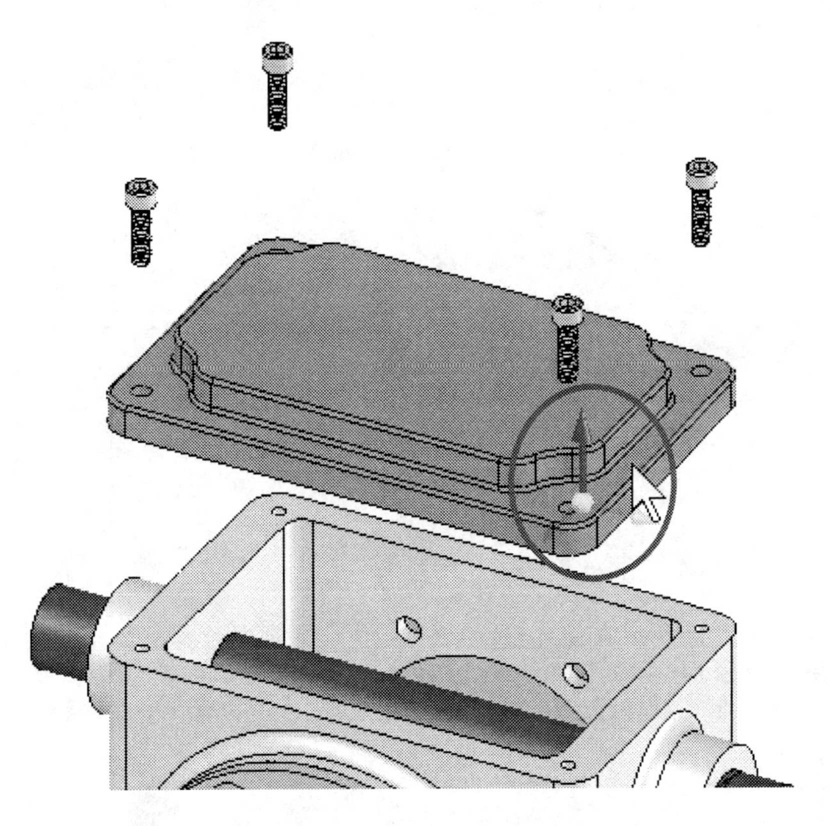

268. - For the third step, select the six Screws of one of the Side Covers and drag them to the left dragging the blue manipulator arrow.

269. - The rest of the explode steps are done the same way, selecting the component(s) and dragging the tip of the manipulator arrow along the explode direction desired. The next step is to explode the Side Cover.

270. - Now explode the Worm Gear Shaft and the Screws from the other side of the gear assembly in one step. Rotate the assembly and select the Screws and the Shaft, then click and drag the manipulator's arrow to explode them.

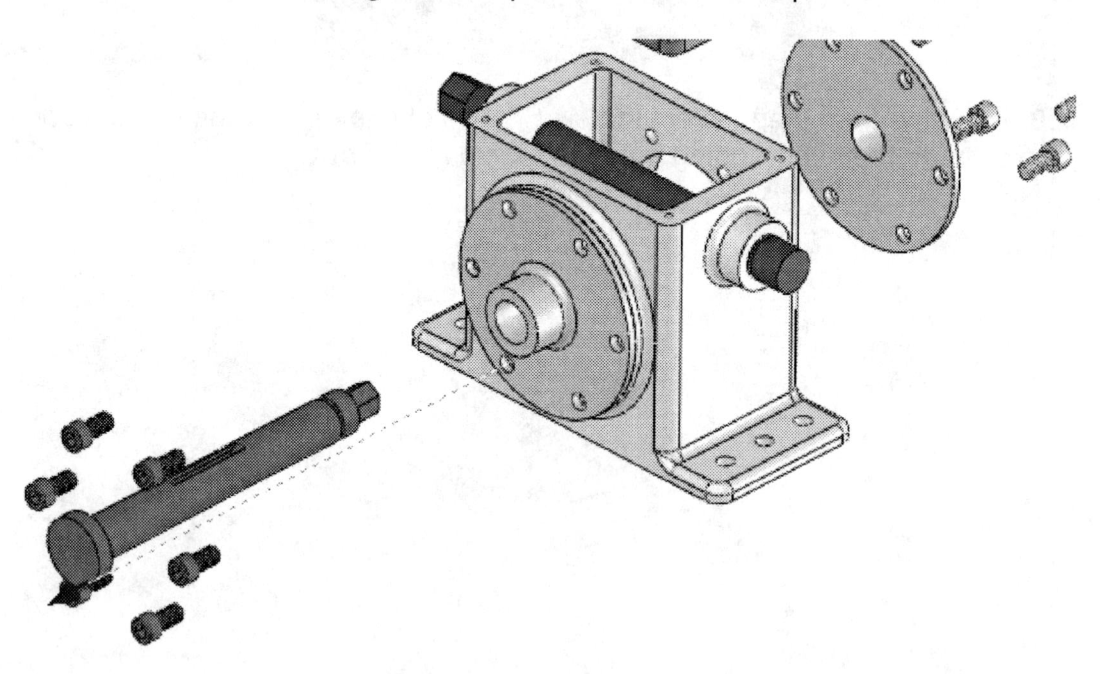

271. - Now explode the second Side Cover.

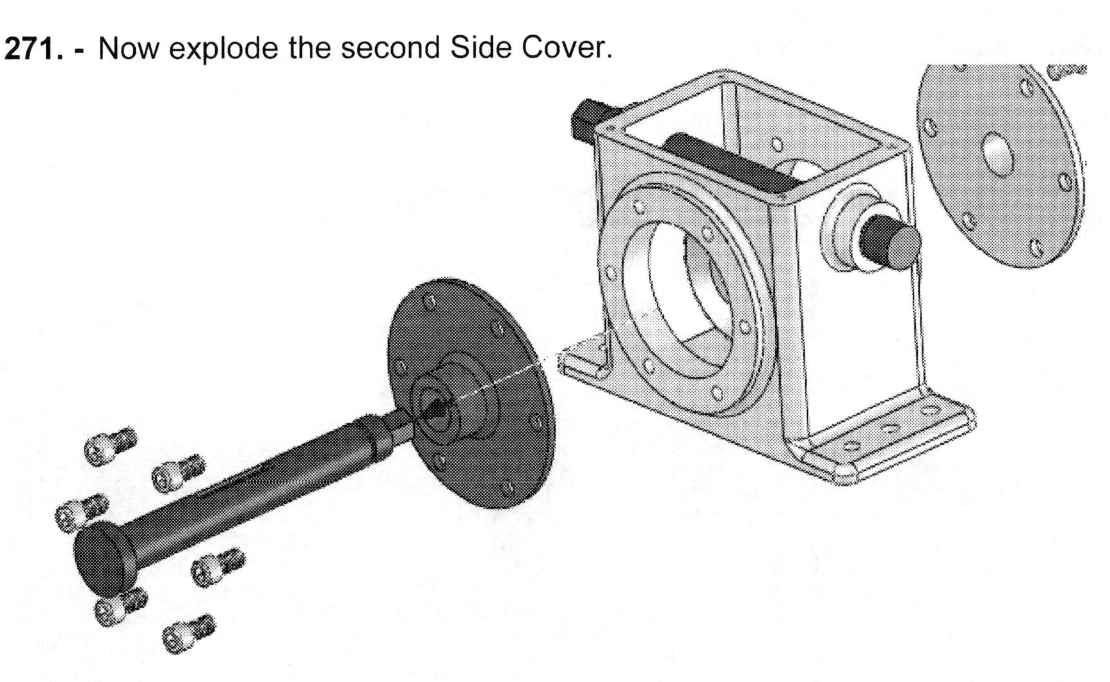

272. - Finally explode the Worm Gear and the Offset Shaft, one in each step, to get the exploded view in the next image. Try to group parts that will be exploded the same distance and direction in the same step; this way we'll reduce the number of steps required to document our design.

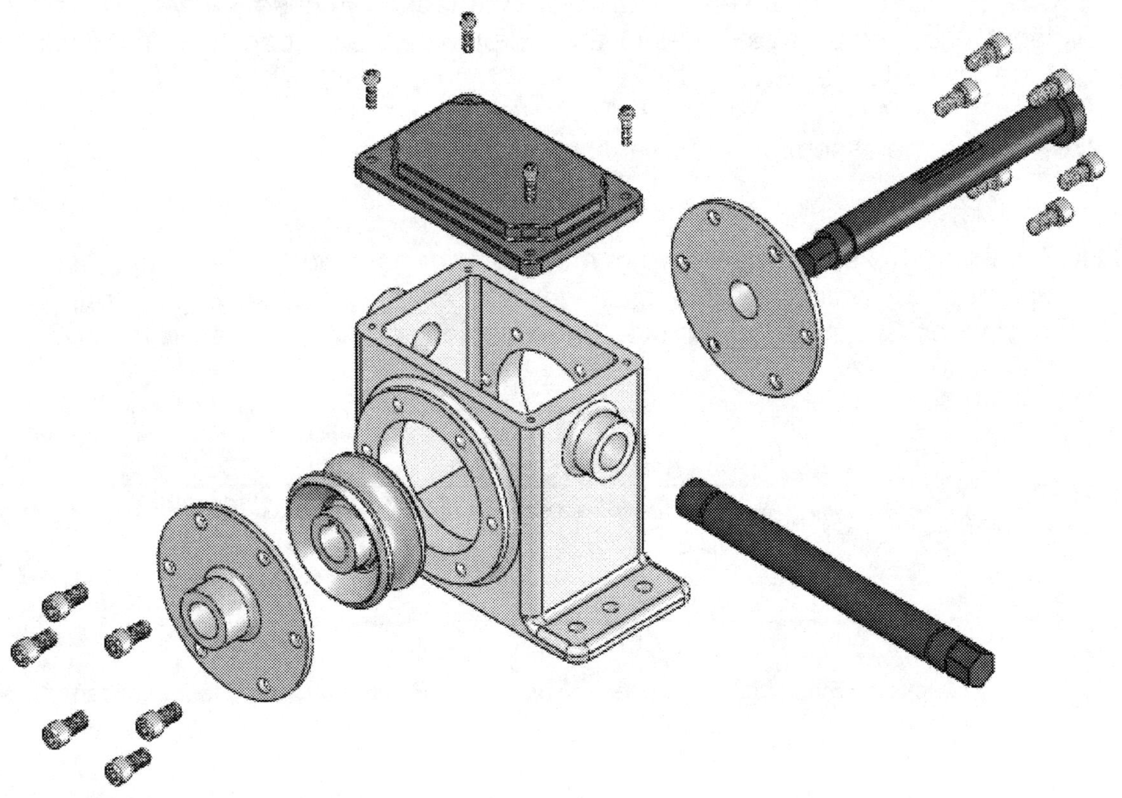

NOTE: If an explode step needs to be edited or "tweaked", simply select it from the "Explode Steps" list and drag the blue arrow to adjust the distance. When done adding explode steps click OK to finish the Exploded view.

273. - After exploding the assembly, we may want to collapse the assembly. In order to do this, click with the Right Mouse Button at the top of the Feature Manager at the Assembly name and select "**Collapse**" from the pop-up menu.

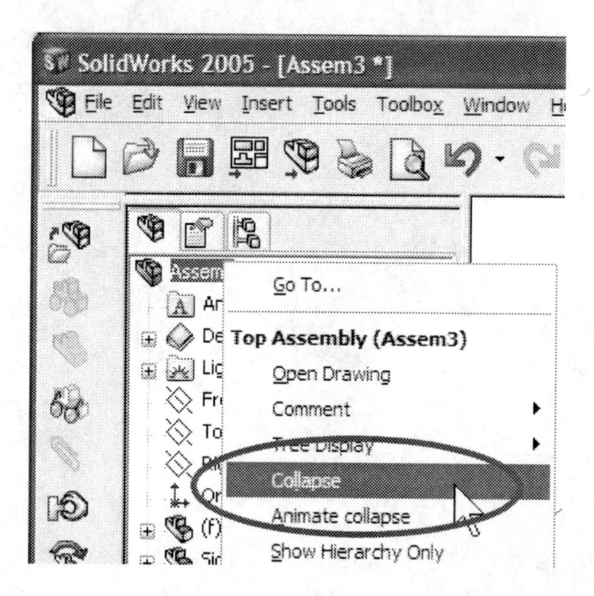

274. - To explode the assembly again, click with the Right Mouse Button at the top of the Feature Manager as we just did, and select "Explode". This option will only be available if the assembly has been exploded. In case we need to edit the exploded view steps again, select the "Exploded View" icon to bring back the explode Property Manager.

275. - Save the assembly as "Gear Housing" and close the file.

NOTE: From this same menu, we can select "**Animate collapse**" if the assembly is exploded, or "Animate explode" if the assembly is collapsed. This brings up the Animation Controller to animate the explosion using VCR like controls.

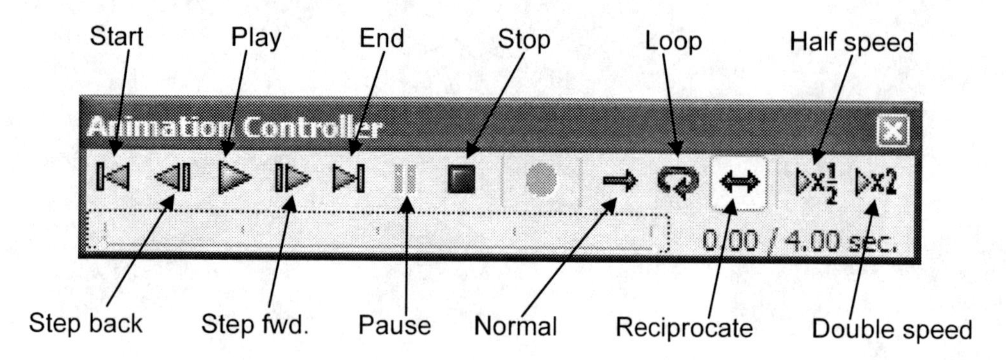

Assembly Drawing Section

ITEM NO.	PART NUMBER	DESCRIPTION	QTY.
1	Housing		1
2	Side Cover		2
3	Worm GearShaft		1
4	Worm Gear		1
5	Offset Shaft		1
6	Top Cover		1
7	HX-SHCS 0.138-32x0.5x0.5-S		4
8	HX-SHCS 0.25-20x0.5x0.5-S		12

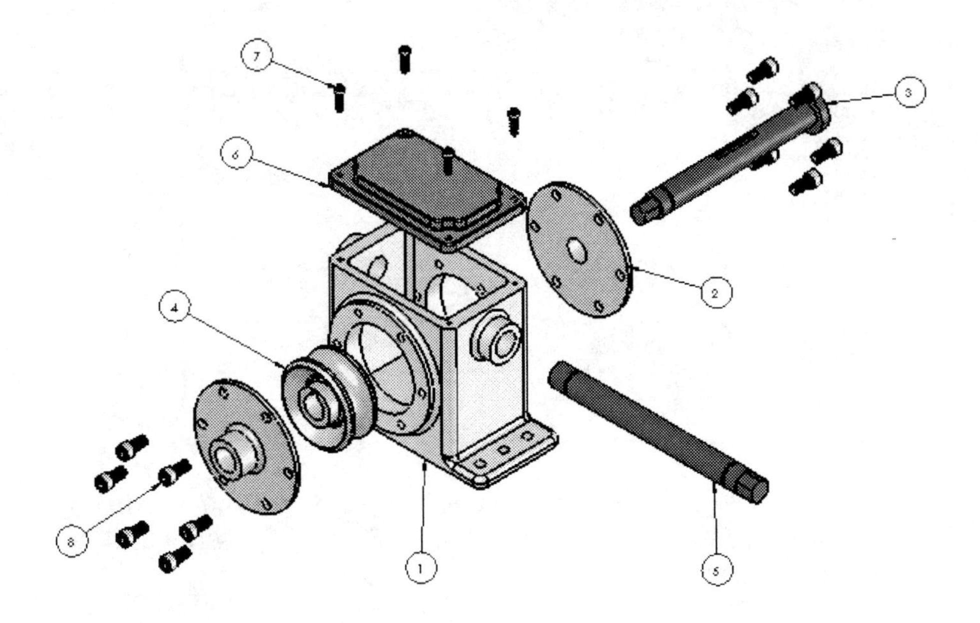

276. - After making the assembly it is often required to make an assembly drawing with a Bill of Materials for assembly instructions and/or documentation. In this case we will make a drawing with an exploded view, Bill of Materials (BOM) and identification balloons. To make the drawing, open the assembly "Gear Housing" and make sure it is in an exploded view. (Go to step 274 to explode the assembly if needed.) Select the "Make drawing from part/assembly" icon as before, make a new drawing using the "B-Landscape" sheet size, and turn off the "Display sheet format" options.

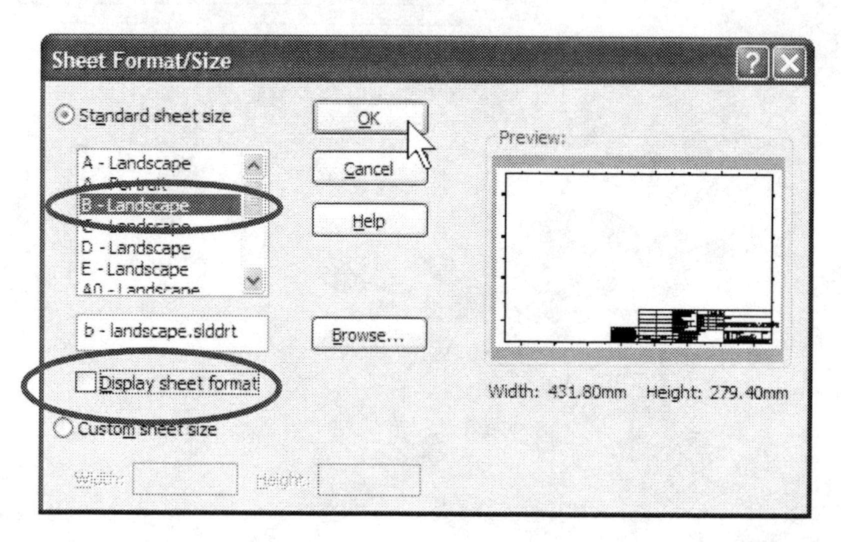

277. - When the "Model View" dialog is displayed, select "Isometric" from the "Orientation" box, "Shaded with Edges" from "Display Style", "Use custom scale, 1:2" from the "Scale" options box, and locate the view on the sheet.

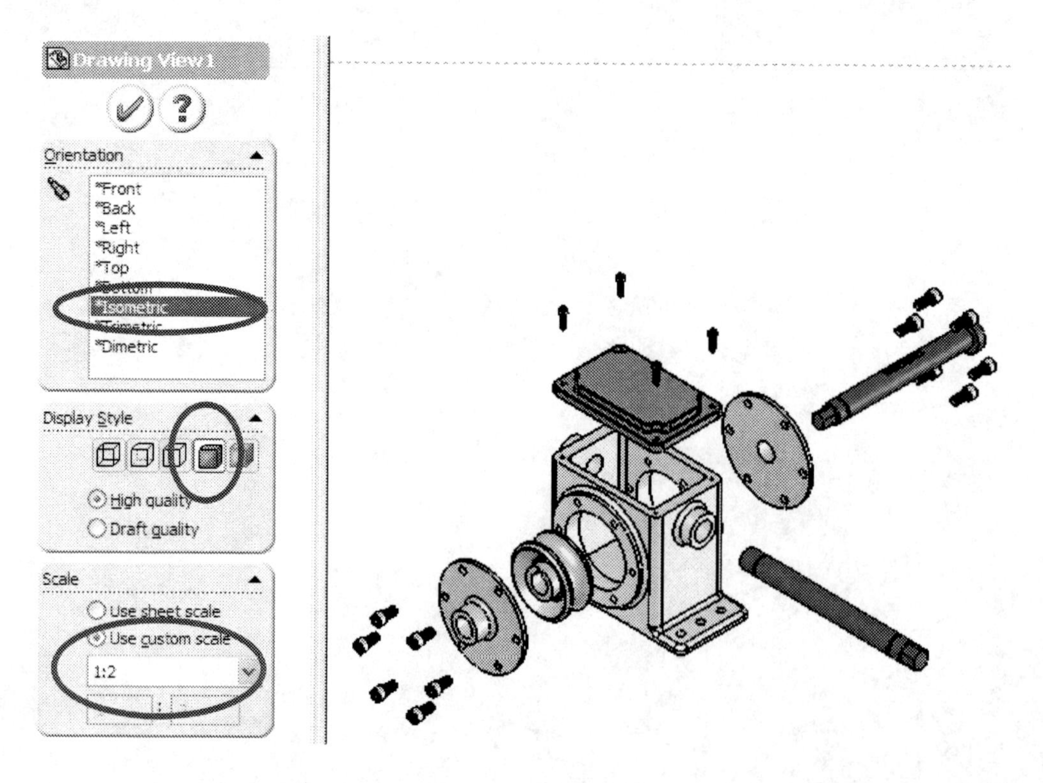

278. - Once we have the Isometric view in the drawing, we'll add the **Bill of Materials** (BOM). Select the isometric view from the graphics area by clicking on it, and from the "Insert" menu, select "Tables, Bill of Materials".

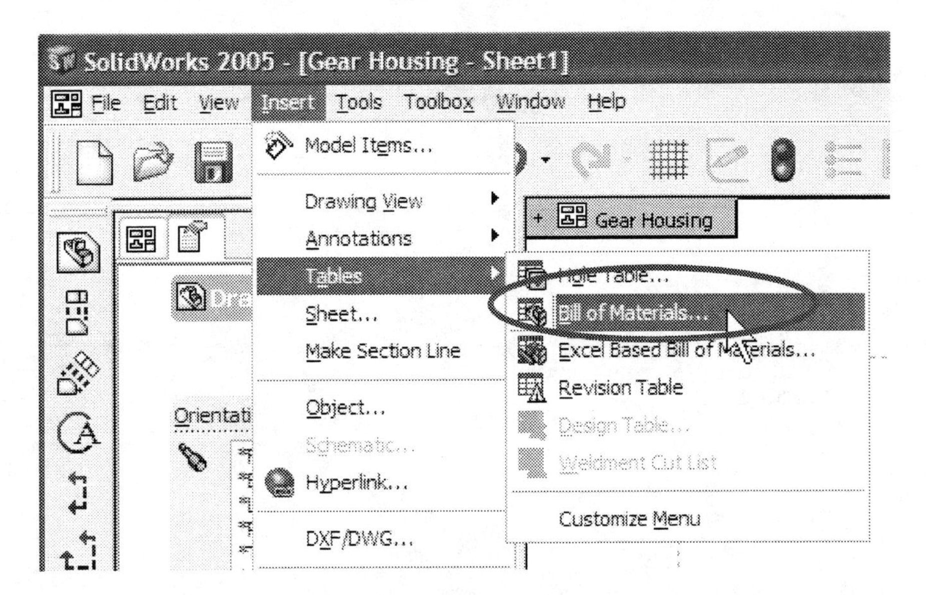

279. - Click OK in the Bill of Materials Property Manager to accept the default options. For more options and BOM customization go to the "Help" menu and search for "Bill of Materials" to modify the template and include properties such as weight, volume, vendor, cost, etc.

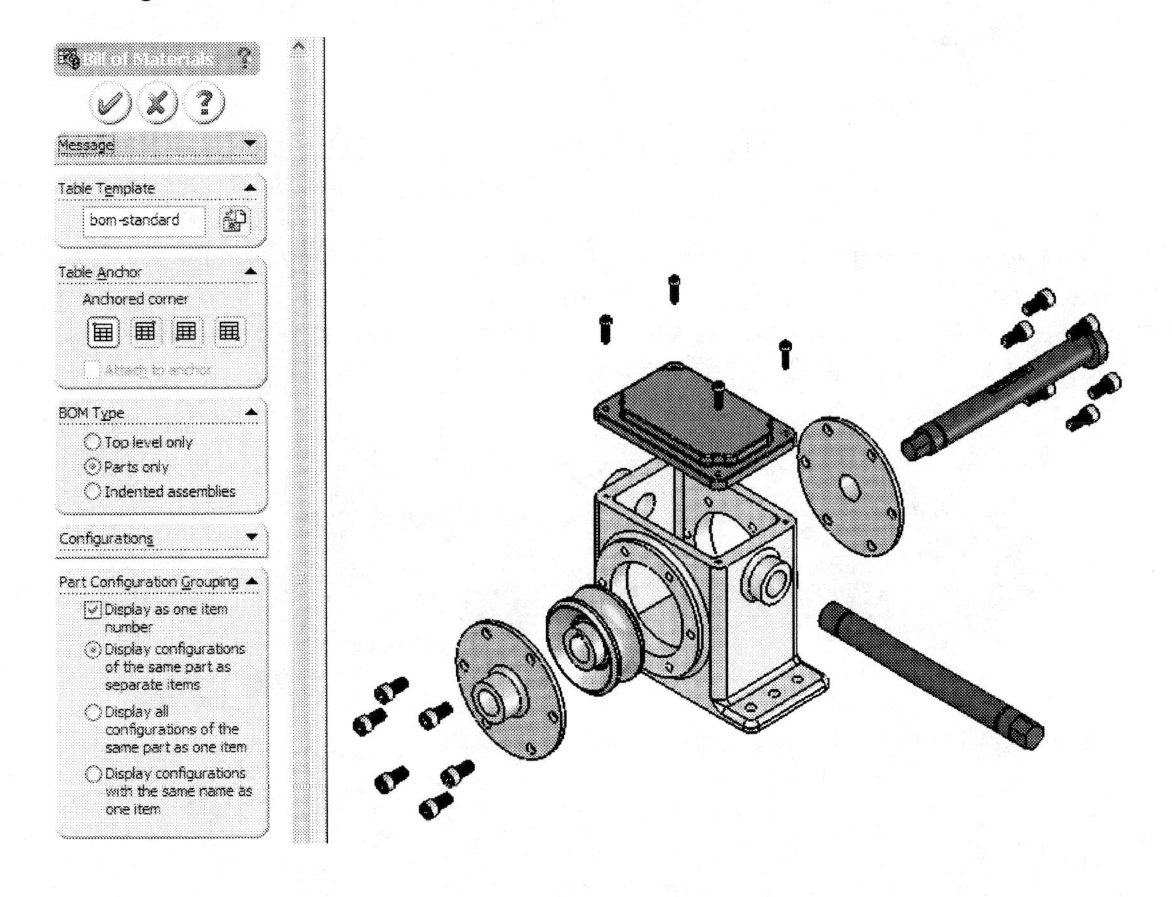

280. - After clicking on OK we can locate the Bill of Materials on the sheet. Move the mouse and locate the BOM in the top right part of the drawing; be sure to allow some space for the identification balloons. If needed, move the assembly view.

ITEM NO.	PART NUMBER	DESCRIPTION	QTY.
1	Housing		1
2	Side Cover		2
3	Worm Gear Shaft		1
4	Worm Gear		1
5	Offset Shaft		1
6	Top Cover		1
7	HX-SHCS 0.138-32x0.5x0.5-S		4
8	HX-SHCS 0.25-20x0.5x0.5-S		12

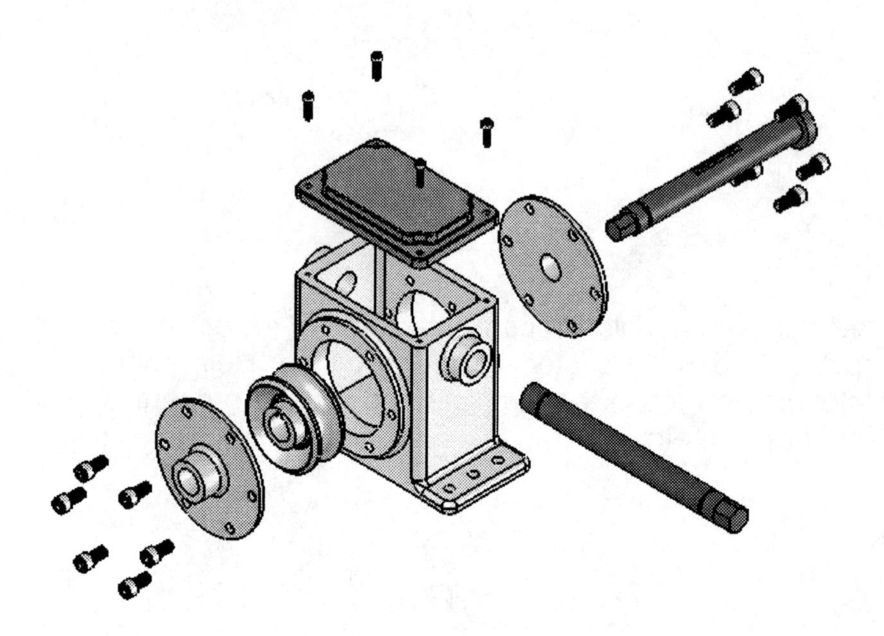

281. - Now let's add the identification balloons to our assembly drawing. Click in the Assembly drawing view with the Right Mouse Button, and from the pop-up menu select "Annotations, **AutoBalloon**". Another option is to go to the "Insert" menu, "Annotations, AutoBalloon".

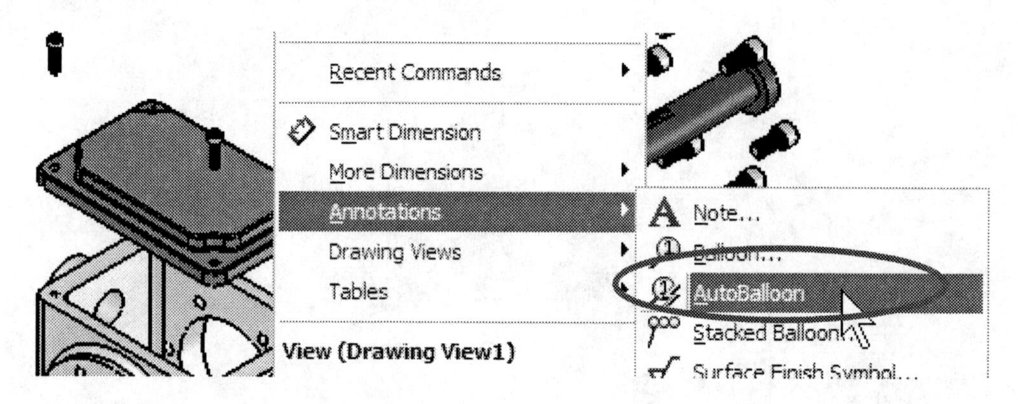

282. - When the AutoBalloon Property Manager is displayed select the options "Square" for the "Balloon Layout", "Ignore multiple instances" to avoid adding balloons to repeated components, and "Circular, 2 Characters, Item Number" from the "Balloon Settings" options box for the size and style of the balloons. Click OK to add the balloons to the drawing view. After we add the balloons we can modify their positions as needed by simply dragging them individually.

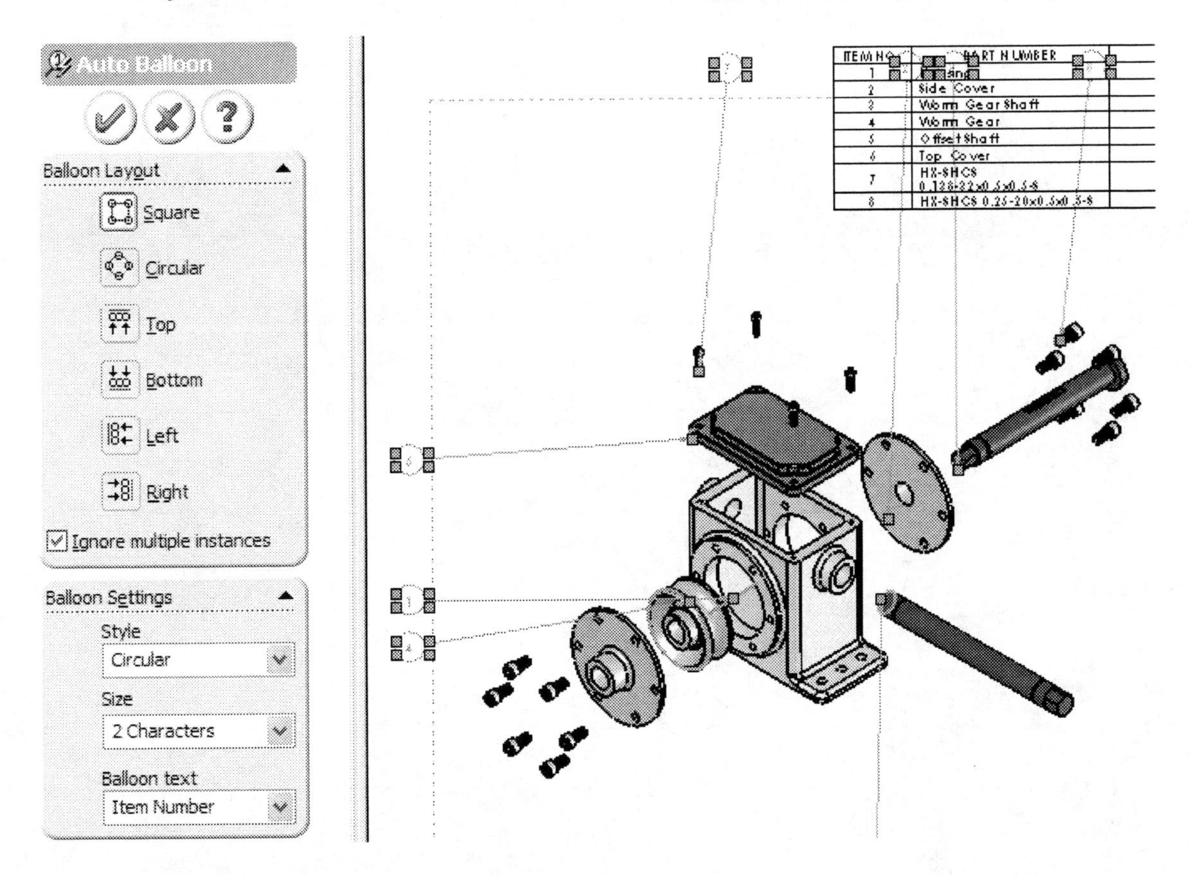

283. - A balloon's arrow tip can be dragged to a different area of the part for visibility and is also smart; if the arrow tip is dragged to a different component the item number displayed will change to reflect the item number of the part that the balloon is attached to.

284. - Arrange the balloons as needed for readability. Save the drawing and close the file.

ITEM NO.	PART NUMBER	DESCRIPTION	QTY.
1	Housing		1
2	Side Cover		2
3	Worm GearShaft		1
4	Worm Gear		1
5	Offset Shaft		1
6	Top Cover		1
7	HX-SHCS 0.138-32x0.5x0.5-S		4
8	HX-SHCS 0.25-20x0.5x0.5-S		12

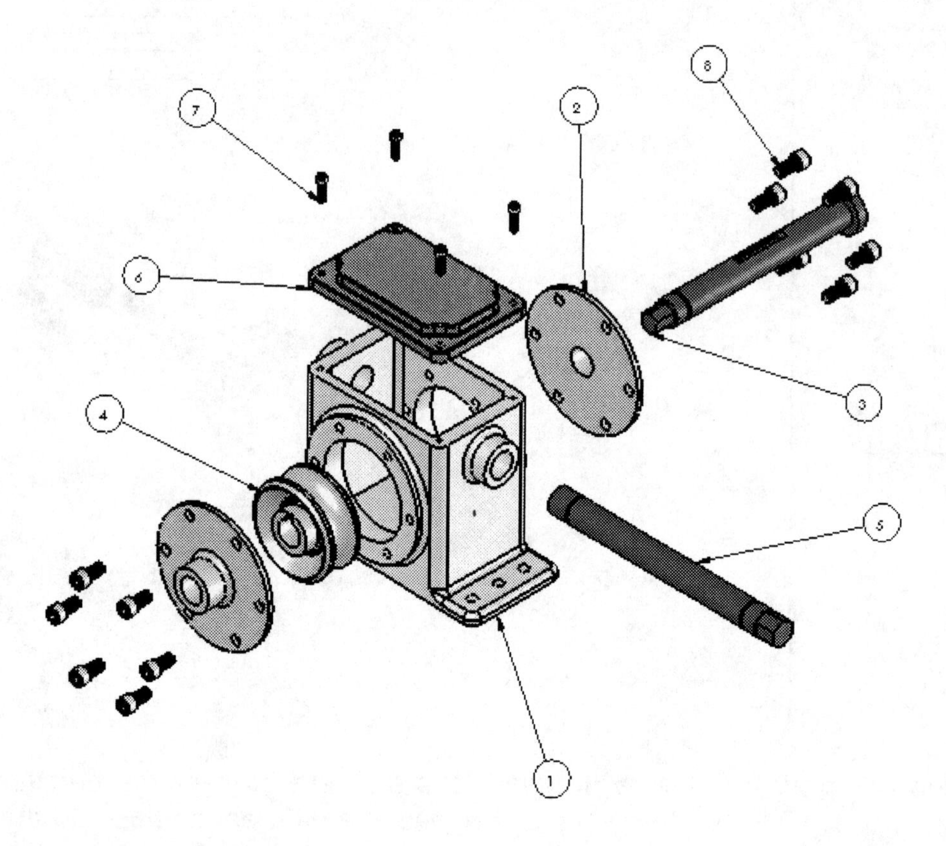

Final Comments

By completing the exercises presented in this book, we learned how to apply many different SolidWorks features and its various options to common design tasks. As we stated at the beginning, this book is meant to be an introduction to SolidWorks, and as the reader was able to see, the breadth of options and possibilities available to the user are enough for any design task at hand. After completing this book the reader is capable of accomplishing many different tasks from part and assembly modeling to detailing in a very short period of time, using the most commonly used commands in SolidWorks.

We hope this book serves as a stepping stone for the reader to learn more. A curious reader will be able to venture into more advanced features, and take advantage of the similarities and consistency of the user interface to his/her advantage. We tried very hard to make the content as understandable and easy to follow as possible, as well as to get the reader working on SolidWorks almost immediately, maximizing the "hands-on" time.

After completing a total of six different parts, their drawings, an assembly including the fasteners, an exploded view and the assembly drawing complete with Bill of Materials and identification balloons, the reader is ready to apply the learned concepts in different design applications.

We'd love to hear from you, your experience with this book and any comments or ideas on how we can make it better for you by sending an email to: areyes@mechanicad.com.

One final tip: if you get lost, or can't find what you are looking for while working in SolidWorks, click with the Right Mouse Button. Chances are, what you are looking for is in that pop-up menu. ☺

Notes:

Appendix

Document templates

One of the main reasons to have multiple templates is to have different settings, especially units; we can have millimeter and inch templates, different materials, dimensioning standards, etc. and every time we make a new part based on that template, the new document will have the same settings of the template. One good idea is to have a folder to store our templates, and add it to the SolidWorks list of templates.

Using the Windows Explorer, make a new folder to store our new templates. For this example we'll make a new folder in the Desktop called *"MySWTemplates"*. In SolidWorks, go to the "Tools" menu, "Options" and in the "System Options" tab, select "File Locations". From the drop down menu select "Document Templates".

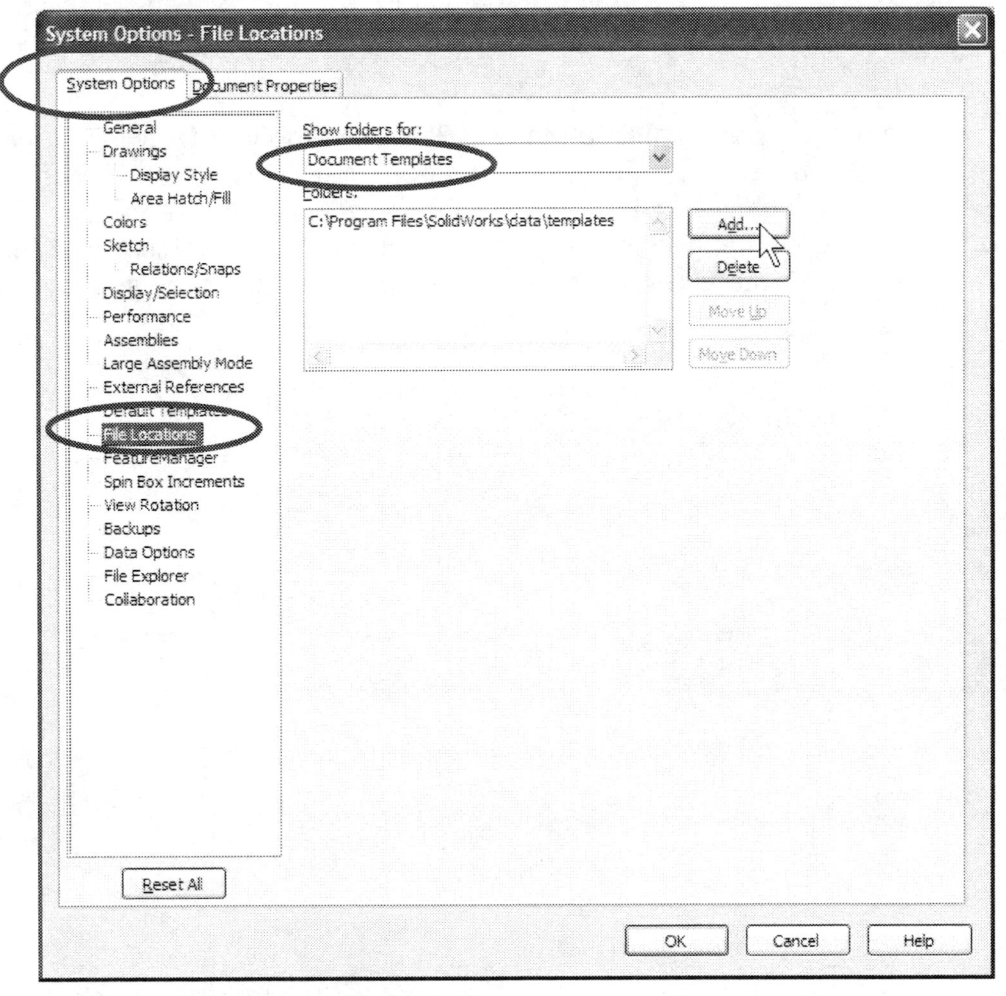

Click the "Add" button and locate the folder where you want to save your templates and click on OK. You will now see two template folders listed.

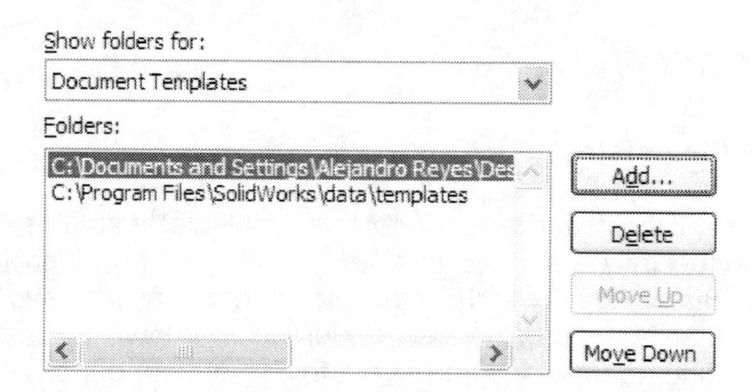

After adding new templates to this folder, we'll see an extra tab when we select the "New Document" icon. In our example, this is how it will look:

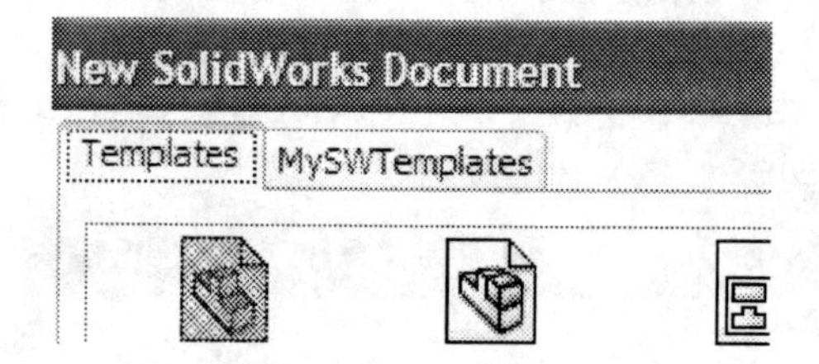

To make a new template make a Part, Assembly or Drawing file, and go to the "Tools" menu, "Options" and select the "Document Properties" tab. Everything we change in this tab will be saved with the template. The most common options changed in a template are Units, Grid, detailing standards and Annotation Fonts. Let's review them in the order that they appear in the options.

The first one is "Detailing"; here we can change the Dimensioning standard to ANSI, ISO, DIN, JIS, etc.; by changing the standard, all the necessary changes will be made to arrows, dimensions, annotations, etc. according to the standard.

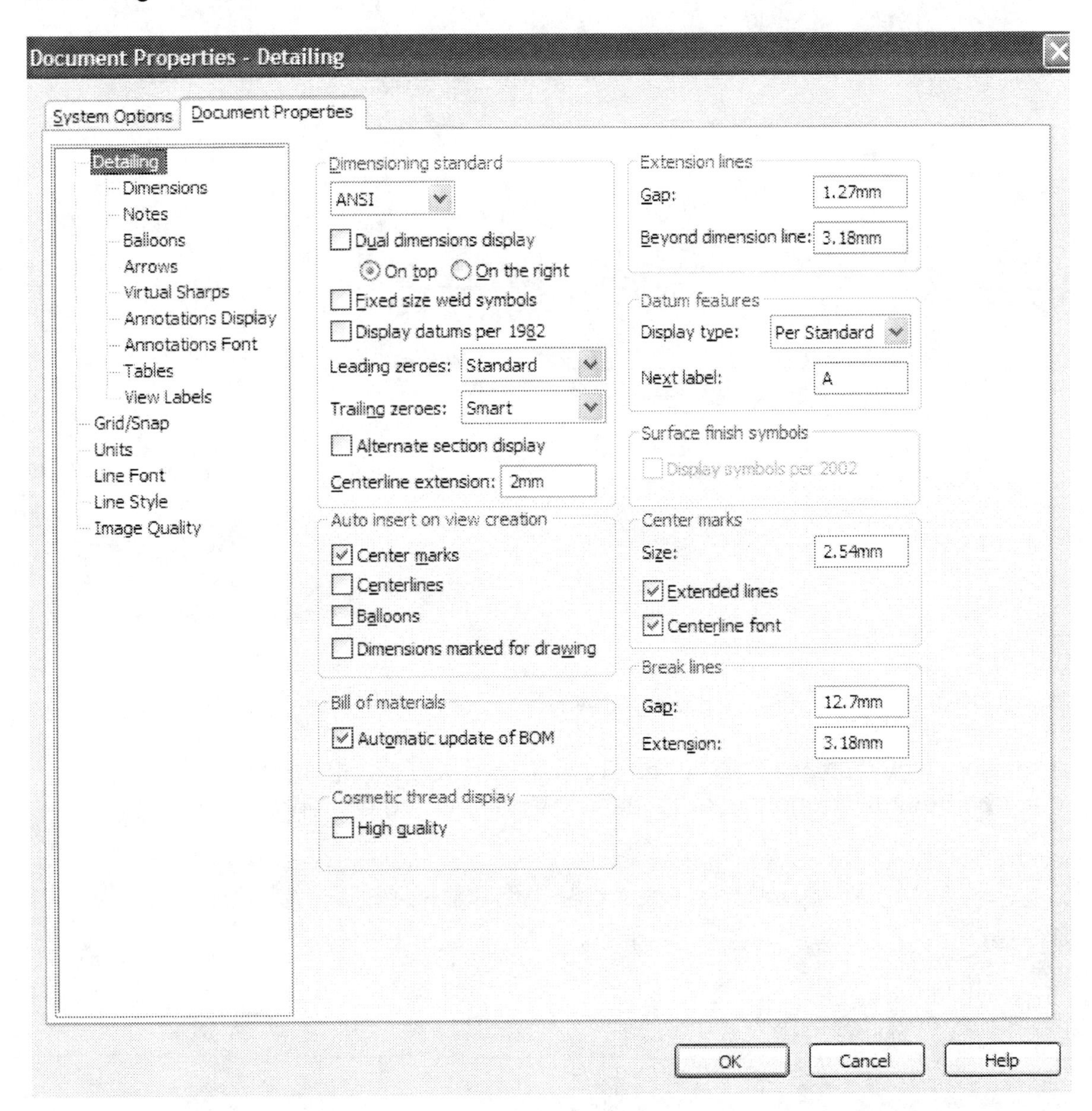

The options listed under "Auto insert on view creation", "Center marks", "Bill of Materials", "Break lines" and "Cosmetic thread display" are used only in drawing templates. All other options are used for Part, Assembly and Drawing templates.

In the "Annotations Font" section we can change the font used for all of the items listed. Select the annotation type that we want to change, and the standard Windows Font selection box will be presented to change the font. Select the desired font parameters and click on OK.

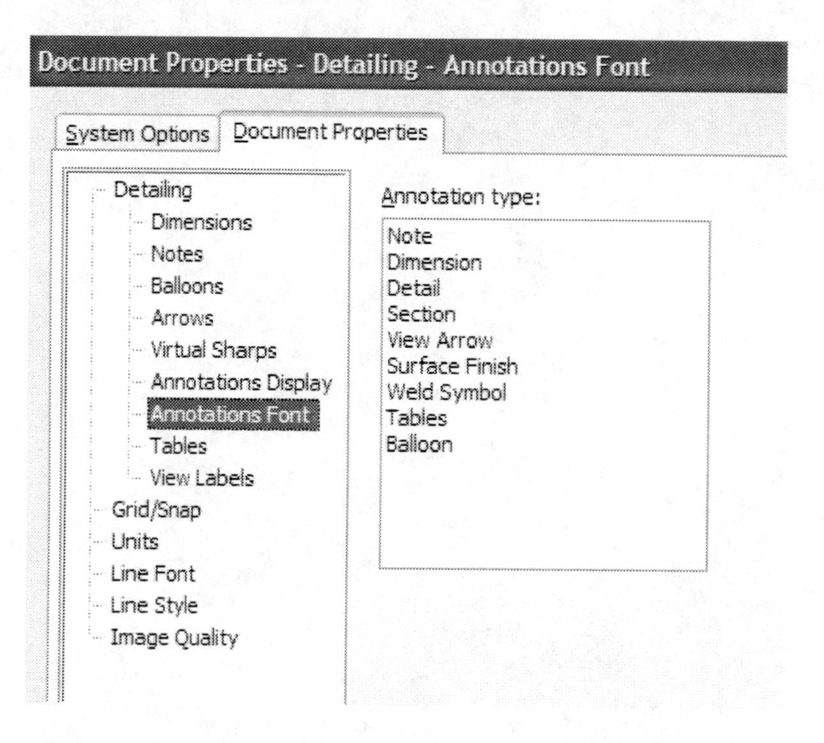

For the "Grid/Snap" section, we can define if we want to have a grid and its settings when we work in a sketch. Most users don't use the grid, but that's precisely why it's called an option. Turn it on if you like it; turn it off if you don't. It may be helpful to a new user to identify when we are working in a sketch.

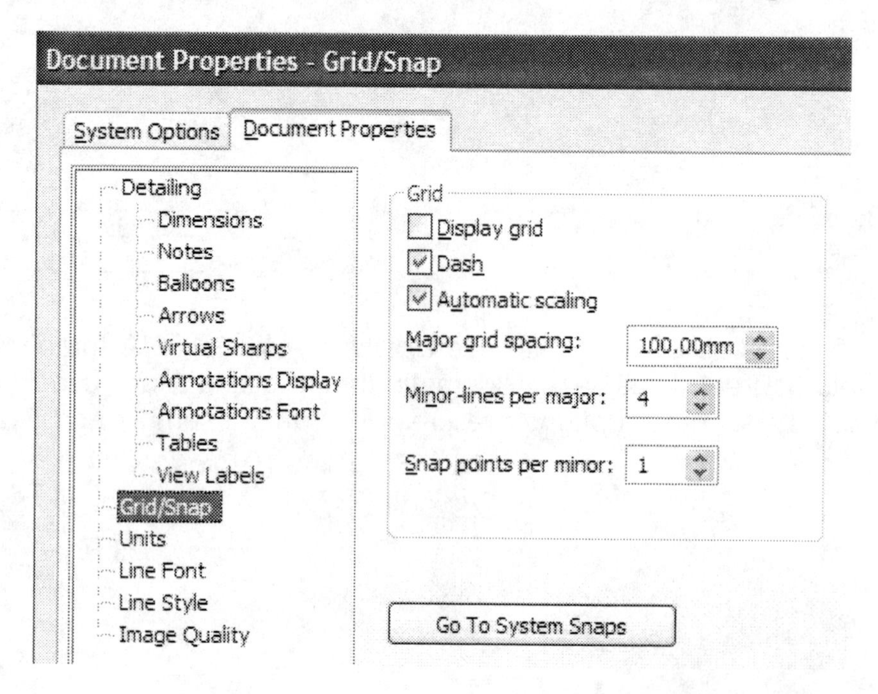

For the "Units" section, simply select the units that we want to use in the template, number of decimal places, angular units, length, mass, volume and force. Selecting a unit system will change all the corresponding units to that system, saving us time. Selecting "Custom" will allow us to mix unit systems if so desired.

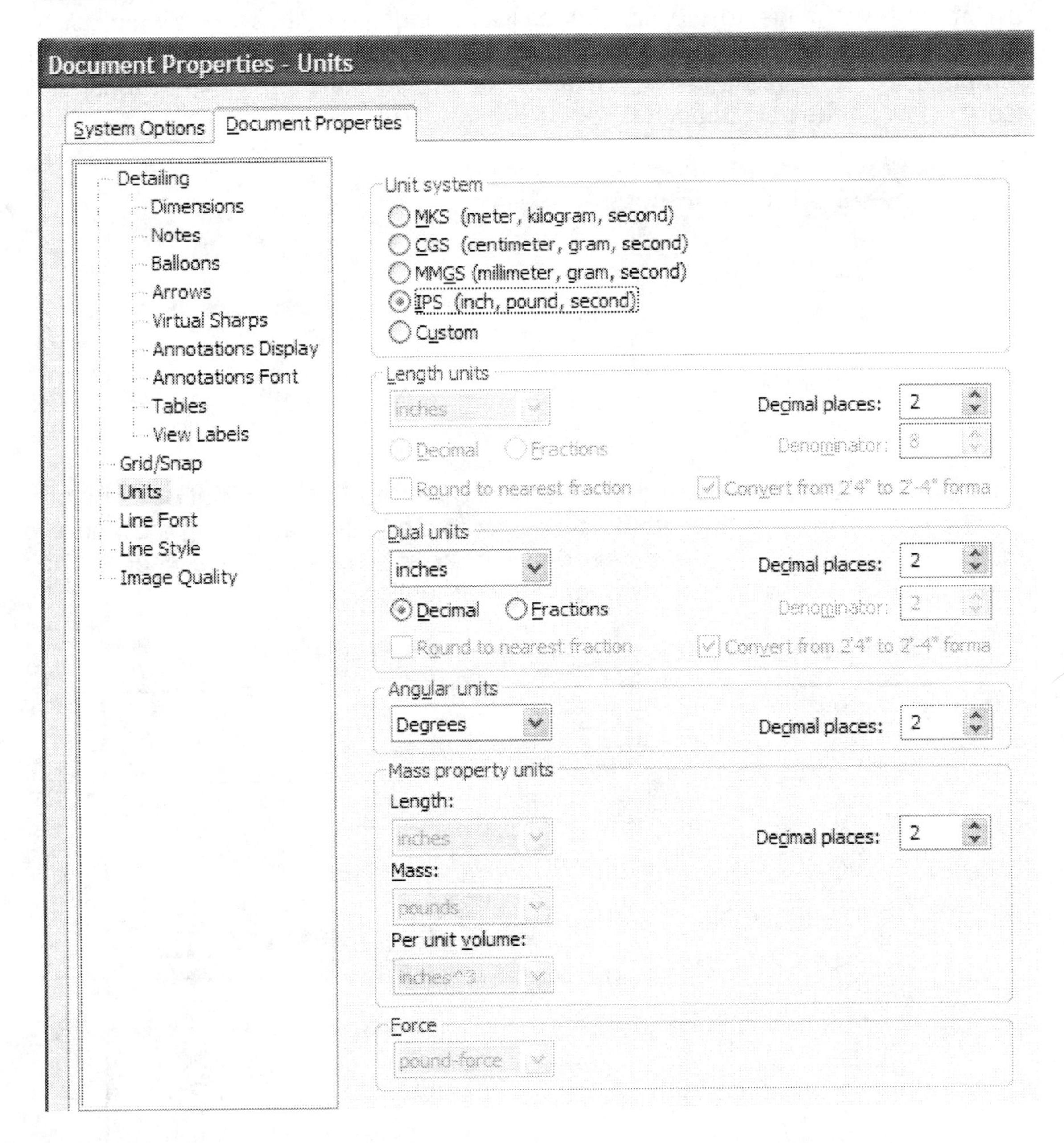

Click OK to change the document properties to the options selected. To save the documents as a new template go to the "File" menu, "Save As". From the "Save As Type" drop down box, select "Part Template (*.prtdot)", "Assembly Template (*.asmdot)" or "Drawing Template (*.drwdot)" (depending on the type of document we are working on we'll only see one option). SolidWorks will automatically change to the first folder listed under the "File Locations" list for templates; if needed, browse to the folder where we want to save it. Give your template a new name and click "Save". When we select the "New Document" icon, we'll see the new templates tab with our new template.

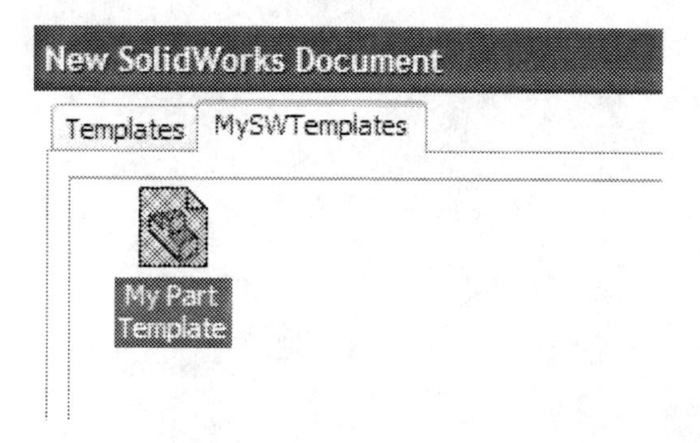

For the Drawing templates we can also set the Sheet format or title block. To change a drawing's **title block**, it's easier to start with an existing drawing and modify it. To do this, make a new drawing, select the sheet size needed, and make sure we have the "Display Sheet Format" option on.

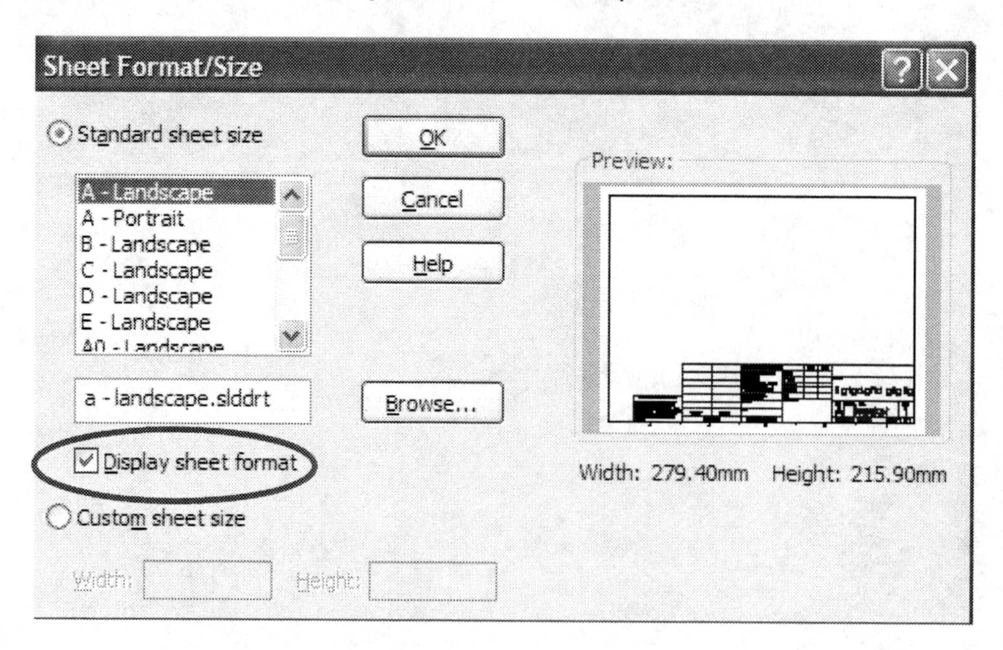

To change the title block, click with the Right Mouse Button in the drawing area and from the pop-up menu select "**Edit Sheet Format**".

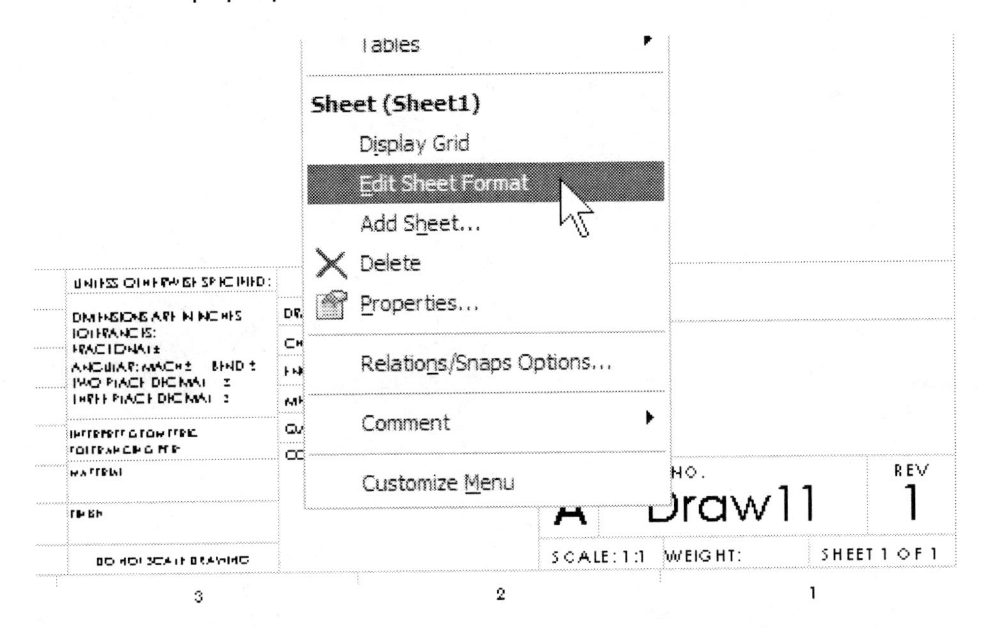

We are now editing the title block and can change it; we have all the sketch tools available to modify it as needed. Some of the notes in the drawing are parametric, meaning that they change automatically, like the scale, revision, drawing name, etc. Search the SolidWorks help file for "Link to Property" for instructions on how to link a note to a document property.

After modifying the title block to our liking, exit the "sheet format" clicking with the Right Mouse Button in the graphics area and selecting "**Edit Sheet**" from the pop-up menu.

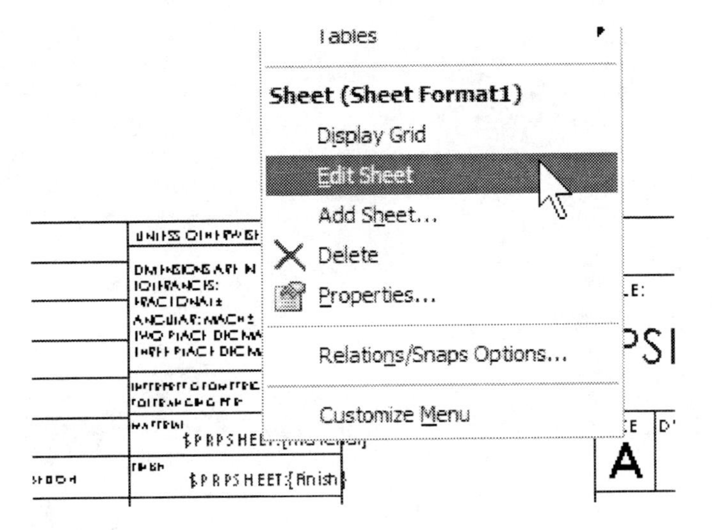

If we save this drawing as a template the title block changes will be saved with the template.

Notes:

Index